THE RAINBOW
From Myth to Mathematics

THE
RAINBOW

From Myth to Mathematics

by

CARL B. BOYER

With New Color Illustrations
and Commentary by
Robert Greenler

Princeton University Press
Princeton, New Jersey 35

Published by Princeton University Press, 41 William Street,
Princeton, New Jersey 08540

First Princeton Paperback printing, 1987

LCC 87-2288
ISBN 0-691-08457-2
ISBN 0-691-02405-7 (pbk.)

This edition replaces the black and white illustrations
of the original edition with eight pages of new color illustrations

Clothbound editions of Princeton University Press books are printed
on acid-free paper, and binding materials are chosen for strength
and durability. Paperbacks, while satisfactory for personal collections,
are not usually suitable for library rebinding

Printed in the United States of America by Princeton University
Press, Princeton, New Jersey

To
Marjorie

Preface

THE RAINBOW HAS HAD HOSTS OF ADMIRERS—MORE, PERHAPS, THAN ANY OTHER natural phenomenon can boast—but it has had few biographers. Virtually every volume on mythology contains legends connected with the rainbow, and practically all modern textbooks of physics include some exposition of the optical principles which account for the bow; yet historians of science have not taken advantage of the opportunity which the rainbow affords of bridging the gap between the sciences and the humanities. Mankind has been thinking, talking, and writing about the rainbow for thousands of years; but there is not a single up-to-date volume devoted to man's recorded ideas on the subject. In 1699 there appeared in Latin a substantial volume by Volkamer, entitled *Thaumantiados, Sive Iridis Admiranda*, which recounted, in a discursive and disorganized manner, man's notions on the rainbow; but neither the history nor the theory of the bow has stood still during the last two and a half centuries. Almost a hundred years ago three German scholars, Just, Kunze, and Reclam, each published an historical treatment of the rainbow; but these less ambitious projects, comprising roughly a score of pages each, scarcely could do justice to the rich scholarly tradition surrounding the bow. Within the past few years Crombie's *Robert Grosseteste*, with practically two full chapters devoted to the rainbow, has furnished a stimulating record of some of man's thoughts during the medieval period; but this leaves untold the beginning and the end of man's story of the bow. Preparation of the present book was undertaken long before the appearance of Crombie's erudite volume; but it has been possible to incorporate some suggestions which have been derived from the *Robert Grosseteste*, and these are gratefully acknowledged. There are a great many other sources, primary and secondary, which have been used; and it is hoped that citation of these in the notes and the bibliography will be taken as an expression of appreciation to the authors. Considerable attention has been paid to the accurate identification of sources in the hope that readers may be tempted to study further some aspects of the history of the theory of the rainbow.

The author owes a very special debt of gratitude to the John Simon Guggenheim Memorial Foundation for the award during the academic year 1954-

1955 of a fellowship which made possible the completion of this project and other studies. Assistance is acknowledged also from numerous other sources. Many libraries—including the New York Public Library, the Library of Congress, the John Crerar Library, the Burndy Library, the library of the American Philosophical Society, and the libraries of Brooklyn College, the University of Chicago, Columbia University, Harvard University, the University of Michigan, the University of Oklahoma, the University of Pennsylvania, Princeton University and the Institute for Advanced Studies, the University of Virginia, the University of Wisconsin, and Yale University—have aided through interlibrary loans of rare books, through photostats and microfilm, or through other services. The typing of the manuscript was capably and cheerfully done by Mrs. Thelma K. Langer; and to my wife I am especially grateful for neverfailing patience and encouragement during the long days of study and writing. Dr. Pearl Kibre of Hunter College and Dr. I. Bernard Cohen of Harvard University generously gave of their time to read the manuscript and to make valuable suggestions. Other colleagues and friends who helped to speed the work on its way must not think that because they are unnamed their assistance is unappreciated. *Nil sine multis sociis.*

C. B. B.

Contents

Figures

Plates

The following plates appear as a group after page 142

PAINTINGS

Namirrgi, Danbon Group, Arnhem Land Plateau, Australia, Bark
painting: *Rainbow Serpent*
Thirteenth-century mosaic, Cathedral of San Marco, Venice: *Noah,
His Family, and the Animals Leaving the Ark*
Jacob van Ruisdael: *The Cemetery*
Frederic Church: *Rainy Season in the Tropics*
Edouard Manet: *Fishing in Saint-Ouen, near Paris*
Prince Eugen of Sweden: *The Rainbow*
Henry Mosler: *Above the Rainbow*
Awa-tsireh, Hopi: *The Rainbow*

PHOTOGRAPHS

Wide-angle rainbow
Rainbow with supernumerary arcs
Saltwater bow over the Pacific Ocean
Reflected rainbow seen in a pond next to the Atlantic Ocean
Reflected-light rainbows
White rainbow or fogbow with supernumerary bow
Red rainbow
Infrared rainbow

THE RAINBOW
From Myth to Mathematics

I

The Beginnings

MAN'S STORY OF THE RAINBOW, LIKE OTHER ASPECTS OF THE HISTORY OF SCIENCE, has no inescapable origin and no discernible end. There is no record, oral or written, of the precise date at which a rainbow was first noticed; and even now, in the middle of the twentieth century, it is not possible to boast that the formation of the bow is understood in all details. The theory of the rainbow evidently arose in man's sense of wonder; and now, thousands of years later, it has become enmeshed with the intricacies of advanced mathematics. To trace the gradual development of this theory from early primitive conjectures to sophisticated contemporary formulations will be to tread the pathway to human knowledge—to live again the life of science.

Men of every age seem to have felt a deep sense of awe in gazing upon a beautiful rainbow. The poet Wordsworth wrote for all when he penned the movingly simple lines expressing joy in the present and hope for the future.

> My heart leaps up when I behold
> A rainbow in the sky:
> So was it when my life began;
> So is it now I am a man;
> So be it when I shall grow old,
> Or let me die! [1]

The deep yearning in these lines is akin to man's search for God, and one need not be surprised to find the rainbow early associated with divinity. Hebrew and Greek literature both point to the religious significance of the bow. In the Bible one reads:

And God said, This is the token of the covenant which I make between me and you and every living creature that is with you, for perpetual generations: I do set my bow in the cloud, and it shall be for a token of a covenant between me and the earth. And it shall come to pass, when I bring a cloud over the earth, that the bow shall be seen in the cloud, and I will remember my covenant, which is between me and you and every living creature of all flesh; and the waters shall no more become a flood to destroy all flesh. And the bow shall be in the cloud; and I will look upon it, that I may remember the everlasting covenant between God and every living creature of all flesh that is upon the earth. And God said unto Noah, This is the token of the covenant which I have established between me and all flesh that is upon the earth.[2]

Homer in the *Iliad* wrote that Aphrodite, wounded by the warrior Diomedes, fled from the field of battle to Olympus along the rainbow route, carried swiftly as the wind by the goddess Iris.[3] The rainbow as a mark of deity has ever since held its place in literature. In Dante's *Paradiso*, for example, the Greek and Hebrew traditions appear together in a passage in which the poet looks on the bow as a symbol of the handmaid of Juno and as a sign of God's covenant with Noah.[4]

Were Noah and Homer among the first to view the rainbow? It is not unlikely that both the Greek and the Hebrew accounts were derived from older Mesopotamian traditions. The rainbow legend is not found in the Gilgamesh epic; but in an early Sumerian hymn the rainbow is called "the bow of the deluge," and a Chaldean story of the flood tells how, at the moment of her coming, "the great goddess (Ishtar) lifted up the mighty bow which Anu had made according to his wish." [5] Then, too, the Biblical passage called forth considerable controversy as to whether or not there had been pre-Noachean rainbows, for here there seemed to be a conflict between science and religion. A narrow interpretation of the passage can lead to the inference that no bow was seen before the time of the flood; but uniformitarian principles of geology made it appear most unlikely that atmospheric conditions had undergone some catastrophic change in the days of Noah. Disinclined to believe that the laws of nature were changed after the creation, writers of the Talmud classed the rainbow among the half-dozen things created at twilight on the last day of creation just before the seventh day.[6]

Most Christian exegetes interpreted the passage broadly as indicating that God here gave to the already familiar beauty of the rainbow a new signification, causing it to become a symbol of divine promise. The choice which one makes between the interpretations causes an enormous difference in the chronology of the rainbow. Under the narrower view the bow is a matter of a few thousand years old, for the Biblical flood is said to have taken place in 2348 B.C. when Noah was 600 years old. (He died at an age of 950 years!) The broader point of view would leave room for a prehistory of the rainbow —unrecorded, but reconstructible with reasonable plausibility by extrapolating backward in terms of scientific evidence. Rain, sunlight, and cloud formations would appear to have been sufficiently similar (throughout temperate regions and during the period of man's existence) to those familiar today to justify the assumption that rainbows were observed by our most primitive ancestors.

Modern paleontology indicates that the age of man may well extend as far back as one or two million years ago. Is this, then, a *terminus a quo*—a limit in time beyond which one can not hope to penetrate? Not necessarily. Man may indeed represent the highest form of life; but, even in the Biblical account of creation, he was by no means the first. Beyond him, counting backward through the dim eons of the past, lies a vast sea of sentient life. A sense of vision may have been present in early animal forms of half a billion years ago, and it is to be presumed that many of these looked upon rainbows similar to our own. The size and shape of the bow must have been much the same; but the colors were non-existent for the vast majority of early living creatures, and they undoubtedly varied widely for those animals so fortunate as to enjoy color.[7] Man is not alone in his possession of the priceless gift of color-vision, but he is one of the exceptions in this respect. Fishes, reptiles, birds, and primates, not directly interrelated in the evolutionary family tree, are isolated islands of color vision in the great ocean of animal species. These categories all are among the diurnal vertebrates; and a sense of color seems to be associated especially with a cone-rich retina, cones being more sensitive to red. Nocturnal vertebrates (and they are a majority) do not distinguish clearly between hues, for in their case rods predominate. Thus owls' and cats' eyes are sensitive to blue, which predominates at night. Many cone-bearing animals are known not to have color sense, but there are no color-seeing animals whose retinae contain rods alone. One may envy the feline ability (due to a preponderance of retinal rods) to distinguish objects in semi-darkness, but few would willingly trade acuity of nocturnal vision for the rich aesthetic satisfaction of color sense. Hue discrimination among the lower animals has served for protection, but man's intelligence has made this function largely superfluous and now it contributes primarily to his enjoyment. Does man, then, see the most beautiful rainbow? This question is difficult to answer, for sensitivity to light and color varies considerably from species to species. Insects, for example, presumably see colors in the rainbow which for us are not there, and they probably fail to note some of those familiar to us. The range of visibility for bees is shifted toward the short-wave side; they are not sensitive to pure red or to blue green, hues which occur rarely as flower colors in our flora. For bees the blue and purple-red flowers stand out clearly from the foliage.[8] Birds, on the other hand, are much more sensitive to red than are men, but they are subsensitive to green, blue, and violet. If fish pay any attention to the rainbow, they are more attracted by the inner colors, for the short-wave brightness is higher for them than it is for us. However, it is probably safe to say that no form of life gets greater aesthetic enjoyment from the rainbow than does man, and there is as yet no evidence that subhuman species share in the satisfaction of intellectual curiosity which we have known. In this respect an anthropocentric view of the rainbow is readily justified.

Granted that the semicircular form of the rainbow may have been observed by early forms of life endowed with a sense of vision, does this mark the beginning of the story of the rainbow? Were there not bows formed when, possibly a billion or more years ago, rain fell upon a hot earth from steaming clouds? Does the rainbow, in fact, exist independently of the observer? This question leads one to the familiar philosophical problem of distinguishing between subjective and objective, or between primary and secondary, qualities; and the question here has unusual pertinence. The rainbow is just about the most subtle phenomenon that everyday nature presents to us; the creation of the familiar sun and rain, it is nevertheless as unapproachable as a spirit. One can not say of the rainbow, "Lo, here," or "Lo, there." The rain producing the bow may be miles away, or part of it may be so close that by stretching out one's hand it would appear possible to grasp the end of the arc. It seems to have been many generations before mankind became fully aware of the fact that the rainbow is not a "thing" in the ordinary sense of the word, for there are at least as many rainbows as there are observers. One refers ordinarily to the rainbow; but one properly should speak rather of his rainbow, inasmuch as the appearance of the bow depends not only upon atmospheric conditions but also upon the position (and physiological idiosyncracies) of the one who views it. (In fact, even the two eyes of the same observer do not see precisely the same rainbow.) Were the bow a unique objective circular band, the theory of conic sections shows that it would appear elliptical to those who view it in oblique perspective. And primitive man noticed that as he moved, the bow followed him like his shadow. Flee from it as one will, the bow nevertheless pursues relentlessly, and one can not by his utmost effort escape from its presence. The intangible nature of the bow gives added significance to the problem of its existence prior to the advent of sentient beings; but the question pales into insignificance in comparison to the absorbing but unanswerable one of the possibility of life and rainbows on planets other than our own.

The descendants of Noah rejoiced at sight of the rainbow because it was a symbol of God's beneficence, but for early man the bow had also a sinister aspect as a hostile force. The anthropomorphic Iris, messenger of the gods, appears in the *Iliad* as a harbinger of war and turbulence,[9] in contrast to Hermes, messenger of peace. She was the daughter of Elektra and sister of the Harpies.

> But Thaumas chose Elektra for his spouse,
> Of the deep-flowing Oceanus child;
> She bore swift Iris, fair-haired Harpies too.[10]

Fig. 1.—Two ancient representations of Iris, one from a bronze plate, the other from a poly-chrome lecythus. (Reproduced from Roscher.)

In the story of the Argonauts, when the sons of Boreas are wrathfully pursuing the Harpies, Iris leapt from the sky and warned the brothers not to strike the Harpies, the hounds of mighty Zeus.[11]

When Hera and Athena were ready to intervene to help the Greeks against the Trojans, Zeus sent Iris to warn them that he would smite their chariot if they pursued their intent; and Hera sent Iris to stir up trouble among the Trojan women in Aeneas' band during the funeral games for Anchises.[12] In Virgil's Aeneid, too, the arc is portrayed malevolently as a precursor of storm; and it was Iris who cut the last tie holding the dying Queen Dido to life, the morning rainbow thus foreshadowing the tempest that cast Aeneas on the shores of Sicily. (There is a somewhat similar legend that Hercules languished between life and death until Iris came and cut his hair, breaking the spell and permitting him to die.) [13] By command of Zeus, the goddess Iris carried a ewer of water from the Styx with which she put to sleep all who perjured themselves. Ovid, when describing the deluge of Deucalion, introduces Iris as significative of the disaster that was to befall; [14] and tradition ascribes to Iris the role of the bearer of ill tidings to Menelaus on the flight of Helen and Alexandrus. So frequently was the significance of Iris malignant that it has been conjectured that the words *iris* and *ire* are related. Commonly, however, the name of Iris is thought to be derived from the Greek word for "speaker" or "messenger." [15]

Iris was a goddess of the elements who also worked to man's advantage. The granddaughter of Oceanus (and sometimes the spouse of Zephyr), she seems to have been at first a wind and rain goddess. (In Babylon the rainbow was the fiancée of the rain; in far South America the moon is the wife of the rainbow; at other places the bow is the companion or servant of the sun, heralding its reappearance, or sometimes the great spirit which covers the rain with a mantle.) It was Iris who charged the clouds with water from lakes and rivers in order that rain might fall again upon the earth. The farmer, therefore, paid honor to the rainbow goddess who helped thus to quicken his crops, and he looked to the bow for signs of their success: a predominance of red signified a good year for the vine, green presaged an abundance of olive oil, and white announced a bumper wheat crop. Even in recent times in France, near the Spanish border, the agricultural implications of the rainbow are preserved in the names *arc de sedo* or *pont de sedo* (*de soie* in some parts of France). Such phrases probably are suggested less by the vari-colored hues themselves than by the belief that the colors of the bow are a prognos-

FIG. 2.—Iris, the rainbow goddess, carrying a child. From an ancient vase. (Reproduced from Roscher.)

tication of a good or bad harvest of silkworm cocoons, a prominent yellow band presaging an abundant yield.

The goddess Iris was associated with meteorological phenomena in general, and so it was she who, at the entreaty of Achilles, sped to the home of the Zephyrs and to Boreas to urge them to consume with fire the pyre of Patroclus. And when the grieving Demeter would have destroyed the race of man with cruel famine, Zeus sent golden-winged Iris to summon rich-haired Demeter. The speed of Iris is described poetically by Aristophanes:

> And Iris, says Homer, shoots straight through the skies
> With the ease of a terrified dove.[16]

Just about two thousand years later Shakespeare also looked to Iris as a swift and faithful messenger. He has Queen Margaret, wife of Henry VI, say to the exiled Duke of Suffolk:

> Let me hear from thee;
> For wheresoe'er thou art in this world's globe,
> I'll have an Iris that shall find thee out.[17]

The swiftness of Iris may well have been suggested to the ancient poets by the rapidity with which the rainbow in the heavens vanishes. In the very earliest representations Iris is without wings, but later she generally was pictured as a beautiful virgin with wings of gold and robed in bright colors of varied hue. (Sometimes, on the other hand, she was said to be the mother of Eros.) Often she was depicted, as in a recent Greek postage stamp, riding a rainbow, sometimes with a nimbus on her head, and with the colors of the bow reflected in the cloud. (In Finnish mythology a divine maiden sits on the rainbow and weaves golden raiment.) Her genealogy supports the opinion that originally she was the personification of the rainbow; but later poets say that the rainbow merely indicates the route she takes, appearing when she wants it and vanishing when she no longer has need of it. No statues of Iris

Fig. 3.—Iris and the rainbow as depicted on a Greek airmail stamp (No. 773) of 1935–1937.

have been preserved, but she appeared frequently on vases and bas-reliefs. In works of art she can easily be confused with the winged Eros and Nike, which may be why the talaria, the caduceus, and a vase came to be recognized as her attributes. Often associated with Hermes, her role in art and literature declined as his increased. Her subordinate position among the gods may be taken as an indication of the small importance attached by Greek fancy to personifications of natural phenomena. The poet-philosopher Xenophanes, for example, said of Iris that "she is but a cloud tinted in diverse colors."

Meteorology and hagiology have been richly associated in Christianity, and names of saints have been freely attached to the rainbow. St. Martin's day is especially important in the farm almanac because it is supposed to determine the weather for the latter part of autumn. Hence, and also because St. Martin is the patron saint of the wine and fruit crop, the bow sometimes is called the Arc of St. Martin. Occasionally it is known as Arc of St. John, or Gate of St. John, because of the role of this saint in the thunderstorm and the return of Christ.

In China, also, the rainbow served in prognostications, not only in husbandry but also in political affairs and marital fidelity. Chinese literature includes classifications of rainbows into such categories as white, red, green, or gray, curved or straight (rain galls?), male or female. The formation of the rainbow was in part ascribed to some kind of combination of yin and yang, the male and female principles, in part to the breaking up of certain stars. The double rainbow was recognized at least by 1010 in China, for there is a quotation before that time asserting that when a pair of rainbows appear simultaneously, the brighter one is the male rainbow, the fainter one is female. When the cloud is thin and when the solar rays are reflected from the raindrops, then the male rainbow appears.[18]

Rainbow prognosticative notions have been remarkably viable, and some of these survive today in folk-rhymes in many languages:

> Regenbogen am Morgen
> Macht dem Schafer sorgen;
> Regenbogen am Abend
> Ist dem Schafer labend.

or,

> A rainbow in the morning
> Is the shepherd's warning;
> A rainbow at night
> Is the shepherd's delight.

Such popular rules are dependent upon the fact that the rainbow generally implies a local passing shower, soon over, and that changes in the weather in many parts of Europe are likely to pass from west to east. With more fanciful abandon, the Mojave Indians of Arizona look upon the rainbow as a succession of charms needed by the Creator to bring the rain to an end. To terminate a violent storm, the entire sequence of colors is needed.

The rainbow was not, of course, the only optical phenomenon which served as a weather forecaster. Halos and parhelia generally were regarded as signs of either fair weather or light rain; and there was a category of appearances, known in Latin as *virgae*, which, like the rainbow, were believed to presage storms. Virgae are literally "streaks" or "columns" of light, and the generic term was used rather loosely to include a number of things not too closely related as to cause. In the narrower sense the word referred to what now usually are called raingalls or watergalls or windgalls. These are imperfect rainbows (in Germany sometimes called Wasserkalb)—the lower portions of bows which, rising but a short distance upward, streak the leaden mass of clouds with partial gleams of faded red and purple. (Such broken fragments of rainbows are known also as weathergaws or as sundogs, although the latter phrase today sometimes is applied also to a mock-sun or parhelion.) Light-pillars or sun-pillars, shafts of light sometimes tinted and generally nearly vertical, were grouped also among the virgae, the study of which was coupled in antiquity with that of the rainbow.

Many primitive peoples viewed the rainbow with fear and misgiving, sharing with Homer the comparison of Iris to a serpent. In African mythology the rainbow is thought of as a giant snake that comes out after rainfall to graze; and the hapless person upon whom he falls will be devoured. (Some people still believe that a home upon which the bow falls will suffer misfortune.) The Zulus, and also tribes in South America, likewise regarded the snake bow as cannibalistic; and there are rainbow figures in numerous myths involving the swallowing of cattle. Among Eastern Europeans the giant serpent was believed to suck water from seas, lakes, and rivers, sprinkling it anew as rain. Sometimes the bow was believed to draw children, as well as water, on high. Tartar tribes call the bow "The thunder drinks water"; while Esthonians held that the rainbow was the head of an ox which is lowered to a river in drinking. The Greek poet Hesiod shares a touch of the water-drinking myths in his representation of Zeus as dispatching Iris to bring, in a golden jug, the eternal and primeval water of the Styx, [19] while other legends interpret the rainbow as a stream which souls in heaven can drink. Plutarch records the legend in which Iris has the head of a bull and consumes the water of streams; and Ovid, too, speaks of her as drinking water. Among the Shoshoni Indians in America the firmament is a dome of ice against which the rainbow, a giant serpent, rubs his back. Particles of ice thus are rubbed off and fall to the earth, in winter as snow, in summer as rain. (This reminds one somewhat of the fable of the bow as the razor which shaves the heavens.) Southern Tartar tribes call the rainbow the "rainbelt." In Central Australia, on the other hand, the Kaitish tribe believe that the rainbow is the son of the rain, and that with filial regard he is anxious to prevent his father from falling to the earth. Hence if the rainbow appears in the sky at a time when rain is wanted, they try to enchant the bow away.[20]

The Kirghis name for the bow is "the old woman's sheep-halter," a consequence of a fable of a man who had two quarrelling wives. The mother-in-law cursed the elder, who with her sons fled to the heavens with the cattle which she now tethers to the rainbow.

The shape of the bow, as well as its inescapability, caused primitive peoples of Nias to fear it as a huge net spread by a powerful spirit to catch their shadows or souls. Among Finns and Lapps it was the sickle or bow of the Thunder God, a skillful archer whose arrow is the lightning. Some held it to be the god's crossbow and said that ancient stone weapons found in the ground are the bolts which he had used to kill the Forest Spirit hiding among the trees. In middle and northern Asia it is related that the rainbow is a camel with three persons on its back: the first beats a drum (thunder); a second waves a scarf (lightning); and the third draws reins causing water (rain) to run from the camel's mouth. The Blackfoot Indians called the rainbow "Rain's Hat" or "The Old Man's Fish Line," or "The Lariat." Among Germanic myths is one which looks on the rainbow as the bowl which God used at the time of the creation in tinting the birds. To the Greenlander the rainbow has been the hem of a god's garment, to the ancient Welsh, the chair of the goddess Ceridwen.

In the Hebraic-Christian tradition, unlike many others, the rainbow generally was looked upon with hope and confidence, but there are vestiges also of fear. The Bible compares the rainbow with the brightness of the throne of God, and by some it was interpreted as a sign of censure. Following the example of Ezekiel before the throne,[21] certain of the Hebrews believed that one should prostrate himself before the rainbow; but others enjoined the custom as savoring of heathenism, and approved instead the repetition of a benediction:

> Praised be the Lord our God, the King of the
> Universe, who remembereth the covenant and is faithful
> in His covenant, and maintaineth His word.[22]

A Medieval Germanic tradition held that for forty years before the end of the world there would be no rainbow, and from this belief men drew comfort whenever they saw a bow silhouetted against the dark sky:

> So the rainbow appear,
> The world hath no fear,
> Until thereafter forty year.[23]

A Hebrew belief asserted that if Yahweh lays aside his bow and hangs it in the clouds, this is a sign that his anger has subsided.[24] Other peoples have had similar ideas, based upon the tradition that an archer carries his bow with the ends pointing downward when he wishes to indicate his peaceful intentions.

In ancient classical literature the rainbow sometimes was deified as Iris; at other times it was regarded merely as the route traversed by the messenger of Hera.[25] The conception of the rainbow as a pathway or bridge has been widespread. For some it has been the best of all bridges, built out of three colors; for others the phrase "building on the rainbow" has meant a bootless enterprise.[26] North American Indians were among those who thought of the rainbow as the Pathway of Souls, an interpretation found in many other places. Among the Japanese the rainbow is identified as the "Floating Bridge of Heaven"; and Hawaiian and Polynesian myths allude to the bow as the path to the upper world. In the Austrian Alps the souls of the righteous are said to ascend the bow to heaven; and in New Zealand the dead chieftains are believed to pass along it to reach their new home. In parts of France the rainbow is called the *pont du St. Esprit*, and in many places it is the bridge of St. Bernard or of St. Martin or of St. Peter. Basque pilgrims knew it as the "puente de Roma." Sometimes it is called instead the *Croix de St. Denis* (or of St. Leonard or of St. Bernard or of St. Martin). In Italy the name *arcu de Santa Marina* is relatively familiar.

Associations of the rainbow and the milky way are frequent.[27] The Arabic name for the milky way is equivalent to Gate of Heaven, and in Russia the analogous role was played by the rainbow. Elsewhere also the bow has been called the Gate of Paradise; and by some the rainbow has been thought to be a ray of light which falls on the earth when Peter opens the heavenly gate. In parts of France the rainbow is known as the *porte de St. Jacques*, while the milky way is called *chemin de St. Jacques*. In Swabia and Bavaria saints pass by the rainbow from heaven to earth; while in Polynesia this is the route of the gods themselves.

In Eddic literature the bow served as a link between the gods and man— the Bifrost bridge, guarded by Heimdel, over which the gods passed daily. At the time of the Götterdämerung the sons of Muspell will cross the bridge and then demolish it. Sometimes also in the *Eddas* the rainbow is interpreted as a necklace worn by Freyja, the "necklace of the Brisings," alluded to in *Beowulf*; again it is the bow of Thor from which he shoots arrows at evil spirits.[28] Among the Finns it has been an arc which hurls arrows of fire; [29] in Mozambique it is the arm of a conquering god. In the Japanese Ko-Ji-Ki (or Records of Ancient Matters),[30] compiled presumably in 712, the creation of the island of Onogoro is related to the rainbow. Deities, standing upon the "floating bridge of heaven," thrust down a jeweled spear into the brine and stirred with it. When the spear was withdrawn, the brine that dripped down from the end was piled up in the form of the island.

In myth and legend the rainbow has been regarded variously as a harbinger of misfortune and as a sign of good luck. Some have held it to be a bad sign if the feet of the bow rest on water, whereas a rainbow arching from dry land to dry land is a good augury. Dreambooks held that when one dreams of seeing a rainbow, he will give or receive a gift according as the bow is seen in the west or the east. The Crown-prince Frederick August took it as a good omen when, upon his receiving the kingdom from Napoleon in 1806, a rainbow appeared; but others interpreted it as boding ill, a view confirmed by the war and destruction of Saxony which ensued. By many, a rainbow appearing at the birth of a child is taken to be a favorable sign; but in Slavonic accounts a glance from the fay who sits at the foot of the rainbow, combing herself, brings death. In other tales the bow is associated with "The Old One" (either male or female) who enters so frequently in folklore concerning atmospheric phenomena. Primitive Peruvians held the rainbow in such awe that they remained silent during its duration. Conflicting views are implied by names given to the rainbow, even within the same country. In France the bow sometimes is known as Raie de St. Martin, or again it is called Raie du diable. In Arabia the rainbow also sometimes is known as the devil's arc; and in Germany the paler secondary bow is called Teufelsregenbogen. There has been a widespread belief that the exterior rainbow arc represents an unsuccessful attempt on the part of Satan to outdo the Architect of the rainbow. The colorless lunar rainbow has been regarded as particularly ill-omened.

The rainbow inevitably has shared in numerological and theological fancies. Among Buddhists the colors of the rainbow were related to the seven planets and the seven regions of the earth. In Christianity the colors were linked sometimes to the seven sacraments; and the bow served as a sign of the Trinity when from the element earth at one end it passed through the element air to the other foot resting on a third element, water.

Alchemy, too, found the rainbow symbolism appropriate to its theories. The alchemists' so-called "philosophic rainbow," the efflorescence of metallic colors which heralded the recovery of pure gold, was, like the meteorological bow, a sign that the struggle between the elements was over and that peace reigned.

One of the most persistent facets in rainbow lore is the belief, common to many lands, that it is wrong to point at the bow. One who points the finger at a rainbow may be struck by lightning, develop an ulcer, or lose the finger. Widespread also is the idea that one who crosses, or passes directly beneath, the rainbow will change sex. (In Bohemia the operation of this rule is thought to be limited to girls who are not yet seven.)

The inaccessibility of the ends of the rainbow has encouraged the growth of countless legends. Some gypsies believe that one who at Whitsuntide finds the end of a rainbow and mounts it will win eternal health and beauty. Some in France say that where the rainbow touches the earth a fay has placed a magic pearl; while according to other legends one finds there prized beads which cannot now be manufactured. More prevalent are stories about golden rainbow vessels. It is widely held that angels who mount the bow let fall little dishes which are especially effective in easing the labor of a pregnant woman and in curing the fevers of obedient children who find them and drink out of them.[31] On the other hand, one who dares sell one of these little golden bowls may be struck dumb. These rainbow dishes are also popularly reported to be found in a boot, shoe, or hat which has been thrown over the bow. (Other versions have it that the boot, shoe, or hat is turned to gold, or is filled with gold doubloons, as it falls on the other side of the bow.) In Swabia it has been thought that the ends of the rainbow rest upon two such *Regenbogenschüsselchen*, and that he who comes upon the end of the bow which can be seen the longer will there find the coveted patina. One form of the legend holds that the rainbow draws water by means of these golden vessels; another explains the rainbow as the brightness produced by reflection of the sun's rays in these golden dishes; still a third asserts that from each rainbow the sun causes a golden paten to fall.[32] Until a century ago there was considerable discussion as to whether certain concavo-convex objects, resembling patens (See fig. 4.) and called *patellae iridis*, had fallen from heaven along the rainbow arch or were in reality ancient Roman or Celtic coins. Large numbers of these were found south of the upper Danube, between the Isar and Inn Rivers; and although they have a unity of design which indicates a common origin, none of the objects carries a date or place, nor is there anything on them which resembles script. Finally the numismatist and archaeologist Franz Seraph Streber (1805–1864) showed that these "rainbow" coins were of Celtic origin. Some of them were proved to be gold pieces struck before the year 400 by the Vindelici, a people who lived in Switzerland before their Germanization and the conquest by Rome.[33] The scholarship of Streber and others has dispersed some of the superstition associated with the rainbow; but even today the longing for the elusive pot of

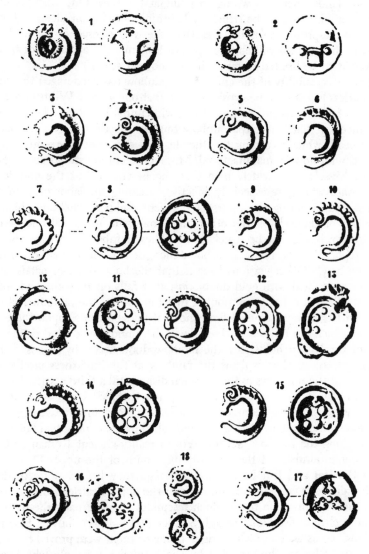

Fig. 4.—Examples of "rainbow patellae" struck by Celtic tribes before the Roman conquest. (Reproduced from Streber.)

gold at the end of the rainbow illustrates the lively persistence of some legends. The word *iris* (and its derivative, *iridescent*) today has been taken over by the work-a-day world; but to our ancestors it was a vivid reminder that the colors of the rainbow are divine. And thaumaturgy, a word which now refers to the mundane science of wonderworking, once deified primitive man's sense of awe. Thaumas, Greek god of wonder, was, quite appropriately, regarded as the father of Iris; and either the daughter or her atmospheric harbinger, therefore, was known also as Thaumantias. Man's story of the rainbow seems to be no exception to the principle expressed by Comte that a branch of natural science arises first in mythology and religion before it achieves its ultimate status as a positivistic science. A study of folklore shows that prescientific legends have persisted to our day in the wide variety of names by which the rainbow is known. In France the name *iris* for the bow was usual until the seventeenth century, since when it has been displaced, except in certain well circumscribed regions, by the more prosaic *arc-en-ciel*, which had been used occasionally as early as the thirteenth century. But in Spain the phrase *arco iris* still survives in common usage. Besides these, there are at least a couple of hundred other names by which the rainbow is known in various localities of France and Spain alone [34]—many more than are attached to other meteorological phenomena. Some of these, such as belt, girdle, or sash, garter, crown, wheel, are simply variations suggested by the circularity of the arc; others, including bridge, door, or gate are derived from the semicircular form. But such single words are seldom used by themselves; they invariably are linked with other associations, generally of a local nature. Reason, in most cases, seems to have been less the determining factor than local language customs and the association of ideas. Some are simple marks of piety, such as Arc de Dieu, or "God's sword," or "Dove of God," or "God's foot"; others, like Arc du temps (in isolated use in Picardy), or "little gold bridge," or "black cow," probably are accidents of habit. Nor is there any apparent significance in the fact that Romance peoples have tended toward names which are more ethereal, the Teutonic toward those more mundane. The story of man's science of the rainbow has been the result of contributions from all races. History shows that in any age and in every land science and myth grow side by side. At all times the number of scientists has been very small as compared with the total population [35]—and much more so in the past than now. While the vast majority have been satisfied to look on the rainbow with awe or fear or hope, the small minority were, and still are, seeking to unravel the mysteries of its formation. To some non-scientists, the search for the causes of the rainbow has appeared to be a profanation of God's handiwork; to others it has been a pious effort to know God better. To the scientist it has been an exhilarating, at times a perversely tantalizing pursuit, inspired by his innate desire to know, of one of the most elusive problems afforded by nature. The result of his search has been a succession of theories, each in its way as beautiful as the phenomenon it would describe.

II

Greece and Rome

THERE IS NO PRECISE DATE AT WHICH MYTHOLOGY GAVE WAY TO SCIENCE IN the theory of the rainbow, nor did the transition take place at the same time or at the same rate in all cultures. The origins of science can be traced back far before the days of Homer and Noah, but there were then no studies equivalent to what now would be called optics or physics or meteorology. Roughly six thousand years ago the Egyptians had introduced a calendar based upon a year of 365 days, showing that astronomy had developed to a respectable level. Some four thousand years ago the Mesopotamian civilization had surpassed the Egyptian in mathematical knowledge, for the Pythagorean theorem and the solution of the quadratic equation had been discovered and were used in connection with numerous mensurational problems. It often has been asserted that these early civilizations were entirely utilitarian in attitude; but it now is clear that such judgments have been exaggerated. Egyptian papyri and Babylonian cuneiform tablets contain problems in elementary number theory and aspects of mathematical recreations which were not even remotely connected with every-day affairs. Scientific curiosity was not lacking in the Tigris-Euphrates and the Nile valleys, as the fragmentary documents show. Rudiments of an astrometeorology are seen in cuneiform tablets of the Sumerian-Babylonian culture of four or five thousand years ago, in which names are given to the smaller halo radius of 22° (tarbasu) and the larger one of 45° radius (supuru).[1] The search for regularities in atmospheric phenomena led the Mesopotamian peoples to the plausible assumption that the weather was influenced by the stars.

There were two other river valleys, besides those of the Nile and Mesopotamia, in which important early civilizations developed—the Indus and the Yangtze. Chronology for both of these latter regions remains highly uncertain, but it appears unlikely that the level of scientific achievement in India or China exceeded, or even equalled, that of the same time in Egypt and Mesopotamia. It is probable, nevertheless, that the idea of a correlation between the weather of the "Twelve Nights" and the twelve months of the year, as it is found in Bede, for example, is of Indo-Germanic origin before the advent of Christianity.[2] With four potamic civilizations flourishing sev-

33

eral thousand years ago, one might expect to find some theory of the rainbow; yet there is no evidence of an attempt at that time at a scientific explanation. Among the Buddhists the seven colors of the rainbow were related to the seven planets and the seven regions of the earth,[3] but such numerological ideas had no scientific basis.

In a very broad and over-simplified sense, one can recognize, in the development of civilizations, three general stages which may be designated respectively as potamic, thalassic, and oceanic, according as the dominant cultures centered about rivers, seas, or oceans. The first of the stages has left nothing scientific on the rainbow, even though it persisted at least as late as the time of Noah and Homer. With the advent of the thalassic civilizations, which thrived throughout the whole Mediterranean area during the first millennium B.C., the situation changed.

Among the peoples pressing down from the north were the Hellenes who occupied the peninsula between the Adriatic and Aegean Seas and then spread east and west to colonize the shores of Asia Minor and the tip of Italy (Magna Graecia). They acquired with amazing alacrity all the knowledge that the potamic civilizations had accumulated, and then they looked about for new intellectual fields to conquer. Unencumbered by hoary traditions and relatively unhampered by political and cultural authoritarianism, Greek scholars investigated nature with an exhilarating freedom and ingenuousness. With them the scientific point of view became a dominant characteristic, for they sought to coordinate observations of natural phenomena into a consistent theoretical structure. The earliest mathematician, philosopher, and scientist known by name was Thales of Miletus (fl.c.600 B.C.), one of the seven wise men of Greece. He became the leader of a group of scholars, known as the Ionian School, who tried to find some basic unity in the multiplicity of natural phenomena. Thales believed that water constituted the unifying basis of all things; but Anaximenes (fl.c.575 B.C.), one of his students, held instead that air was the basic element. With such an inquiring attitude toward events, it is no surprise to find from the members of this school in Asia Minor the first naturalistic statements on the rainbow. Whereas Homer and Hesiod had thought of the bow as dependent upon celestial powers, Anaximenes pointed out the obvious relation of the rainbow to the appearance of the sun. The colors he explained as resulting from the admixture of sunlight with the blackness of the cloud. He said that the rays of the sun fall upon a cloud which is so thick that the rays cannot penetrate it, and the light, therefore, is bent back toward the eye.[4] Anaximander, contemporary of Anaximenes and likewise a follower of Thales, defined winds as a flowing of air, in contrast to Homer's reference to them as gods. As Cicero later pointed out, it was difficult to draw a sharp line between that part of the universe which was thought to be divine and that regarded as human. If the heavenly bodies are divine, why not the next in order, the rainbow? And if this is divine, why not the clouds? Or why not the rain? [5] Greek scientists took the step of including all visible phenomena within the scope of natural philosophy.

One could wish for far more light on Ionian views on the rainbow, for only scraps of information are available. No documents of the period have survived the ravages of time, and the little that is known of the Milesian school is reported by others who lived long afterward.

The school which Thales founded was destined to exert a wide influence, for students went out from it to other regions of the Greek world. In Magna Graecia (the lower part of Italy) the intellectual milieu showed strong Ionian influence, especially through the disciples of Pythagoras, a protégé of Thales. There Empedocles of Acragas (c. 490–c. 435 b.c.), a poet, scientist, philosopher and social reformer, propounded the pervasive doctrine of the four elements. According to this extraordinarily viable doctrine, all forms of matter are made up, in varying proportions, of earth, air, fire, and water. This theory dominated all of physics and strongly influenced early ideas on the rainbow.

Greek science in Italy was more tinged with mysticism than was that in Asia Minor (possibly as a result of the oriental travels of Pythagoras), and hence the only statement on the rainbow attributed to the Italian peninsula at that time comes close to ancient mythology. Empedocles is reported to have believed that the rainbow carried wind and rain from the sea,[6] an idea that may have come down from legends of the bow as a drinker of water. On another of the fundamental problems of optics, Empedocles opposed the erroneous view of his day. It was generally believed among the Greeks that the transmission of light (or sight) is instantaneous, an assumption that may have been encouraged by the fact that, upon opening one's eyes, distant objects become visible immediately. Empedocles held, nevertheless, that light does indeed travel in time; [7] but a demonstration of this fact was awaited even longer than was a satisfactory explanation of the rainbow.

As Pythagoras had brought Ionian science to the Italian peninsula, so did Anaxagoras of Clazomenae (born c. 500 b.c.), during the fifth century b.c. carry this influence to the Attic peninsula. Anaxagoras, who became the tutor and friend of Pericles, found at Athens a popular atmosphere which was hostile to natural science; and when he asserted that the sun, far from being divine, was nothing but a huge white-hot stone, he was jailed for impiety. (While in prison he attempted to square the circle, a problem which was destined to lead to remarkable developments in mathematics.) Ultimately Anaxagoras was released through the influence of Pericles; but his impiety seems to have continued, for he questioned the divinity of Iris. In one of the earliest conflicts between science and folklore, he declared that the rainbow is but a reflection of the sun from a spherical cloud, as from a mirror. There are no extant scientific treatises by Anaxagoras (and virtually none by his contemporaries), and his views have been reported only incidentally—and probably incompletely—by others; [8] but his theory that the rainbow is caused by reflection persisted, in variously elaborated forms, for about two thousand years.

The loss of ancient scientific treatises is to be regretted also in the case of another scientist of this period—Democritus of Abdera (fl.c. 460). He was one of the founders of the atomic theory, and he speculated on mathematical ideas ultimately leading to the infinitesimal calculus. Moreover, he is said to have written two books on optics; and Albertus Magnus ascribed to him the idea that the colors of the rainbow are due to the positions from which it is viewed. But it is utterly impossible to reconstruct his thought beyond such hearsay. One can not even determine whether or not either Anaxagoras or Democritus was aware of the optical law of reflection, or if it had been applied in those days to a geometrical demonstration of the formation of the rainbow. The law may have been discovered during, or very shortly after, the Periclean Age, for Plato (427–347 B.C.) seems to have been aware of some uniformity in the angles in optical reflection. In the Timaeus he wrote:

And now there is no longer any difficulty in understanding the creation of images in mirrors and in all smooth and bright surfaces. The fires from within and from without communicate about the smooth surface, and form one image which is variously refracted. All which phenomena necessarily arise by reason of the fire or light about the eye combining with the fire or ray of light about the smooth and bright surfaces. And when the parts of the light within and the light without meet and touch in a manner contrary to the usual mode of meeting, then the right appears to be left and the left right; but the right again appears right, and the left left, when the position of one of the two concurring lights is inverted; and this happens when the smooth surface of the mirror, which is concave, repels the right stream of vision to the left side, and the left to the right. Or if the mirror be turned vertically, then the face appears upside down, and the upper part of the rays are driven downwards, and the lower upwards.[9]

This passage does indeed prove that Plato was acquainted with the general properties of mirrors, both plane and curved; but there is in it no allusion to the essence of the law of reflection, the strict equality of angles. Consequently any inference from this quotation that Plato knew the law is largely gratuitous. That he may well have had such knowledge is perhaps better shown in the fact that Aristotle, "the mind of his school," all but gives formal expression to the mathematical facts of reflection. Plato was especially interested in mathematics, believing that "God ever geometrizes," and hence one might well expect of him some application of geometry to the theory of the rainbow. However, such expectations are rudely dashed, for Plato mentions the bow only in connection with the mythological paternity of Iris. He has Socrates say [10] that he was not a bad genealogist who said that Iris is the child of Thaumas (wonder); for wonder is the feeling of a philosopher and philosophy begins in wonder. In another Platonic passage [11] it is suggested that the name Iris was derived from εἴρειν (to tell) because she was a messenger of heaven.

But if Plato himself made no great scientific contribution, his school, nevertheless, was a center in the enthusiastic analysis of natural phenomena. It is reported, for example, that Philip, an associate of Plato, attempted to justify the belief that the rainbow is caused by reflection. He called attention to the fact that as an observer moves from side to side, so also does the position of the bow shift from side to side in the same sense. (The behavior of the rainbow is much the same as that of one's shadow.) But this is precisely what takes place in the case of an image seen in a mirror, from which Philip concluded that the phenomena are produced in the same manner.[12]

The *Timaeus* of Plato is the earliest extant Greek treatise dealing with scientific material, and it enjoyed an immense popularity during the Middle Ages; but to the modern reader it is fantastic and repulsive. Biological and physical science are so thoroughly blended with numerology and metaphysics that it is difficult to find in the dialogue anything of lasting scientific value. Astronomy and mathematics were practically the only scientific disciplines which had achieved intellectual independence, but not a single primary work in these fields has survived from the Greek period up to the days of Plato. The theory of the rainbow did not fall within the scope of these subjects, and hence it had remained a part of a crude matrix of natural philosophy.

There was among Plato's students at the Academy one who had a penchant for the proper organization and classification of knowledge. Aristotle (384–322 B.C.) had grown up under the Greek medical tradition, and this may account for the fact that observation and orderliness were more pronounced in his work than in the soaring flights of imagination of his master. The knowledge and experience of Aristotle were so encyclopedic that some arrangement into subdivisions became inevitable. Among the new sciences which thus came into being was meteorology. Plato seems first to have used the word meteorology, but with Aristotle it took on a special meaning. It was coined from three Greek words: *meta*, "beyond"; *area*, "air"; and *logos*, "word" or "reason"; and it was used to designate natural phenomena taking place within the sphere of the moon's orbit. The subject did not suddenly spring into being, for rudiments of such a natural science can be seen in the *Clouds* of Aristophanes, probably through the influence of Anaxagoras. Subsequent tendencies in Greek thought, however, had thwarted its development. A younger contemporary of Anaxagoras, Diogenes of Apollonia, may have written a work on *Meteorologia*; but with him the aura of phantasy surrounding meteorological phenomena became heightened. The Sophists took advantage of this situation to advance their Weltanschaung. Then, too, the attitude of Socrates was not far from that of the later medieval world—if science does not teach the good life, it is worthless.

The beginnings of meteorology had come into some disrepute because in the popular mind the subject was associated, partly through the opposition of Anaxagoras to the Athenian theodicy, with atheism.[13] Moreover, the material in this new field, between Anaximander and Aristotle, had not been clearly differentiated out of the crude matrix of ideas making up the content of the early poetic works *On Nature* or found in the Pythagorean scientific trivium dealing with celestial, sublunar, and terrestrial phenomena. The work of Diogenes, for example, included both supralunar and sublunar spheres. Aristotle, however, made a clear-cut distinction between the realm of astronomy (including the sphere of the moon and everything beyond it—what now might be called astrophysics) and that of his new subject, meteorology. The latter was restricted to phenomena thought to be sublunar, with particular reference to whatever takes place in the atmosphere. Although the upper bound for meteorological phenomena (the moon's orbit about the earth) was well defined, the lower limit remained so vague that much of modern geology and chemistry were subsumed under the broad general heading of meteorology. The oldest comprehensive treatise on the subject is Aristotle's *Meteorologica*, the broad scope of which is evidenced by the fact that the fourth (and last) book of this is tantamount to one of the earliest chemical textbooks. This last book, however, forms so inappropriate a portion of the work that its genuineness was brought into question early, but modern scholarship tends to support its authenticity.

In the narrower sense there were three types of "meteors" in antiquity: aerial (winds), aqueous (rain, hail, sleet, and snow), and luminous (halos, northern lights, igneous meteors). In the last category Aristotle specifically included the rainbow (He also mistakenly included comets, in the belief that they were sublunar phenomena.); and his treatise *De Meteorologica* devoted a substantial amount of space to a penetrating discussion of the causes of the bow. This section may, in fact, be taken as the first truly systematic theory of the rainbow which has come down to us.[14]

Aristotle appears to have had a very definite interest both in optics in general—which he includes, along with harmonics and astronomy, among "the more physical of the branches of mathematics"[15]—and in reflection in particular. It would seem, he says, that in the time of Empedocles "there was no scientific knowledge of the general subject of the formation of images and the phenomena of reflection"; [16] but in his own day Aristotle could write of these as a part of "the theory of optics." [17] Aristotle's view here (as elsewhere) was dominated by common sense rather than abstract idealizations, and hence he was led to make a plausible but unfortunate—from the modern point of view—distinction between two types of reflection:

In some mirrors the forms of things are reflected, in others only their colours. Of the latter kind are those mirrors which are so small as to be indivisible for sense.[18]

Rainbows and haloes fell naturally into the latter category, for Aristotle, confusing reflection and refraction, erroneously regarded these phenomena as caused by "reflection of the sight to the sun" from drops of water and air respectively.[19] Although he went into considerable detail in seeking to justify this viewpoint mathematically,[20] Aristotle does not refer in this connection to the correct quantitative law of reflection. In fact, his geometrical explanation of the rainbow will be seen to be incompatible with the equality of angles, and this has led a recent commentator to conclude that Aristotle "knows that there is a law according to which reflection takes place, but he does not know, and also, perhaps does not feel the need of knowing the specific form of the law itself." [21] However, the apparent inconsistency is more readily understandable if one keeps in mind Aristotle's clear-cut distinction between two types of reflection. The rainbow is explained as a reflection, not from a smooth mirror, but from a cloud. Hence the light of the sun is "seen on the uneven mirror," [22] and the reflection in this case is of the type that causes color rather than images. Probably he thought the strict mathematical law held only for images. Even now one scarcely thinks of the equality of angles when reflection results in color, as in looking at the blue sky or a red rug. The law of reflection is suggested by a study of images rather than colors. Unfortunately, the extant works of Aristotle contain no mathematical treatment—comparable to the discussion of the rainbow—of mirrors in which "the forms of things are reflected." The quantitative principle in such cases therefore is not explicitly stated; but considerations in analogous acoustic and mechanical situations make it clear that the optical law was indeed known.

Aristotle on several occasions pointed out analogies between the phenomena of sight and sound,[23] and hence the following passage on echoes may, by implication, be regarded as directly relevant:

Why is it that the voice, which is air that has taken a certain form and is carried along, often loses its form by dissolution, but an echo, which is caused by such air striking on something hard, does not become dissolved, but we hear it distinctly? Is it because in an echo refraction takes place and not dispersion? This being so, the whole continues to exist and there are two parts of it of similar form; for refraction takes place at the same angle. So the voice of the echo is similar to the original voice.[24]

The representation of acoustics given here undoubtedly was inspired by optical phenomena. The "dissolution" or "dispersion" of the voice is a striking parallel to the optical reflection from small uneven mirrors in which form is not preserved; the "two parts of it of similar form" [the incident voice and its reflection or echo] are obviously analogous to the "visual" and "reflected" rays in connection with mirrors in which "the forms of things are reflected." There can be but little doubt that the equality of angles specifically mentioned in the acoustic case was carried over by analogy from the optical situation in which the phenomenon is far more apparent to the senses. In fact, in another place [25] Aristotle specifically states, in connection with the causes of echo, of reflection, and of the rainbow, "The connections to be proved which these questions embody are identical generically, because all three are forms of repercussion." It should be remarked that the recognition of the acoustic law of reflection in itself constitutes a significant achievement, quite probably of the Aristotelian school, which is not generally appreciated [26] and which probably would not have been made without a knowledge of the corresponding optical principle. In the case of refraction there was a lag of more than two centuries between the discovery by Snell in 1621 of the optical law and the demonstration by Soundhauss in 1852 of its acoustical counterpart. Here again the principle for acoustics resulted directly from the thought of analogies between light and sound.[27]

In another passage in the *Problemata* the reflection of light or vision is compared in its turn to a mechanical reflection in which the preservation of angles is perhaps even more palpable to common sense. In this connection Aristotle (or one of his students) asks, "Why is it that objects which are travelling along, when they come into collision with anything, rebound in a direction opposite to that in which they are naturally travelling, and at similar angles?", and in answering the question he points aptly to optical reflection as an illustration.

Now every object rebounds at similar angles, because it is travelling to the point to which it is carried by the impetus which was imparted by the person who threw it; and at that point it must be travelling at an acute angle or a right angle. Since then the repelling object stops the movement in a straight line, it stops alike the moving object and its impetus. As then in a mirror the image appears at the end of the line along which the sight travels, so the opposite occurs in moving objects, for they are repelled at an angle of the same magnitude as the angle at the apex (for it must be observed that both the angle and the impetus are changed), and in these circumstances it is clear that moving objects must rebound at similar angles.[28]

Here the author probably has in mind the Platonic doctrine of visual rays emanating from the eye. The visual ray proceeds to the mirror, at which point it meets the ray of light from the object. Both rays therefore travel toward the mirror, whereas in the case of a rebounding object one motion is toward the wall while the other is away from it. In this sense, then, the behavior in the mechanical situation is opposite to that in the optical reflection; but the definite implication is that the cases are alike in other respects, namely, in the rectilinearity of the motions and in the equality of the angles. The rectilinear propagation of light is sometimes ascribed to Euclid,[29] but the quotation above (as well as other passages which are even more explicit) [30] makes it clear that this phenomenon was taught at the Lyceum, if not still earlier at the Academy.

The above excerpts from Aristotelian works do not constitute explicit formulations of the optical law of reflection, but they do serve to make almost self-evident the inference that the equality of angles, at least in the case of plane mirrors, was familiar to the Aristotelian school. Moreover, doubt as to the authenticity of the *Problemata* does not essentially alter the situation inasmuch as the excerpts quoted are in close agreement with relevant passages cited from the more genuinely Aristotelian *Meteorologica* and *Analytica posteriora*.[31] In view of this, and unless further evidence to the contrary is presented, one may with reasonable assurance regard the time of Aristotle as a *terminus ante quem* for this discovery. How long the law had been known can only be conjectured. The passage from Plato cited above does not clearly establish such anterior knowledge. On the other hand, Aristotle touches upon the quantitative aspect of reflection in a manner so perversely adventitious as to imply that in his day the discovery was by no means a recent one.

This is in contrast to the Aristotelian treatment of the rainbow, where the ample explanation leaves the definite impression that here the author was contributing something new, in spite of his reference to statements on the rainbow by the "ancients."

Aristotle's explanation of the rainbow may be thought of as made up of four parts: (1) the physical agents involved; (2) the shape of the bow; (3) the size of the bow; and (4) the origin of the colors. In answering the first part, Aristotle called attention to what he regarded as the three fundamental elements: (a) a source of light, generally the sun; (b) a dark rain-cloud; and (c) the eye of an observer. He believed with his predecessors that the bow is formed when rays of sunlight strike the surface of the cloud and are reflected to the eye of an observer, or, in the Platonic language of extramission to which

Aristotle here conforms, when visual rays from the eye are reflected toward the sun by the cloud. (Geometrically it makes no difference which theory of vision is adopted, and Aristotle in other contexts adopted the quasi-modern theory that light is an impulse transmitted through a pellucid medium. Even today a remnant of the older extramission theory is apparent in the phrase, "Cast your eye upon it.") In finding the essence of the rainbow in the reflection of rays of light or of vision, Aristotle added nothing new; he simply lent the weight of his authority to a dogma which went unquestioned for another millennium or more. It was in his answer to the second and third parts that he showed his greatest originality. Earlier writers (including possibly Anaxagoras) seem to have made the plausible elementary assumption that the circularity of the bow is due to the sphericity of the reflecting cloud. Aristotle rightly rejected this, perhaps because it conflicted with even the most casual observations of cloud formations. He had resort instead to a geometrical argument which begged many questions, but which in at least one essential does not differ from the explanation of circularity given today. He realized that the circularity of the bow is due to some basic geometrical relationship between the positions of the sun, the cloud, and the observer. He began with an idea which is quite foreign to modern meteorology, although its analogue is seen in the "celestial sphere" in astronomy—the notion that the objects one sights can be regarded as lying on a hemisphere the center of which is at the eye of the observer and the base of which is the plane of the horizon (Fig. 5). That is, Aristotle's explanation assumes that the sun (or at least the points from which the rays are regarded as emanating) and a portion R of the cloud which causes the reflection are equidistant from the eye of the observer at O. This probably is not to be taken literally, for Aristotle himself assumed in other connections that the sun lies well beyond the moon, and that clouds very definitely are sublunar. Just how much reality is to be ascribed to this hypothetical sphere (which we shall hereafter refer to as the Aristotelian meteorological sphere) is never made quite clear. Perhaps Aristotle here was following the advice Plato had given to contemporary astronomers to seek a geometric construction which should "save the phenomena." In this sense, then, as a schematic geometrical representation accounting for the circularity of the bow, rather than as a physical theory, Aristotle's explanation is generally acceptable. For the first stage of his explanation, Aristotle assumed that the sun is on the horizon at S, and he drew the semicircle SRA (the upper half of that great circle of the meteorological sphere which passes through S and is perpendicular to the plane of the horizon). He drew also RC perpendicular to SA, and the lines SR and OR. By a principle of sufficient reason, there must be some uniform relationship among the relative positions of the points S, R, A, C, and O. But if the entire configuration made up of these points is

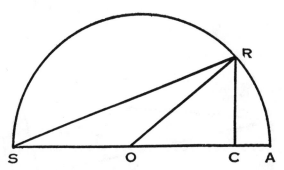

F<small>IG</small>. 5.—Aristotle's idea of the formation of the rainbow when the sun is on the horizon at S, the cloud at R, and the eye at O.

revolved about the line SA, this relationship remains invariant; and hence, if R is a point on the rainbow, every point through which R passes during the revolution of the diagram will also be a point on the bow.[32] Obviously, the line RC (extended) sweeps out, during the motion, the plane through C which is perpendicular to the horizontal line SA; and this plane cuts the meteorological sphere in a circle, the center of the circle or bow being at C, and the radius of the bow (or the greatest altitude of a point on the rainbow) being the distance between C and R. That only half of this circle on the meteorological sphere can be seen is due to the obvious fact that only the portion lying above the plane of the horizon is visible from O. During the rotation of the figure above, the line SR sweeps out a conical surface which may be designated the cone of solar rays, and the line OR generates a surface which may, for purposes of exposition, be called the cone of visual rays.

The Aristotelian explanation of the circularity of the rainbow is applicable also, with only slight modification, when the sun is above the horizon. As before, let O be the eye of the observer, S the position of the sun, R a point on the rainbow, let RC be perpendicular to the line SOCA, and let DE be the line in which the plane RSOCA cuts the horizon (Fig. 6). Then again, as the plane RSOCA rotates about the line SOCA, the relative positions of S, R, and O remain the same. Hence, "since the reflection takes place in the same way from every point," [33] every point on the path of rotation of R will lie on the rainbow, and this path is a circle on the meteorological sphere. Less than half of this circle will be visible to an observer at O, for more than half of the circle lies below the plane of the horizon. Aristotle was aware of the fact that the greater the elevation of the sun above the horizon, the smaller is the apparent altitude of the rainbow. This should have suggested to him that the angle ROC is independent of the solar elevation—that is, that the angular radius of the rainbow is constant. But here optical illusion led him

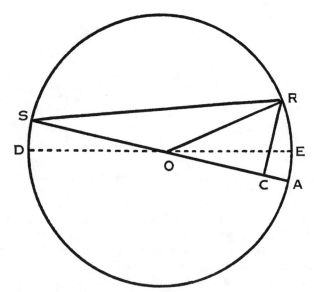

Fɪɢ. 6.—Aristotle's idea of the formation of the rainbow when the sun S is above the horizon DOE.

into one of the most persistent of all erroneous beliefs attaching themselves to the story of the rainbow. At sunrise and sunset the rainbow is a full semicircle; but it is half of a circle the radius of which Aristotle believed to be smaller than the radius of the circle of which the bow forms an arc when the sun is above the horizon. That is, he lent his authority to the popular belief that the higher the sun, the smaller is the arc of the rainbow but the larger is the radius of the circle of which it is a part.[34] Aristotle did, nevertheless, point out that there is a degree of solar elevation beyond which the formation of a rainbow is impossible; and this elevation is exceeded, for the latitude of Athens (38° N, or roughly the same as that of Washington, D.C.), toward midday during the summer season. Moreover, he did call attention to the important fact that the sun S, the eye of the observer O, and the center (or pole) of the rainbow C are collinear, an element in his theory which was not always kept sight of by his successors.

The Aristotelian explanation of the circularity of the bow does not differ essentially from that which would be acceptable today, granted the postulate that there is a unique point R on the azimuthal circle SRA which is appropriate for the formation of the bow. The soundness of the geometrical argument does credit to the Philosopher, the more so when one recalls that his interests lay closer to biology than to mathematics. Had his successors con-

tinued his work on the same high level, the story of the rainbow might not
have been such a tale of frustration as it was destined to be.

The third of the basic questions on the rainbow which Aristotle raised
concerned the size of the bow. Not all points on the circle SRA are illuminated with the brightness of the rainbow, but only a narrow portion about
the point R. What accounts for the unique property of this point? Rays from
the sun not only make up the surface of the cone of solar rays, described
above; they fill up the entire volume within and without the cone. Why are
not these other rays equally effective? Why is not the entire surface of the
meteorological sphere illuminated? The facile answer that one might have
expected, in the light of Aristotle's knowledge of the law of reflection, is that
only at R is the angle of incidence equal to the angle of reflection. But elementary geometry shows that this can not be the case if one thinks of the
meteorological sphere as the reflecting surface. The visual ray RO, being a
radius of the sphere, is perpendicular to the surface at R; the solar ray SR is
not a radius and hence can not be perpendicular to the sphere (unless, of
course, R coincides with A, in which case the bow would consist of one point
only). The reflecting surface, therefore, can not be the meteorological hemisphere itself, for the law of reflection for continuous surfaces would be violated. Aristotle assumed instead that the reflection is of the second type—it
takes place from tiny mirrors on the hemisphere which are so small that color
and not form is the result. And yet reflection does not take place at all points
on the cloud or sphere. There must be some law of uniformity involved—the
uniformity upon which the argument on circularity hinged. Why are only certain portions of the cloud functional in a reflection of the second type? If the
meteorological hemisphere consists of a myriad of tiny dispersively reflecting
surfaces or "Aristotelian mirrors," diversely oriented as in a rough wall or the
earth's atmosphere, then all points on the hemisphere should be effective in
reflection. In this case the rainbow should not be a band only, but a bright
inverted bowl similar to the blue vault of the sky. To answer this critical point,
Aristotle made a bold assumption. He held that the efficacy of the reflection
depends on the distance from the sun at which the reflection takes place.
More precisely, he assumes that, to be effective, the reflection must be such
that the solar ray SR bears to the visual ray RO a very special ratio. Following
through the implications of this postulate (which we shall call Aristotle's
second law of reflection), he showed how to determine the point R geometrically. Every schoolboy today knows that the locus of points equidistant from
the two points S and O is the perpendicular bisector of the segment SO; but
few know the more general locus of points R for which the ratio of the distances RS and RO is a constant K different from unity. This latter locus is
named for, and generally is attributed to, Apollonius, the Great Geometer;

yet Aristotle, who did not claim the discovery as his own, was familiar with the locus a century earlier. The locus of points R such that RS : RO = K is a circle with center on the line SO. Hence to find the point R on the meteorological great circle SA (Figs. 5, 6, and 7) satisfying the Aristotelian requirement, one need only find on the line SA the two points B and F such that

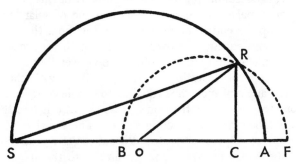

FIG. 7.—Aristotle's determination of the point R at which light producing the rainbow is reflected to the eye at O.

BS : BO = K and FS : FO = K and construct the Apollonian semicircle on the segment BF (Fig. 7). The desired point on the rainbow will be the point of intersection R of this latter semicircle with the meteorological semicircle on SA. Other points on the rainbow are then obtained through the rotation, explained above, of the entire configuration about the line SBOCAF. Aristotle's explanation thus "saves the phenomena."

Aristotle's ingeniously contrived explanation of the size of the rainbow is quite erroneous in the light of modern science. Nevertheless, it should be recognized that for an age in which astronomy was the only science (except for some rudimentary ideas in music and optics) to which mathematics had as yet been applied, this earliest geometrical meteorological theory was remarkably sophisticated. It was the very first quantitative attempt on what has turned out to be an extraordinarily difficult problem, and it had no successors in this respect for nearly two thousand years. This long interval will be seen to have been a period of intense interest and speculation on the rainbow, during which Aristotelian ideas on the agencies causing the bow were to be frequently questioned and occasionally improved. On the mathematical explanation of the magnitude of the rainbow arc, however, Aristotle had no peer until the sixteenth century.[35] On the other hand, the chief qualitative characteristics of his theory, with the emphasis on the meteorological sphere and the role of the cloud as a whole, were to hamper the study of the rainbow for many centuries to come.

The reference by Aristotle to "mirrors which are so small as to be indi-visible to sense" contains profound implications for the explanation of the rainbow. In the first place, it ruled out a speciously simple theory of the cir-cularity of the bow which nevertheless found wide acceptance in Greco-Roman circles—the idea that the circular form of the bow, being an image of the sun, was due to the sphericity of the sun. Then, too, Aristotle's statement makes it clear that the surface of the cloud is to be thought of as made up of a myriad of tiny little mirrors. What could be more natural than to think in terms of reflection by tiny droplets? Aristotle, however, did not hit upon the idea of making a geometrical study of what might happen to light which strikes a single drop of rain. He thought of the dewy cloud as a medium which, while not smooth like a mirror, was nevertheless continuous in its parts. It was not an aggregate of discrete droplets, but a unified mass, the component parts of which were cohesive and contiguous. One wonders to what extent this view may have been due to the intense Peripatetic opposition to the atomic theory of matter which Democritus had so warmly espoused over a century before.

The fourth of the Aristotelian questions on the rainbow, concerning the manner in which the colors are formed, has been the favorite topic from that day to this. Aristotle already had committed himself to the idea that the re-flection causing the bow was not of the image-forming type but of the sort which results in color. But why did the hues appear in such a variegated pat-tern, unlike the blue of the sky or the red of the setting sun? Knowing that when the sun sets, it often is viewed through mist and clouds, Aristotle be-lieved that the red coloration of the setting (or rising) sun results from a weakening of the light as it passes through the obstructing media—mist, smoke, or haze. In the same way rays which pass through water are weakened; but, the obstruction being greater than that of mist, the light is so unfeebled that it here takes on a greenish-blue tinge. In other words, the Aristotelian theory was derived from the basic assumption, derived from Anaximenes, that light and darkness, or white and black, produce through mixture the other colors—from red, which contains more light than darkness, to blue (or pur-ple), in which blackness predominates over whiteness.[36] The light and dark must, of course, be thoroughly intermingled so that the minute parts are indi-vidually indistinguishable, the resulting impression being a sort of compound of the two. The multiplicity of colors is the result of differences in the pro-portions of the component parts. As was the case in musical harmony, so also with respect to colors, those compounded in simple ratios of small whole numbers, such as purple and scarlet, are most pleasant. (This extension of the Pythagorean laws of music to the field of color is an early recognition of analo-gies between light and sound which later were to encourage the use of the wave theory of light and a new theory of the rainbow.) Hues corresponding to all conceivable proportions, commensurable or incommensurable, are possible; but Aristotle believed that there are three primary colors. Arranged from the strongest to the weakest—i.e., from that containing the largest proportion of white to that containing the smallest—these are red, green, and blue. Aristotle's trichromatic theory, an anticipation of modern color theories, was not entirely new. Aristophanes also had distinguished three colors in the rainbow —purple, green, and yellow [37]—possibly because the number three, even before the advent of Christianity, was a sacred number. Empedocles naturally preferred four basic colors, associating them with his four elements: white with fire, black with water, red with air, and yellow-green with earth. Democritus, too, accepted this quadrivium of colors; but he assumed that color was a subjective phenomenon depending on the shapes of atoms and on their arrangement, white being due to smooth round atoms, black to rough crooked ones. Plato in the Timaeus inclined to a tetrachromatic theory and suggested that color is produced by the reciprocal action between a fluid penetrating the eye and the fiery fluid of the eye; but his attitude toward the colors of the rainbow was more one of awe than of scientific curiosity. Metrodorus of Chios, according to Plutarch, held that the red of the rainbow is from the sun's rays, the blue from the cloud, the other shades being mixtures.[38] But the color theory which achieved the status of orthodoxy in ancient science was that of Aristotle. The arguments he cited in support of this view are, it must be admitted, far from convincing. The step-by-step change from white to black is described as follows:

When sight is relatively strong the change is to red; the next stage is green, and a further degree of weakness gives violet. No further change is visible, but three completes the series of colors (as we find three does in most things).[39]

In another place one finds listed among the things "three completes" the number of dimensions in geometry.[40] The number three in Greek days often was regarded as the first true number, and Aristotle points significantly to the fact that it is the smallest of the number of objects to which one applies the term all. Even in English one never says "all two"; and in Greek grammar there are three "numbers": singular, double, and plural. Another argument for the primacy of red, green, and violet-blue is Aristotle's belief that these are the three colors which painters can not create by mixing.[41] Aristotle suggests that other colors result from the juxtaposition of primary colors through contrast, just as "black beside black makes that which is in some degree white look quite white." [42]

Aristotle's application of his color-theory to the rainbow led to a dogma which dominated thought for many centuries. "If the principles we laid down about the appearance of colors are true," he wrote, "the rainbow necessarily has three colors, and these three and no others." So persistent has been the belief in a tricolored rainbow that in a popular French song of the nineteenth century the bow was called "rainbow of liberty." It should be noted, however, that Aristotle did not deny the fact that there appear to be other hues in the rainbow:

The appearance of yellow is due to contrast, for the red is whitened by its juxtaposition with green. We can see this from the fact that the rainbow is purest when the cloud is blackest; and then the red shows most yellow. So the whole of the red shows white by contrast with the blackness of the cloud around: for it is white compared to the cloud and the green.[43]

But the rainbow is not a fortuitous congeries of varied hues; the colors appear always in the same clearly discernible arrangement. The red band is always outermost, the blue innermost, and the green betwixt the other two. Such uniformity indicated the operation of a scientific principle, and this Aristotle was perhaps the first to seek to explain.

Color, according to Aristotle, is caused by weakening of either light or sight, and such a weakening can result (1) from a reflection or refraction of light, (2) from the distance through which light or sight is transmitted, or (3) from weakness of vision. It was known, for example, that persons with weak or moist eyes see about lamplights ringed bands of color, for light which is weakened becomes tinged as though mixed with darkness. The colors of the rainbow, on the other hand, are caused by the reflection of light or sight and by the weakening of rays with distance of transmission. In the explanation above of the size of the rainbow, it was assumed that reflection occurs at a unique point R; but in reality the reflection is effective for a small interval on the meteorological sphere above and below this point. Above R the ratio SR:OR is somewhat smaller, and below R the ratio is slightly greater. Aristotle, therefore, argued that the light reflected from points just above R is stronger (due to the relative shortness of the ray from the sun), that from points just beneath R is weaker (the ray from the sun is longer), than the light from R itself. The invariable order of the rainbow colors, therefore, is from red above to blue below, with green in the middle. This type of argument he applied in a comparison of the formation of the rainbow with that of the halo. The halo he likewise attributed to reflection; but in this case the reflecting medium is placed between the eye and the sun, whereas in the case of the rainbow the eye is between the sun and the reflecting cloud. In the case of the halo, therefore, the rays travel in a path which is more direct and shorter than that for the rainbow. The rays causing the halo, therefore, are less weakened and hence are not so deeply colored as those producing the rainbow. Moreover, the opacity of the reflector is also a factor, and rays reflected from the moist blackness of the cloud are weaker than those reflected from the lighter medium causing the halo about the sun.[44]

The treatises of the Aristotelian corpus were not necessarily all actually written by the master himself, although they certainly represent his ideas. In spite of an inherent consistency within the whole, there are discrepancies here and there; and one of these concerns the nature of light. Before the days of Aristotle there were several doctrines concerning light and vision. One, sometimes ascribed to the Pythagoreans, sometimes to the atomists, held that luminous objects emit particles which reach the eye and hence stimulate vision. Another, often associated with the name of Euclid but in reality of earlier provenance, argued that the eye emitted visual or exploratory rays, like tentacles, which travelled outward to reach the object perceived. (This theory had the greatest vitality in antiquity.) A reconciliatory hypothesis, sometimes attributed to Plato, compromised by assuming that both eye and object emitted rays the conjunction of which produced vision. Aristotle's view appears far more modern (except for his belief in the instantaneity of light), for he introduced the postulate that between the eye and the object there exists a pellucid medium in which light impulses are transmitted from the object to the eye.[45] Nevertheless, in some of the Aristotelian treatises the language of visual rays continued to be used, possibly in deference to views current at the time. It should be noted that the Aristotelian geometry of the rainbow is independent of the doctrine of vision in terms of which it is expressed.

A certain degree of inconsistency creeps into Aristotle's exposition of the rainbow when he explains the double form of the bow. It is customary for one to speak of *the* rainbow in the singular; yet, as most people are aware, frequently two rainbows appear. The inner or brighter bow is known as the primary rainbow, the outer and less bright is called the secondary.[46] Who first noticed the latter is not known, but the first recorded account of it is that given by Aristotle in his treatise *On Meteorology*. This secondary or exterior rainbow Aristotle believed to be caused "in the same manner" as the primary, through the reflection of rays by a cloud at a point above that at which the rays producing the primary arc are reflected. None of the original Aristotelian diagrams for the primary bow are extant, and those appearing in modern treatises are reconstructions by medieval commentators. For the secondary bow we do not have even such reconstructed diagrams, and so the precise meaning of Aristotle can not be determined. His principle of geometric regularity in the second type of reflection would be violated unless one assumes that for the proportion $SR:OR = K$, encountered above, there are at least two values of K for which the reflection is effective, so that there are two points R and R' (Fig. 8) on the great circle SA of the meteorological sphere at which appropriate reflections take place, one producing the primary bow, the other the secondary. So far there is no inconsistency with earlier ideas in such an *ad hoc* explanation (although one can well raise the pertinent ques-

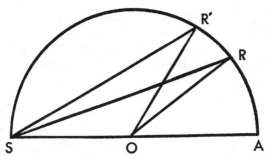

Fig. 8.—Aristotle believed that the primary and secondary rainbows are formed in the same way.

tion, Why can K take on only two values?). But observation discloses that the bow at R′ is much paler than that at R, even though the ratio SR′:R′O, which determines the weakening of the rays, is smaller than the ratio SR:RO (Fig. 7). Moreover, one cannot place R′ farther from O than R without upsetting the entire Aristotelian foundation, for the distance SO, from the sun to the observer, must be the same for both bows. And yet Aristotle argues that "the reflection from the outer rainbow is weaker because it takes place from a greater distance and less of it reaches the sun, and so the colors seen are fainter." [47]

Another more serious difficulty arises in connection with the ordering of the colors. As the ratio SR′:OR′ decreases, the rays are presumed to become stronger, and yet in the secondary rainbow the colors are arranged in just the opposite order: the red band is innermost, the blue on the outside, and the green falls in between. Aristotle therefore abandons, for both the primary and the secondary bow, the color argument given above. Notwithstanding his assertion that the bows are formed in the same manner, Aristotle adduced two entirely different explanations for the color arrangements. For the primary bow he makes the dominant factor the total area on the meteorological sphere illuminated by the rays of sight. The radius of the outermost of the three colored bands being greater than that of the innermost, the area of the outer semicircular ring is greater than that of the inner ring. Color being a function of the sum total of the reflected sight, the outer band naturally should be brighter than the inner; and this accounts for the fact that the former is red, the latter is blue. This argument can not of course be applied to the secondary bow, for the colors are arranged in an order contrary to that in the case of the primary. Aristotle has recourse to an entirely different line of reasoning, one which dispenses with the meteorological sphere and substitutes for it a vertical plane containing the bow. Let B, G, and R be the highest points, respectively, of the blue, green, and red bands in the secondary bow, arranged on a vertical line; and let S be the sun and O the observer (Fig. 9). Inasmuch as the ray SBO is longer and is reflected more obliquely than the ray SRO, the former appears blue, the latter red, with green intermediate.[48] Thus the difficulty presented by the inverted ordering of the colors is resolved; but Aristotle,

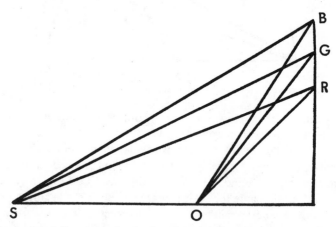

Fig. 9.—An Aristotelian explanation for the ordering of the colors in the secondary rainbow.

who believed that nature does everything in the simplest and best way, could scarcely have taken much pride in such tortuous reasoning. (Incidentally, this argument again shows that reflection in the case of the rainbow is of the second type, for the optical law of reflection is violated.)

Aristotle referred to the secondary rainbow as though its existence were common knowledge. What could be more natural, then, than to inquire whether a third bow is possible. Posing this problem in the *Meteorologica*, Aristotle answered in the negative.

Three rainbows or more are not found because even the second is fainter, so that the third reflection can have no strength whatever and can not reach the sun at all.[49]

Presumably the third bow would have to be formed in the same manner as for the other two; but the reflection would take place at so great an altitude, and with such an obliquity, that the rays would be so weakened as to be ineffectual. It was natural, therefore, for Aristotle's successors to seek the elusive tertiary rainbow in the region of the sky above the secondary bow.

From the arguments on the ordering of the colors in the first two bows it is impossible to predict what arrangement of colors one should anticipate in the hypothetical tertiary rainbow; and Aristotle does not commit himself on this point.

That the rainbow often consists of a double arc must have been common knowledge in ancient Greece; but the possibility of a lunar bow seems not to have been generally accepted.[50] Aristotle, in the *Meteorologica*, wrote:

> The rainbow is seen by day, and it was formerly thought that it never appeared by night as a moon rainbow. This opinion was due to the rarity of the occurrence: it was not observed, for though it does happen it does so rarely. The reason is that the colors are not so easy to see in the dark and that many other conditions must coincide, and all that in a single day in the month. For if there is to be one it must be at full moon, and then, as the moon is either rising or setting. So we have only met with two instances of a moon rainbow in more than fifty years.[51]

The moon rainbow Aristotle believed to afford an apt instance of his theory of contrast in color production. The lunar bow appears to be quite white, and this he explained as the result of contrast with the blackness of the dark night cloud upon which it appears.[52] His last sentence in the passage above was twisted by later careless transcribers into an assertion that not more than two lunar bows within a given fifty-year interval are possible. As a matter of fact, one is fortunate indeed if he sees as many as this, for factors beyond those given by Aristotle make such observations rare. For one thing, the large majority of local showers, over land areas where most people live, occur in the summer afternoon when the moon has no chance to produce a bow. Then, too, the full moon would appear to the east of the observer during the early hours of the evening, whereas the prevailing winds in many regions are from the west and hence make the eastern sky too cloudy for a bow. When one realizes that lunar rainbows occur about a hundredth as frequently as solar bows, and if one recollects that fully three-fourths of these will be missed through sleep, he will better appreciate the rarity of observations of lunar bows.[53]

In Aristotle's day the theory of the rainbow was not far removed in time from the mythological stage, and so the Philosopher took the trouble to refute or explain some popular superstitions related to it. For example, Pliny and Plutarch, long afterward, reported that priests preferred, in connection with offerings, wood on which the rainbow had touched while it was damp, imagining that it had a more agreeable odor than other wood. Yet Aristotle had held this to be contrary to the fact that the bow is only an affection of sight—it does not really exist objectively in nature. The scent supposedly associated with the occurrence of the rainbow he explained as due perhaps to the warm dampness experienced during a light shower. The fact that a rainbow may be accompanied by very inconsiderable rain seems to be responsible for the mistaken ascription of the scent to the bow instead of the dampness.

The Aristotelian theory of the rainbow is, of course, today untenable. The most serious deficiency undoubtedly is the ascription of the bow to reflection alone, with no role accorded to the essential phenomenon of refraction. The word refraction was used in his works as synonymous with reflection; but the context shows that it was reflection which he had in mind. Then, too, the Aristotelian distinction between two types of reflection, while quite understandable in the light of his physics, was unfortunate insofar as the rainbow theory is concerned. A characteristic of his explanation which was perhaps even more obstructive was the macrocosmic approach—the concentration of attention on the cloud and the meteorological sphere, rather than on the "little mirrors" in the cloud, where the key to the problem was in the end to be found. There is also, in Aristotle's ingenious theory, the utter incongruity in equating the distances from the observer's eye of the sun and the cloud. Finally, one misses in his account any mensurational element, for Aristotle neither suggested any numerical value for the critical ratio SR:RO (Fig. 5), nor did he estimate the angular radius of the bow from which this ratio can be deduced. The trigonometric law of cosines was known in geometric form to Euclid, and it is probable that Aristotle was familiar with it. If θ is the angular radius of the rainbow, (i.e., the angle ROA in figure 5), the law of cosines leads immediately to the result SR:RO $= \sqrt{2 + 2 \cos \theta}$. Conversely, from the postulation of a value for the ratio on the left side of this equation, the angle of the rainbow could have been inferred. But then Aristotle was a product of the Platonic Academy, where the motto had been "God ever geometrizes," rather than the modern counterpart, "To measure is to know," and Aristotle's geometry, even without measurement, was more sophisticated than that of any successor for well over a millenium. Moreover, his work includes at least two very significant advances in theory which became a permanent part of our scientific heritage on the rainbow. The first of these was the idea that the size of the rainbow could be explained geometrically in terms of the relative positions of the sun, the rain cloud, and the eye of the observer. Another was the correct explanation of the semicircular form of the bow—the observation that the triangle SRO is unchanged if rotated about the side SO.

Whatever may be one's judgment of the place of Aristotle in the history of science, criticism must be in terms of the status of scientific knowledge at that time. The rainbow problem concerns one of the most elusive portions of science; and when one compares the idiosyncrasies of the atmosphere with the regularity of the heavens, it is easier to appreciate why the rainbow appeared so enigmatical to the ancients. In view of the fact that Aristotle placed the rainbow not in optics, but in meteorology, along with hydrology, seismology, geology, and other portions of natural philosophy, it is greatly to his credit that he gave so thoroughly mathematical a treatment of the bow. Among many of his successors this mathematical element was omitted altogether.

Mathematics and astronomy, the oldest of sciences, had advanced rapidly during the days when Athens became the center of "the glory that was Greece." There seems also to have been an incipient science of optics, although the earliest treatise which survives is one attributed to Euclid, a younger contemporary of Aristotle. In any case, the study of optics would have included only very elementary propositions involving the rectilinear propagation of light, such as occur in the theory of perspective. Later this was supplemented by a division, known as catoptrics, devoted to the reflection of light by plane and curvilinear mirrors; and ultimately a third branch, dioptrics (sometimes also mesoptrics or anaclastic or diaclastic), covered the phenomena of refraction. Sometimes still another division, scenography (the study of optical illusions and perspective drawing) was recognized.

Aristotle undoubtedly was acquainted with the colors formed when sunlight passes through a glass prism, but he seems not to have associated these with the rainbow. Among the many works attributed to him is one entitled De Coloribus, but this contains nothing further on the rainbow. He was the author also of a long treatise (in eight "books" or chapters) called De Physica, but this is far removed from what is known today as physics. The Greek word physis was more akin to the Latin phrase de rerum natura or to the modern word "essence;" and hence Aristotle's Physica is concerned with ontology rather than with the mathematical laws governing natural phenomena. Aristotle was, in fact, primarily a philosopher and biologist; and hence it is all the more surprising that the first mathematical theory of the rainbow should have come from him. The surprise deepens into admiration when one realizes that no superior explanation was proposed for a period of more than fifteen hundred years. Archimedes (287–212 b.c.), the greatest mathematical scientist of antiquity, was especially interested in optical phenomena; yet, so far as one knows, he left the problem of the rainbow quite untouched. Here one sees the sharp difference in approach of the two outstanding scientists of ancient times. Aristotle gave answers—often times rough-and-ready, occasionally more sophisticated—to all questions that turned up; and hence many of his answers have not stood the test of time. Archimedes concentrated attention upon a few aspects of mechanics and optics, and his treatises are as impeccable today as when they were written.

The most celebrated student of Aristotle was, unquestionably, Alexander the Great; and while the master was busily engaged in building an intellectual empire, the pupil was rapidly acquiring the territorial kingdoms of the world. But how short-lived was the temporal power of the latter in comparison with the cultural hegemony of the former! Within a few years of his death the empire of Alexander crumbled. The territory was divided among his generals and, disorganized and distrustful of each other, the Greek states succumbed to the relentless expansion of Rome. In 146 B.C. Corinth, as well as Carthage, was utterly destroyed, and Rome became master of western civiliza-, tion. But, chiefly through the influence of Alexander's teacher, it may be said that "captured Greece took captive her captor, Rome." Politically the Mediterranean had become Roman; intellectually it remained Hellenic. When Aristotle died, one year after Alexander, his school, the Lyceum, continued as a rival of the Academy; but meanwhile there had grown up two other philosophical schools, the Stoic and Epicurean. From these four intellectual groups in Greece came the men who taught the Roman leaders, and hence Athens remained the cultural center of the western world until in 529 the schools were closed by the emperor Justinian and the scholars dispersed. Hence the period from 146 B.C. to 529 A.D. often is referred to as the Greco-Roman age. During this time there were outstanding developments in astronomy and mathematics, but physical science failed to make much progress. Theophrastus, the successor of Aristotle as head of the Lyceum, had continued his master's work in meteorology; but his interests were directed especially toward terrestrial meteors, and his speculations on the origin of metals were of more concern to alchemy than to what now is called meteorology. In any case his works have largely been lost. It would appear that Theophrastus, and after him Posidonius, the Stoic, sought to give meteorology a new name, "metarsiology," because of some terminological confusion which continued to subsume sidereal phenomena under the traditional name. This alias survived in commentaries as late as the sixth century; [54] but even the strong influence on classical culture exerted by Posidonius could not carry the field against the entrenched terminology of Aristotle.

It is curious that the later Greek physicists, even those within the Lyceum itself, paid little attention to the Aristotelian geometrical theory of the rainbow. The little that we know of their ideas, and also of the fragments on pre-Aristotelian views, comes largely from a scholar far removed in time—from Aetius, a physician to the Byzantine court under Justinian. Some of the physicists are reported to have rejected Aristotle's view that the bow is a composite of reflections taking place in the individual drops. The rapidity with which it vanishes they felt suggested that the reflection takes place in the cloud itself, and that a displacement in the particular cloud formation involved causes it to lose its efficacy in mirroring the form of the sun. That the image of the sun is distorted should occasion no surprise, for a mirror often produces distortion, especially when, as in the case of the cloud, it is not clear and serene.[55] There were thus, in the centuries just before our era, three chief theories to account for the circularity of the bow: most ascribed this to the shape of the cloud; some to the form of the sun; and a few (among whom Aristotle is of course to be counted) to the relative positions of the sun, the cloud, and the spectator.

Among the later Greek physicists one must include Posidonius of Rhodes (c. 130–50 B.C.), a philosopher of the Stoic school who nevertheless was strongly influenced by Plato and Aristotle. In science he is known for his determination of the relative distances of the sun and moon (superior to those of Aristarchus, Hipparchus, and Ptolemy) as well as for his measurement of the earth (which probably was inferior to that of Eratosthenes); but it is known that he wrote also on meteorology. Unfortunately, all of his works have been lost; but they served as a mine of information for later writers, and hence fragments of his ideas have survived. It appears that he disagreed with Aristotle that there are only three colors in the bow and that he held the colors of the bow to be illusory only, having no reality. Judging from reports which sometimes are mutually contradictory, Posidonius accepted the bow as a composite reflection from a concave cloud as from a multitude of tiny mirrors; but just how the individual drops were related to the total image is not clear,[56] for it is difficult to reconstruct his thought. (He is also one of the most controverted figures of antiquity in philosophy.) Even today such a question cannot be lightly brushed aside, for the interaction of light with the atoms making up the surface of a smooth mirror is very complicated. Did Posidonius consider the geometry of the bow with respect to the cloud as a whole or in terms of each drop? It is tantalizing not to know. Two centuries after he died, Posidonius was praised by Galen, physician to Marcus Aurelius, as the greatest scientist of his time by virtue of his knowledge of geometry; yet his comments on the rainbow have been reported on by men who were not themselves geometers, and all traces of any mathematical treatment which he may have given have vanished. It is known that of the three traditional divisions, optics, catoptrics, and dioptrics, Posidonius placed the study of the rainbow in the second. The atomistic approach seems to have been stronger in his case than it was in Aristotle, leading to increased emphasis on reflections in little drops; but since he postulates a cloud in the form of a sphere, any geometry he may have had in mind probably concerned the spherical cloud rather than the sphericity of the drops. Betraying Platonic leanings, he preferred a four-color theory, holding that, under the influence of the sun on the cloud, there were two types of color in the rainbow, the blue-green and the yellow-red.[57]

The background of Posidonius, who was born at Apamea in Syria and probably died at Rome, was thoroughly Greek, but his ideas were freely carried over into the Latin world. He was one of those who did most to spread Stoicism among the Romans; he was a teacher of Pompey, and Cicero was numbered among his friends and admirers. Although his works are now lost, it is evident that they were known at Rome for many centuries. In particular his work on meteorology was familiar to Seneca the Younger (3 B.C.– 65 A.D.), one of the earliest Latin scholars to write on the rainbow. Seneca had been appointed tutor to the eleven-year-old Nero and served as confidential adviser to Agrippina, his mother. Seneca became enormously wealthy and rose in political influence until in the year 57 he became consul; but the emperor Nero came to dislike and suspect him, and finally compelled him to commit suicide. The numerous writings of Seneca are mostly of a literary or moral nature, but he composed also a scientific work, the *Quaestiones Naturales*, which enjoyed a considerable vogue during the earlier Middle Ages.

The *Quaestiones* is a popular treatise on geology, meteorology, and astronomy which contains some shrewd passages of significant value. In one of these Seneca espoused, quite contrary to the doctrine of Aristotle, the idea that comets resemble planets in that they revolve in fixed orbits. His long discussion of the rainbow includes nothing of comparable value, but it is significant as an indication of the views on the subject current in his day. In fact, one can not tell to what extent his views are original with him, for virtually all material on the rainbow in antiquity has been lost, except for the works of Aristotle and Seneca. The ideas of Seneca can not be regarded as superior to those of Aristotle, but they show that Peripatetic theories were not accepted without question. Aristotle had held that there were only three true rainbow colors, other putative hues being due to optical or physiological illusion; but Seneca asserted that there were indefinitely many colors—a thousand diverse hues blending into each other. In the *Aeneid* the bow similarly is described as an arc of a thousand colors.

> As when the rainbow, opposite the sun
> A thousand intermingled colors throws.
>
> With saffron wings then dewy Iris flies
> Through heaven's expanse, a thousand varied dyes
> Extracting from the sun, opposed in place.[58]

Seneca believed the varied colors are due to the peculiarities of clouds. Some parts are swollen, others hollow; some are too dense to transmit sunlight, others are too rare to exclude it. The differences in consistency cause alternations of light and shade, producing color. Thus Seneca was confident that the variegation of color is due to a combination of the brightness of the sun with the dullness of the moist cloud. As to the reality of the colors, however, Seneca unhesitatingly sided with the Peripatetics and firmly rejected the belief of Posidonius, the Stoic, that they are illusory.

With respect to the shape of the bow Seneca admitted that there is great uncertainty as to the causes. He accepted the basic idea that the rainbow is a reflection of sunlight by a cloud, but where Aristotle had categorically rejected the roundness of the bow as resulting from the sphericity of the sun, Seneca toys with this ingenuous view and suggests that the circle of the rainbow is much larger, in apparent size, than that of the sun for the same reason that images in a sphere of water appear enlarged. He suggested also a totally different explanation—that the circularity of the bow (as well as of the halo) is due to motion communicated to the atmosphere (and the vapors it contains) by the impulse of the light. The effect of such a motion must necessarily be circular, as in the case of waves spreading out from the place where a stone is thrown into a still lake.[59]

Seneca seems to be unable to make up his mind whether the cloud serves as one large spherical reflector or whether each little spherical raindrop mirrors the form of the sun. He is inclined to reject the latter because the drops fall too quickly to be able to snatch up and reflect the form of the sun; but he feels nevertheless that the total picture of the rainbow is the composite reflection from many component parts. Mathematical proofs, Seneca believed, leave no doubt that the bow is an image of the sun, but one that does not resemble it. After all, objects are not always faithfully represented in mirrors. Seneca compared the prismatic spectral colors with those of the rainbow (as Nero did with those on a dove's neck), but he could make little of this observation. Analogy with a prism suggested that rays must strike the cloud at a proper angle to produce color, "just as Tyrian purple must be held high to reveal its full blaze."

Seneca agreed with Aristotle that in winter a bow can be formed at any hour of the day, but in summer it appears only early or late in the day. This he explained as due to the fact that at midday the great heat of the sun dispels the clouds. Also, being too high, the sun looks down too directly on the earth to encounter clouds at the correct angle of obliquity. Extending Virgil's idea of Iris as a harbinger of storms, Seneca associated the rainbow with weather prognostications: If a rainbow appears toward the north (i.e., at noontime), a heavy rainfall is forecast, for the force of the moisture cannot be mastered by the strongest midday sun; if it is in the west (in the morning), there will be only dew or light rain; if it appears in the east (in the evening), this portends fine weather. This formula for the bow as a forecaster goes back to days before Seneca, being ascribed to Aristotle, and it was frequently repeated throughout succeeding ages. Even to this day the role of the rainbow in weather prediction has remained a very popular speculation.

It will be admitted that Seneca added little of permanent value in the theory of the rainbow. His chief contribution lay in his popularization of the rainbow problem and in his emphasis upon the role of the individual raindrops or "mirrors." The most tantalizing aspect of Seneca's explanation is his frequent reference to the possibility of geometrical demonstration, although he himself made no effort to supply this. The practical Romans were ever poor mathematicians, and one looks to them in vain for any improvement over the Aristotelian geometrical theory of the rainbow. Aristotle was interested in man as a creature who by nature desires to know; the Romans were interested in man for what they could get from him by force or persuasion—through the Roman legions or Latin oratory. As the late philosopher A. N. Whitehead put it:

The death of Archimedes at the hands of a Roman soldier is symbolical of a world change of the first magnitude. . . . No Roman lost his life because he was absorbed in the contemplation of a mathematical diagram.

And yet there was one Roman, Pliny (23–79 A.D.), who died through scientific curiosity, for he was overcome at the eruption of Vesuvius. His voluminous *Historia Naturalis* is a vast repository of fact and fable, but it contributed next to nothing to the story of the rainbow. The little one finds is negative, for Pliny denied, against the authority of Aristotle and Seneca, the possibility of lunar rainbows. Lucretius also is a promising Latin source, and he too is a disappointment on the rainbow. The famous didactic poem *De Rerum Natura* includes only the simple statement that the rainbow is caused by rays of the sun shining on rain or heavy aqueous vapor. On the matter of color he wrapped the atomic views of Democritus and Epicurus in a poetic mantle, adding nothing really new.

The Romans referred to the Mediterranean Sea as *Mare Nostrum* (our sea); but it was only in a political sense that the borders of the Mediterranean were Latin. When Caesar took possession of Alexandria, he did not interrupt the tradition of Greek culture. The city no longer boasted a Euclid or an Archimedes; but the spark of science which remained was sufficient to inspire during the second century of the Christian era a resurgence which sometimes is called the late, or second, Alexandrian Age. Among the subjects attracting greatest attention were mathematics and astronomy, and the best-known figure is that of Ptolemy, author of the *Almagest*. Ptolemy wrote treatises on other subjects also, including astrology and optics; and it was in the latter of these that he presented his studies in refraction, later to play a fundamental role in the theory of the rainbow.

Various effects of refraction must have been noted quite early in the history of science. There is a poem, supposedly Orphic and dating back to about 780 B.C. in which the sacred fire is lit by concentrating rays of the sun with the aid of a crystal.[60] An irreverent reference is made to refraction in Aristophanes' *Clouds*, where an old stupid fellow tells Socrates that he had found out an ingenious contrivance against paying his debts. When an attorney brought written action against him, he said he would take a fine transparent stone at the apothecary, with which they kindle fire; and, standing at a distance, he would melt down the wax letters in the record. (Despite the facetiousness of this passage, the greater permanence of Babylonian cuneiform tablets must indeed have been a boon to ancient creditors, as well as to modern archeologists.) Archimedes and Euclid may have been familiar with the penny-in-a-vase experiment (in which an object at the bottom of an empty vessel, just concealed from view by the edge, becomes visible when water is poured into the vessel), for it appears in works sometimes ascribed to them. It certainly was known to Cleomedes (c. 40 B.C.), who suggested that in the same way, through atmospheric refraction, the sun may be visible (and hence able to produce a rainbow), when it is actually somewhat below the horizon.[61] Cleomedes, a disciple of Posidonius, called attention also to some of the elementary qualitative properties of refraction—the reversibility of the ray, and the bending of the ray toward the perpendicular when it passes from the less dense to the more dense medium.

Ptolemy (died after 161), following the suggestion of Cleomedes, made allowances in his work for astronomical refraction (by which the length of the day is increased some five or ten minutes); and, of more importance for the rainbow, he left, in his *Optics*, the earliest surviving table of angles of refraction from air to water. Beginning with an angle of incidence of 10°, and proceeding by intervals of 10° to 80°, the angles of refraction are given in order as 8°, 15½°, 22½°, 29°, 35°, 40½°, 45½°, 50°. This table, quoted and requoted until modern times, has been admired as "the most remarkable experimental research of antiquity."[62] A closer glance at it, however, suggests that there was less experimentation involved in it than originally was thought, for the values of the angles of refraction form an arithmetic progression of second order—that is, their differences form the progression 7½°, 7°, 6½°, 6°, 5½°, 5°, 4½°. As in other portions of Greek science, confidence in mathematics was here greater than that in the evidence of the senses, although the value corresponding to 60° agrees remarkably well with experience.[63]

The *Optics* of Ptolemy has come down to us in a version from which Book I and the end of Book V are missing. The loss of Book I is especially

to be regretted, for it appears that ideas Ptolemy expressed on colors and the
rainbow may well have formed part of it. The earlier Greeks did not distinguish adequately between light and color; but Ptolemy had made a beginning, so later commentators inform us, in the psychology of color sensation.
Although he apparently accepted Aristotle on the fundamental role of black
and white, he seems to have considered colors as having objective reality,
independent of the geometric cone of visual rays. On the number of colors
in the rainbow he also broke with Aristotle, for it was later reported that
Ptolemy (perhaps in Book I of the Optics) placed the number of colors in
the bow at seven.[64] The eclectic nature of his work rouses one's curiosity about
his notions on the fundamental causes of the rainbow. One wonders in particular whether he may not have attributed the bow to refraction; but the
silence of his contemporaries and successors in this connection makes this
appear unlikely. The theory of the rainbow had not been rescued from the
doldrums in which it was caught by philosophical indecision as to whether
the chief agent was a minute drop or a massive cloud. The subject seems to
have been shunned by writers on optics, and as a result it fell into the rut of
an unprogressive peripateticism which failed to appreciate the mathematical
approach of Aristotle himself. One of the signs of the decline in scholarship
which set in during the Greco-Roman period was the tendency to substitute
commentaries for the original treatises of the classical authors. More and
more the Aristotelian theory of the rainbow came to be known at second
or third hand, for certain of the commentators achieved a popularity rivaling
that of The Philosopher. Chief of these in the ancient world was Alexander
of Aphrodisias, the Athenian philosopher who became the head of the Lyceum
between 198 and 211.[65]

Neoplatonic and Stoic tendencies had had no significant effect on the
theory of the rainbow, but at least they afforded some basis for the discussion
of rival views. Alexander, the greatest of Aristotelian commentators, wished
to purge the school of rival influences, with the result that the orthodox
theory of the rainbow became more firmly established. Explanations of the
rainbow all seem to have been in agreement that it is caused by reflection;
differences of opinion arose on more specialized questions. One error which
Alexander found it necessary to refute was the idea, suggested no doubt by
the contrariety in the ordering of the colors, that the secondary rainbow is a
mirror-image of the primary. Nor is the impossibility of the third bow to be
ascribed, he argued, to the fact that it would have to be an image of the
secondary—that is, an image of an image. The paleness of the exterior arc, in
comparison with the interior, and the non-appearance of a third bow are
both due to the same cause—the greater distance at which the reflection takes
place.[66]

The nature of Alexander's commentary is perhaps better illustrated by his assertion that the figure of the bow has no substance, being apparent only; and yet he argued that the colors are real. The reality of the colors in the rainbow was to be one of the most popular of all the questions argued sophistically during the later medieval period.

Alexander gives a generally accurate interpretation of his master's theory, including some of the geometrical reasoning. The diagrams of the original Aristotelian *Meteorologica* have been lost, and so it is to the account of Alexander that we are indebted for the appropriate figures. In connection with a diagram resembling figure 5 above, he proves that, given a value K of the Aristotelian critical effective ratio, there is only one point on the circle SRA for which SR:RO = K. Another figure showing how the halo, like the rainbow, is generated by reflection (according to Aristotle), is reproduced below (Fig. 10) to show how vague the peripatetic meteorological ideas really were. The sun is at b, the eye at a, with the caliginous air producing the

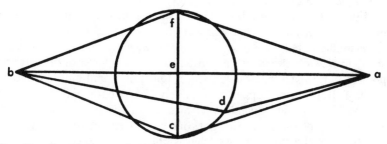

Fig. 10.—Alexander's diagram illustrating the formation of the halo. This was copied and recopied by later Aristotelian commentators.

halo between them. The circle cdf, with center at e, represents the locus of points in the medium at which the solar rays are reflected to the eye. The "proof" that this locus is a circle runs as follows. From the "equality" of the reflection at the points c, d, and f it follows that angles acb, afb, and adb are equal to each other, as are also the lines cb, db, and fb, as well as ac, ad, and af. Hence the triangles acb, adb, and afb are congruent, and the points c, d, and f are equidistant from e, the center of a circle (whose plane is presumably perpendicular to the line ab). This argument is essentially that of Aristotle himself, and Alexander "clarifies" it in his commentary by saying that the reflection involved is made at "equal angles." This presumably is not to be taken to mean that the angle of incidence is equal to the angle of reflection, but only that the reflection takes place in the same way at all points similarly placed with respect to the eye and the sun. The reflection is made from the "small parts" in the medium; but it is not clear whether or not these constitute the surface of a continuous medium analogous to a cloud, or if the rays traverse the medium without refraction. Nor is the law of uniformity for the reflection expressed, although, by analogy with the rainbow, one can assume the equality of the ratios bc:ca = bd:da = bf:fa = k.

The whole elaborate Aristotelian-Alexandrian explanation really boils down to the assertion that the halo is circular because of the uniformities of optics. To this extent the explanation of circularity is correct; but the two assumptions, (1) that the halo is due to reflection and (2) that it is produced in the same manner as the rainbow, both are thoroughly erroneous. The geometry for the halo would appear, on the surface, to be simpler than that for the rainbow; but in the end it was the rainbow, rather than the halo, for which a satisfactory solution was first found. Meanwhile, the Alexandrian diagram for the halo recurred with monotonous regularity in connection with later meteorological commentaries until the seventeenth century, when an adequate account became possible for the first time.

Alexander, a thoroughly loyal Peripatetic, did not seriously question Aristotle's theory of the rainbow, but he put his finger on a particularly vulnerable point. In the assertion that the secondary bow is formed in the same way as the primary, he noted a difficulty—one which we shall call the "Aphrodisian paradox" in view of Alexander's attention to it. In the lower or primary band, he pointed out, the strongest color, red, is uppermost, and blue, the weakest, is on the bottom; in the upper or secondary band, red is nethermost, while blue is on top. By the principle of continuity in the increase and decrease in the intensity of the reflected light, one should expect that the region of the sky between the two rainbows (that is, the space between the two strong red arcs) would be exceptionally bright. In actuality, however, this region between the two bows, which we shall refer to as "Alexander's band," is noticeably darker than any other portion of the entire heavenly dome. Alexander had no easy resolution of the paradox, being satisfied to report that, according to the opinions of others, reflection cannot take place all over [the meteorological sphere], for the positions at which reflection is possible are strictly defined.[67] Later, in the fourteenth century, a more determined effort was to be made to solve the Aphrodisian paradox, but only in 1637 was a satisfactory answer found.

There is no need to analyze in detail the dozen and a half folio pages of Alexander's commentary on the halo, the rainbow, parhelia, and virgae, for they paraphrase faithfully the explanations of Aristotle. All of these were believed to be produced in the same way as reflections accompanied by color; they differed only in the crassitude of the material causing the reflection and in the relative positions of the sun, the reflector, and the observer. So closely does Alexander follow the text of the *Meteorologica* as it appears in our present editions that one can conclude from his work that the Aristotelian treatise, which certainly had not been put in final form by the author himself, had achieved definitive status at least by the second century.[68] Alexander reported Posidonius as agreeing with Aristotle's reflection theory of the halo, although others hold that it is caused by a fraction or disruption of the rays.[69] This would appear to mean that the dissenters intended to replace reflection by refraction, a step in the right direction; but if this was their intention, they failed to convince those who left written documents. So convinced was Alexander of the omnipresence of reflection that he attributed the "horizontal moon" to this agent. That is, he believed that the apparent increase in the size of the moon (or the sun) on rising and setting was a result of the reflection of the rays.[70] He cited with approval the argument for reflection in the rainbow which was ascribed to Philip, the associate of Plato. He reported also an extension of this argument by "others" who held that as one moves toward the bow, it also seems to approach, and as one moves back from the bow, it likewise appears to recede, just as is the case with objects in a looking glass. Alexander added that this is uncertain because of the great distance involved.[71] That attention should be paid to such inaccurate observation and hearsay is in itself a sad commentary on the status at that time of the theory of the rainbow; and the situation was steadily deteriorating. The scientific activity of the philosophical schools was becoming submerged in disputation at a time when pagan learning itself encountered in Christianity a formidable rival for the interests of mankind. The pressure of barbarian invasion, with resulting political chaos, further accelerated the decline in profane learning. Books on philosophy, unintelligible to the invader, were ruined; and the scarcity of parchment led copyists to expunge ancient works to make place for more immediate needs. Perhaps even more important was the fact that in the west fewer and fewer scholars were able to read Greek. When the scribes came to a Greek word or phrase in a manuscript which they were copying, they either left a blank space or else wrote *Grecus est*—which we might translate today as "It is Greek (or incomprehensible) to me." Science in particular was hard hit by the loss of Greek works and the decline in the reading knowledge of Greek. Under Rome the status of science sank to the point where philosophers, astronomers, and charlatan magicians often were grouped together and stigmatized, even outlawed, as *mathematicians*. It is small wonder that what little material on the rainbow there may have been at that time largely disappeared, leaving virtually no surviving fragments from the period of three centuries between Alexander of Aphrodisias and the closing of the Athenian schools by Justinian in 529. Justinian's action was little more than an official recognition of the fact that pagan learning had lost its force in the Greco-Latin world.

III

The Earlier Middle Ages

THE END OF THE ROMAN EMPIRE TRADITIONALLY IS PLACED IN THE YEAR 476 IN
which Odoacer set aside the titular ruler, Romulus Augustulus; but from an
intellectual point of view other dates have greater significance. The Greek
schools of philosophy at Athens had lost much of the vitality they enjoyed
during the days of Aristotle, but they continued to serve as a focus of scholarly
activity until their dissolution in 529. This year marked also the founding
of the monastery at Monte Cassino, a coincidence which may be taken as
symbolic of the shift in interest from secular learning to religious activity.
The attitude of the early church fathers to natural science had been characteristically expressed by Lactantius (c. 260–c. 340), the "Christian Cicero"
and tutor of the son of Constantine, when he wrote:

> To search for the causes of things; to inquire whether the moon is convex
> or concave; whether the stars are fixed in the sky, or float freely in the air; of
> what size and what material are the heavens; whether they be at rest or in
> motion; what is the magnitude of the earth; on what foundations is it sus-
> pended or balanced; to dispute and conjecture upon such matters is just as if
> we chose to discuss what we think of a city in a remote country, of which we
> never heard but the name.[1]

Eusebius (†c. 340), Bishop of Caesarea, adviser to the Emperor Constantine,
historian of the Christian church, and one of the most learned men of the
age, explained that it was not through ignorance of natural philosophy that
he had turned from science, but through contempt for the uselessness of its
activity, devoting himself to the direction of souls to better things.

Is it any wonder that there was no significant contribution to the science
of the rainbow for hundreds of years to come? Comments on the rainbow by
the early Christian fathers generally were jejune or fanciful observations. In
the Clementine literature from the early Christian centuries, for example,
there is a passage ascribed to Clemens Romanus (fl. A.D. 96) in which the
bow is said to be like a ring in wax. The sun shines upon a thin rarescent
cloud, leaving an impression which is carried back to the observer as the rainbow. If the cloud thickens, the bow disappears.[2] St. Ambrose († 397) asserted

that our rainbow is not to be identified with that mentioned in Genesis IX, 13. Here and there more vigorous sparks of scientific inquiry were found during the Middle Ages. It has been well said that the Dark Ages were never so dark as is our knowledge of them, and this is particularly apposite to the theory of the rainbow. One has come to expect little in the way of scientific activity during this period, but during the sixth century there was considerable independent thought in matters of science. Among the earliest of the Christians to compose scholarly commentaries on the scientific works of Aristotle was John Philoponus, known also as Joannes Grammaticus or John of Alexandria, who flourished in the first half of the sixth century. He was a critical thinker who challenged Peripatetic notions on inertia and the void, but he touched only incidentally on the rainbow, for his commentary on the Meteorologica covered only the first book. The little that Philoponus wrote on the rainbow is disappointing, for he followed the Aristotelian explanation in which rays of sight are reflected to the sun by small drops as in mirrors.[3] He and his arch-opponent, the pagan scholar Simplicius Peripateticus, or Simplicius of Cilicia (fl. 533), took opposite sides on the classical controversy as to the reality of the colors of the rainbow, Philoponus holding that these are an optical illusion. Simplicius also was a scholar who did not hesitate to depart from the teachings of the Master, but unfortunately there is no commentary by him on the Meteorologica. It is a pity that of the three chief Aristotelian commentators of the sixth century (Philoponus, Simplicius, and Olympiodorus), the one who left a full-scale work on the Meteorologica showed the least independence of spirit, but even his comments indicate that there were others who disagreed sharply with traditional views.

Olympiodorus, who taught at Alexandria around the year 564, was one of the last of the Aristotelian scholars from that city.[4] His commentary on the Meteorologica is one of the two surviving Greek paraphrases of this Aristotelian treatise, the other being that of Alexander Aphrodisias, composed four centuries before. The commentary of Olympiodorus adheres to Peripatetic doctrines with a slavishness comparable to that of Alexander, but it is on a somewhat higher level. Instead of a prolix repetition of the Aristotelian views, Olympiodorus frequently adduces new arguments in support of old ideas; and his account takes cognizance of opposing theories which are not mentioned in Aphrodisias. It would appear that Olympiodorus made use of some commentator, now completely unknown, who wrote at some time between the second and the sixth centuries.[5]

The section of Olympiodorus on the rainbow [7] has a unity which is wanting in that of other Aristotelian commentaries, for it is built about a single theme. The dozen and a half large folios on the rainbow and related phenomena are concerned with the defence of the Aristotelian doctrine that the halo, the rainbow, parhelia, and sundogs all arise from one and the same cause —the reflection of rays of sight. Olympiodorus therefore writes that he first will investigate the differences between reflection (the Latin translation uses the word *repercussus*) and refraction (*transpectus*). The first distinction he noted was in the relative positions of the object, the eye, and the medium. The second difference lies in the relationships between the angles: mathematics teaches that reflection takes place at equal angles, whereas refraction involves "obtuse" angles. This is illustrated by diagrams which show the author's clear understanding of the phenomena. It is proved in geometrical detail that the equality of the angles of incidence and reflection implies that the distance from the eye to the object is a minimum. In abbreviated form the proof is as follows. Let E be the eye, O the object, and P the point on the surface AB such that $\angle APE = \angle OPB$; and let O′ be the mirror image of O in the surface AB (Fig. 11). Let Q be any point on AB different from

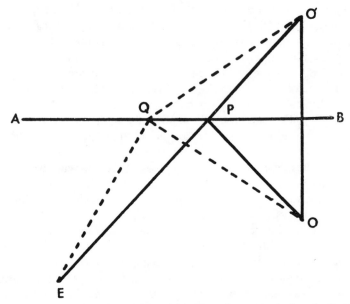

Fɪɢ. 11.—Diagram illustrating the principle of least distance in the law of reflection.

P, and draw the lines EQ, QO, and QO′. Inasmuch as QO = QO′ and PO = PO′, it follows that EQ + QO = EQ + QO′ and EP + PO = EP + PO′. But EP + PO is less than EQ + QO, because, in the $\triangle$EO′Q, the side EO′ is necessarily less than the sum of the other two sides. Hence the path making equal angles, EPO, is shorter than any path such as EQO, for which the angles are unequal. Olympiodorus therefore concluded that if in reflection the angles were not equal, nature would work in vain. This broad principle is essentially Aristotelian, but the proof that it is consonant with the law of reflection is to be ascribed to Heron, who lived at Alexandria about the time of Ptolemy.

In the case of refraction Olympiodorus gave no quantitative law, being satisfied to note that the rays of sight, ga and gb (Fig. 12) are bent toward the perpendicular upon meeting the surface ab. (Throughout the commentary he uses the familiar language of visual rays.) A third difference between reflection

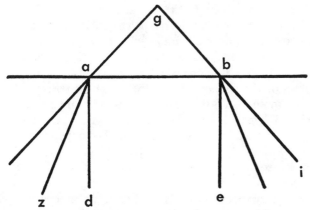

Fig. 12.—Diagram used by Olympiodorus to illustrate the refraction of light.

and refraction which Olympiodorus cited (and which he may have taken from Alexander) was that things seen through the former appear smaller, those seen by the latter appear larger. Thus it is, he thought, that things seen under or through water look larger, and that the sun on rising appears larger because it is seen through vapors. (The latter is, of course, an error, as was later pointed out by Alhazen. Refraction should make the horizontal sun look smaller, not larger.) This observation Olympiodorus turned against those who questioned the Aristotelian reflection theory of the halo and the rainbow. Were the solar rainbow and the lunar halo the result of refraction, he argued, they would be enlarged images of the sun and moon. Inasmuch as this is not the case, the rainbow and the halo cannot be produced by refraction but must be the result of reflection. Moreover, if the rainbow and the halo were caused by refraction, it would be necessary for them to be generated in a medium directly between the sun and the observer, which he held was not the case. Olympiodorus reported that there were men who used Aristotle ill, holding him to be uninformed in mathematical matters and pugnaciously convicting him of error on the corona and the rainbow. He therefore launched into an elaborate defence of the Aristotelian theory, first for the corona and then for the rainbow, in order to free the philosopher of all suspicion of error; but his arguments are, if anything, less cogent than those of Aristotle himself. Olympiodorus made use of three fundamental assumptions: (1) the reflecting body must be sufficiently dense to prevent penetration by the rays of sight; (2) in reflections from tiny mirrors, color alone, and not the form of the object, is preserved; and (3) sight weakened by reflection tends toward darkness, and hence becomes colored.[8] These are the bases from which most ancient and medieval writers argued. His work, therefore, is of interest less as a defence of Aristotle than as a witness to the fact that there apparently were others who entertained the notion that the rainbow is caused by refraction. The fact that it is not possible for us to determine who it was that took such a position is especially to be regretted, inasmuch as the transition from an explanation based entirely on reflection to one which made use of refraction is fundamental in the story of the rainbow. The introduction of refraction in this connection was far more than just casual speculation, if one is to judge from the care with which Olympiodorus seeks to refute it. And that the insurgent view apparently was advanced by mathematically-minded opponents makes the loss of further information still more regrettable, for the key to a satisfactory explanation lay in geometry.

Olympiodorus seems to feel on the defensive in ascribing to the same agent (reflection) two phenomena (the halo and the rainbow) which differ in so many respects. He cites half a dozen points of difference: (1) the shape of the halo is independent of the altitude of the luminous body causing it, whereas the form of the rainbow depends on the altitude of the sun; (2) the halo usually is formed about the moon, the rainbow far more frequently is caused by the sun; (3) the halo generally is seen at night, the rainbow by daytime; (4) the halo rarely is seen near the horizon, the rainbow invariably appears toward the horizon; (5) the halo is seen as a single circle, the rainbow often is a double arc; (6) the halo usually appears in a single color, the rainbow is tricolored. To account for such an array of contrasts Olympiodorus thought it sufficient to assume that the halo is formed by a dark and uniform body of vapor, the rainbow by clouds which are high or low, dense or subtle, and dissimilar in their parts. In much the same way parhelia and virgae differ from each other in that the medium producing the former is smooth and uniform, that causing the latter is difform. Again Olympiodorus takes cognizance of the fact that there are those who hold certain of these phenomena to be the result of refraction, while others are caused by reflection. He believes that these people are deceived by Aristotle's assertion that distorted images are produced in different ways, either by reflection or by refraction. Olympiodorus admits this dual origin of confused likenesses, but again he asserts that for the four meteorological phenomena in question there is but one agency, reflection, the differences arising from the qualities of the media producing them. For the halo Olympiodorus gives a sententious geometrical argument, purporting to justify its circular form, which does not differ in essence from that of Aristotle and Alexander. His account nevertheless is noteworthy in one particular, in that it contains what appears to be the first estimate of the size of the halo. The diameter of the halo, he reported, "contains forty parts." [9] This is a tolerably accurate figure, for the radius usually is given in modern works as about 22°. With so much of the scientific literature of the earlier centuries irretrievably lost, one cannot be certain that the measure of the halo given by Olympiodorus was original with him, but there is no apparent reason to question his priority. That he did not give an analogous estimate for the rainbow probably was the result of his belief that the form of the bow varied with the altitude of the bow.

The commentary of Olympiodorus on the rainbow is not notable for its perspicuity, but it is eminently well organized. It is first argued that reflection can take place from air, as well as from water and the surfaces of mirrors. In support of this contention the author cites the case of Antipherons of Tarentum about whom it was reported that, due to the weakness of his eyes, the visual rays were unable to penetrate the ambient air and, being reflected, caused him to see his own image. This evidently was a case of what is known as "the spectre of the Brocken," one of the many optical marvels which may be experienced while perambulating through mist.[10] One who appreciates the confusing welter in the phenomena of meteorological optics will not lightly condemn the naïveté of the ofttimes premature explanations of the pioneers. The "rainbow rings" seen about candles also were ascribed by the ancients, including Olympiodorus, to the resistance by the air to sight from weak eyes, the rays of which are reflected back to the observer. Inasmuch as rays are reflected more readily by water, or by vapor which is becoming water, than by air, one's eyesight need not be weak in order to observe the solar rainbow in nature.

Olympiodorus believed that it is "clearly apparent" that the bow is caused by the reflection of visual rays to the sun, and he promised to present six points in explanation of this position. (1) He accounts for the unique color of the halo, in contrast to the tricolored rainbow, by repeating the earlier argument on the uniformity and difformity of the reflecting media, as well as by citing the greater proximity of the halo to the quintessence, which cannot be varicolored. (2) The rainbow appears directly opposite, rather than in the direction of the sun, because there the heating effect of the sun is smaller than it is nearer to the sun, so that the vapors producing the bow are not so readily dissipated by the radiation. (3) The rings seen about candles are not tricolored nor semicircular because, as in the case of the halo, there is continuity and uniformity in the parts of the medium producing them. (Later Olympiodorus says that the colored bows seen in water sprays are intermediate between these two types.) (4) Light in darkness produces color, darkness is a weakening of light or sight, and sight is weakened by distance. Hence the rays of sight producing the rainbow are colored. (5) Those rays which are reflected by the nearest superficial portion of the cloud are red; those that penetrate farther, and are further weakened by the greater distance, are green; those that strike the most remote portion of the cloud, and traverse the greatest distance, are blue. (6) As Aristotle says, the colors of the interior bow are determined primarily by the number of rays producing each band, and since the outer band presents the greatest area (and hence the greatest number of rays), it is red. For the exterior rainbow, on the other hand, the dominating factor is not the area of the color band but its distance from the observer and the obliquity of the rays; and hence the inner band in this case is red.[11]

Olympiodorus added to these six main points other typically Peripatetic remarks on the bow. Just as everything has a beginning, a middle and an end, he held, so does the rainbow have precisely three colors. He did not deny that there seemed to be other colors, citing Ptolemy's assertion that the bow appeared septicolored; but the superfluous hues he attributed, following Aristotle closely, to mixture and contrast. The role of optical meteors in weather forecasting is touched on lightly. The halo and virgae, he thought, presaged fair weather, parhelia and the rainbow are harbingers of rain. But most of the last dozen pages of the commentary on the rainbow are made up of a prolix explanation of the calculations required to determine the center or "pole" of the rainbow if the sun is on the horizon and if the value of the Aristotelian effective ratio (the ratio of the distance between sun and cloud to the distance between cloud and observer) is known. Referring to the diagram of Olympiodorus but expressing the work in modern notation, his laborious calculations are in essence equivalent to the following problem. Let c be the eye of the observer, m a point of the cloud at which the iris-producing reflection takes place, i the sun, and p the pole of the bow (Fig. 13). Then if the ratio $r = im/mc$ is known, it is required to find the distance cp in terms of the radius of the meteorological sphere imn of which c is the center. Taking the radius of this sphere as unity, and letting $x = cp$ and $y = mp$, one has, by two applications of the Pythagorean theorem, $x^2 + y^2 = 1$ and $(x + 1)^2 + y^2 = r^2$. Eliminating y from these equations, the required distance cp is found to be $x = (r^2 - 2)/2$. Needless to say, Olympiodorus could make no

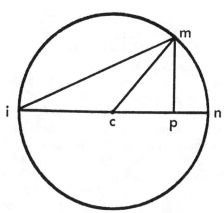

FIG. 13.—Determination by Olympiodorus of the relative distance from c (the position of the eye) to p (the center of the rainbow).

use of the result, for he had not the slightest idea what value was to be assigned to the ratio r.

The commentary of Olympiodorus was the last significant work on the rainbow from the Greco-Roman world. Less than a century after his time, Alexandria was captured by the rising Arabic empire. The story is told that precious manuscripts in the great library were used by the Arabs as fuel to heat the baths of the city; but it is likely that earlier Roman military wantonness and Christian religious fanaticism had destroyed a large proportion of the treatises. At all events it is clear that the first century of Muslim domination was not conducive to scholarship. In the Byzantine empire there was less of a rupture in the Greek scholarly tradition, but a period of stagnation settled over the region which once had been the center of the heroic age of science. Olympiodorus, consequently, may be looked upon as the last of the Greeks to contribute to the story of the rainbow.

The period following Olympiodorus probably was the nadir in man's story of the rainbow, for wherever one turned, whether to Greek or Arabic or Latin civilization, there was nothing of significance to be found. The jejune sixth-century commentary of Olympiodorus, for example, was on a level far above anything which could be found in the West for hundreds of years. Typical of the Latin ideas on the rainbow were those expressed by Archbishop Isidore of Seville (c. 560–636). He was a prodigious writer with a great reputation, and his works cover widely varying fields, from science to theology. His best known work is the Originum sive etymologiarum, but on the rainbow his treatise De natura rerum is of greater pertinence. True to his penchant for reducing things to an etymological basis, Isidore believed that the rainbow was appropriately called iris because it descends through the air (aer, aeris) to the earth. It is caused, he held, by the sun shining upon hollow clouds which receive the rays and fashion them in the form of an arch to which it imparts various colors because tenuous water, clear air, and a dark impenetrable cloud cause such colors. In harmony with the ancient doctrine of the four elements, Isidore of Seville held the bow to be quadri-colored, for it seizes all kinds of elements in it: from the heavens it derives the color of fire; from water it draws the purple; from air it gets its whiteness and from the earth it acquires blackness. The aqueous and igneous colors, some say, are signs of two things—the one that the impious will perish in the flood, and the other that the sinful will burn in hell.[12] Such fanciful observations on the rainbow are a far cry from the mathematically developed views of Aristotle; but if the scientific outlook was dark in Isidore's day, it became even darker during the next few centuries. So feeble did scholarship, especially in science, become in Europe that one reads that nothing could be heard but the scratching of the pen of the Venerable Bede (c. 673–735) at Yarrow in England.

It will be remarked that, by the time of Bede, civilization was passing from the thalassic to the oceanic stage. The Mediterranean was losing its hegemony as the cultural center of the world, for cities were widely established throughout Europe to the Atlantic coastline. The change was, at least at first, of no significance in the theory of the rainbow. In a little work also entitled De natura rerum,[13] Bede could do little but paraphrase the inanities of St. Isidore adding the trivial observations that a rainbow is rarer in summer than in winter, and rarer at night than in the daytime. He held that for forty years before the end of the world the rainbow will not appear in the sky, the absence of the bow indicating a period of dryness. Still later Rabanus Maurus (784–856) echoed similar ideas in a little work De Arcu Coelesti.[14] Maurus, rector of the cloister school at Fulda and later Archbishop of Mainz—and often called praeceptor Germaniae—did for science in Germany what Isidore had done in Spain and Bede in England. Modifying somewhat the Isidorian commentary, Maurus held that the colors of water and fire (blue and red) in the rainbow signify that the earth will not again be destroyed by flood, but by fire instead. While the leading scholars of the Christian world were occupied with such shallow treatments of the rainbow, one must pass on to the Muslim world for more vigorous scientific work.

About the time that Isidore was writing, the hejira of Mohammed (622) touched off a series of conquests rivaling those of Alexander almost a millennium before—and far more lasting. In the early stages the Mohammedan acquisitions were territorial only. Although tales of their burning of the books in the library at Alexandria have been grossly exaggerated, it must be admitted that the conquerors at first displayed little concern for intellectual matters. A century or two later, however, the situation had changed greatly, and Muslim scholars were eagerly seeking out manuscripts of ancient works, mostly from Greece, in science and philosophy, copying them, translating them, preserving them against the ravages of time. It has been well said that the Muslim world put science in cold storage against the day when the Christian world once more was ready to take advantage of it. But such a judgment undervalues the debt of scholars to the Mohammedan world. Granted that Arabic scientific treatises borrowed far more heavily from Greek antecedents than classical antiquity had from the pre-Hellenic civilizations, there were nevertheless sparks of originality to be found in their voluminous works. This is as true of the theory of the rainbow as in other fields.

As in Greece, so also in Arabia early references to the rainbow appear in poetic sources. In one version the rainbow is pictured as the bow of the thunder god who shoots arrows of hail which hang suspended in the clouds; in another it is thought of as a figured tapestry draped by the hands of the south-wind over the dark sphere of air with the tips hanging down to the earth. As in many other lands, it often was called the bow of heaven or bow of the clouds or *arcus daemonis* or Allah's bow.[15] Following the translation into Arabic of the Greek scientific treatises, especially after the middle of the eighth century, less fanciful views prevailed.

The usual tripartite division of the early medieval world into distinct cultural areas (dominated respectively by the Greek, Latin, and Arabic languages) is in a sense an oversimplification. There always was a degree of direct intercommunication between these cultures; and the sharing of ideas was further facilitated by other tongues which served as connecting links. Among the most important of these linking languages were Hebrew and Syriac. When the schools at Athens had been closed by Justinian, a large number of the scholars, including Simplicius, migrated to Syria to form a sort of Greek Academy in exile. There many a classic was translated into Syriac or Hebrew to await the day when Arabic scholars could appreciate it. Syria was far from being a tranquil home for scholarship, strategically situated as it was, but even in times of great tribulation and wars there have been those who devoted themselves to learning. Such a one was Job of Edessa, a Nestorian writer who has left an account of natural science as taught at Bagdad about the year 817. He was learned in Greek, Syriac, and Arabic, and a prolific writer in Syriac, the language in which his *Book of Treasures* survives.[16] This work, containing several pages on the rainbow,[17] may be taken as typical of ideas then current. Although his account of the rainbow is saturated with Aristotelianism, it is not more servile a paraphrase than were contemporary works in Europe and it leaves a far more favorable impression than do the works of Bede and Maurus.

The rainbow appears to us in the shape of a bow, Job explains, because the heavens appear arched.

Indeed the heaven appears to us like a vault, and we imagine to ourselves the colours of the rainbow in accordance with this vaulted shape. Because the side of the sun that is in our direction comes into contact with the colours; because the sun is seen by us in the shape of a sphere, and a sphere has a vaulted shape on any one of its sides; and because it is the sun that is the cause of the rainbow—we see its colours in the shape of a bow.

One wonders whether Job here has in mind the Aristotelian meteorological sphere, but in any case his explanation of the circularity of the bow shows no appreciation of Aristotle's geometric approach. In continuing his explanation, Job departs markedly from the Peripatetic tradition.

After rain has come down and has ceased, and a clear sky has begun, a certain thick, in addition to a thin, even and smooth humidity remains in the air. When the sun shines on the clear and smooth part of that humidity, it causes a reflection on its thick part, and creates in it to our vision the different colours found in the rainbow.

Job compares the formation of the bow to the reflection produced when rays of the sun, striking a smooth and brightly polished utensil of brass or gold, are reflected to other bodies near it. This idea that two media are needed to produce the rainbow was to recur frequently, in other forms, in Arabic and Latin treatises, with Avicenna, the greatest scholar of Islam, and William Gilbert, the great electrician, being numbered among its chief proponents. Job explains that a rainbow is not formed every time there is rain, but only now and then when the humidity that remains in the air is smooth, clear and thin, and when opposite it is a thick and contracted humidity, while the sun is in a place that is fit for reflection, in opposition to the humidity, "for reflection takes place in an opposite direction."

On the colors of the rainbow Job of Edessa again shows his eclecticism With Aristotle he agrees that there are three colors; but his choice of hues is different: date-red, green, and yellow. When the hot and fiery rays of the sun come in contact with the smooth and watery coldness of the air, there is a struggle; and the red color is due to the fire, the green to the part of the humidity which has not been conquered. (Dualistic Persian philosophy probably had a hand in this explanation.) The yellow color is formed through a combination of red and white, as is "the deep saffron dye." Job's explanation of the rainbow, when compared with the work of Aristotle, Seneca, Alexander or Olympiodorus, appears quite primitive; but compared with the roughly contemporary work of Bede, it is a model of independent scientific thought

A century and a half later there was no European work on the rainbow

(Greek or Latin) with which to compare the Arabic. An illustration of the status of science in the Muslim world of the late tenth century can be had in the series of treatises emanating from the secret association formed at Basra and known as The Brethren of Purity.[18] The philosophy of this group was fundamentally Aristotelian and Gnostic, with many other elements also present. The account of the rainbow which was included in their meteorology is essentially Greek science tinged with a bit of folklore. They estimated that the clouds are not more than 16,000 ells above the earth, and taught that the bow is a reflection of the sun's rays from the components of the fresh vapor in the air. (Note that the doctrine of visual rays is here abandoned.) They noted that the higher the sun is, the smaller is the visible portion of the bow; and they pointed out the reversed order of colors in the secondary rainbow. Four colors (rather than the three of Aristotle) were recognized, corresponding to the four qualities (hot, cold, moist, and dry), the four elements (earth, water, air, and fire), the four humours (blood, phlegm, black bile and yellow bile), and the four seasons. It is added that these are the colors found in plants and flowers. And because the bow implies dampness in the air and the consequent abundance of vegetation, so its appearance is a welcome harbinger, sent by nature to man and beast to foretell the year's luxuriant growth. Soothsayers hold that an abundance of a particular color in the rainbow has significance: red foretells slaughter, yellow means sickness, blue signifies want, and green denotes fruitfulness.[19] (Another charming instance of teleology in nature is found in the assertion of the Brethren that the hand of God is evident in the planning of thunder and lightning: if they were nearer, they would damage eyes and ears; if they were farther away, they would not serve to warn us of approaching storms.)

While scientific inquiry in the Latin world lay virtually dormant, scholarly activity in the Arabic civilization burgeoned to a peak about the year 1000. It was then that there appeared the greatest scholar of Islam, Abu Ali al-Husain ibn Abdallah ibn Sina (980–1037). This man, known to the West as Avicenna, was such a voracious reader of works from the fields of mathematics, astronomy, physics, medicine, and philosophy that he has been referred to as the Arabic Aristotle, a true disciple of the father of book-learning.[20] Among the many treatises and commentaries which he composed was one on *Meteorology*, a work which later was translated into Latin and exercised a wide influence in Europe.[21] This work includes a systematic treatment of the rainbow—one which obviously owes much to Aristotle but which by no means follows him slavishly. It opens with a summary of three classical theories of vision: the theory of visual rays (i.e., exploratory rays which go out from the eye to the object perceived); the theory that rays proceed from the object to the eye; and a third compromise theory according to which rays go out both

from the object and from the eye, meeting somewhat as do the rays of incidence and reflection in the case of images appearing in a mirror. Of these
three theories Avicenna, like the Brethren of Purity, prefers the second,[22] a
view which he had supported also in his commentary on Aristotle's *Physica*.
They err, he believed, who hold that sight is something going out from the
eye to the object, for who can believe that the eye has something which extends over half of the celestial sphere? After a long account of perspective
and the formation of images, he turned to the rainbow. With ingenuous
frankness Avicenna admitted that, while many aspects of the problem were
clearly understood, others had not been conclusively investigated. What was
taught about the bow didn't satisfy him; and even the teachings of the Peripatetics, a school to which Avicenna confessed adherence, left him dissatisfied.
He, therefore, presented a modification of the Aristotelian theory which reminds one somewhat of that of Job of Edessa. The rainbow is not located in
or on the cloud, as had generally been held, but in a medium lying directly in
front of the cloud. This was suggested by observations which Avicenna himself had made. Behind a hill there was a cloud which was cut off on one side
by the hill. Yet the rainbow was seen not only on the cloud but also against
the side of the hill. Again he had seen a rainbow above a dark cloud when the
cloud was very low, with a mountain lying behind it. Hence the cloud could
not have been the locus of the bow. The rainbow, he argued, is produced by
the reflection of light by the moist air which is broken up into small transparent particles like dew, rather than by the dark impenetrable cloud.[23] Nevertheless, the particles of moisture do not serve as a mirror unless they have a
dark background, just as rock-crystal will not act as a mirror unless it is covered on the back with a dark substance. Further proof that the bow is not
produced in the dark cloud is found in the fact that the colors of the rainbow
are formed by sprays of water in sunlight and in rings about candle lights in
the early hours of the morning. Moreover, dampness of the eyes can cause
colors, and the colors of the bow appear on the wall of the bath.

   Avicenna here emphasized the important fact that the key to the rainbow
is to be found in the *particles* of moisture, rather than in a cloud continuum.
This should have afforded him the opportunity for a geometrical treatment
of the reflection of light by raindrops, for Avicenna was a capable mathematician.[24] Yet no one at the time saw what now appears so obvious, and consequently no quantitative or geometrical theory was proposed. Avicenna was
content to make a few superficial observations on the form of the bow. It
must, he said, be circular. If the sun is on the horizon, the rainbow is half of
a circle; but the higher the sun, the smaller the bow, and, when the sun
reaches "a great height," no bow appears. He then added the curious Aristotelian error which was to be repeated frequently during the following cen-

turies: the more complete the rainbow (that is, the nearer it comes to being a full semicircle) the smaller is the circle to which the arc of the bow belongs. Thus he rejected the fact of the constancy of the apparent radius. He tried to justify this position by pointing out that the larger the bow, the more nearly perpendicular it is to the horizon. Apparently he had not thought through the geometry of the cone of rays producing the bow. Sections of this cone by a vertical plane are circles when the axis—i.e., the line joining the sun, the eye of the observer, and the center of the bow—is horizontal and perpendicular to the plane. When the sun is above the horizon, vertical sections of the cone of rays become ellipses. In a certain sense, the bow is always circular, for the drops producing it lie on the surface of a right circular cone; but what one thinks he sees is a cross-section of this cone by some surface which serves as a background. In the psychological sense, then, the convexity of the section does indeed vary according to circumstances.

The problem of the colors of the rainbow Avicenna found to be difficult. He accepted the Peripatetic belief that there are three different colors (although he recognized that they blend into each other so that it is impossible to tell where one ends and the next begins), but beyond this he judged the doctrine to be unintelligible. Some say the hues are produced by differences in position of two clouds, or in their intermingling; but Avicenna rejected this on the grounds that two clouds are not necessary for a bow, and, moreover, even when the air is homogeneous, three colors nevertheless appear. It also is false, he said, to hold that the higher regions are nearer the sun and hence reflect the rays more strongly, producing the bright red; for in the case of the secondary bow the ordering of the colors is reversed, thus contradicting the theory. (It will be recalled that Aristotle had used differing hypotheses with respect to the two bows.) Avicenna finally suggested that perhaps the basis is not to be found in the mirror or in the rays producing the image, but rather in the eye. The confusion between the physics and the physiology of color was the basis of much of the controversy, from Posidonius to Goethe, as to whether the colors of the bow are real or illusory. Needless to say, Avicenna was in no position to resolve the difficulty; and this inadequacy led him to conclude with a modesty and suspension of judgment quite appropriate to the true scientific attitude:

My views on the rainbow might appear too uncertain to be put down in this book. It is certain that it is a phantom, and that not more than two can appear at the same time, for the second already is scarcely discernable. How can still a third appear! When here and there in similar connections we say, "It can not be," this simpy means that the possibility is remote, not that it is impossible. This is all that I know about the rainbow. Further clarification must be sought of others.[25]

Among the "others" to whom Avicenna referred, one surely would have hoped to find his contemporary, the mathematician and physicist Abu Ali al-Hasan ibn al-Hasan ibn al-Haitham (c. 965–1039), known generally as Alhazen. He was born in Basra, and there he acquired such a scientific reputation [26] that he was called to Cairo by the caliph to apply his knowledge in the use of the waters of the Nile for the irrigation of lower Egypt. After examining the situation (as well as the failures of his predecessors), he realized that the feat was impossible. Fearing the anger of the caliph, Alhazen feigned madness, for which he was confined and his property confiscated.[27] Upon the death of the caliph, he regained his liberty and property—as well as the opportunity to continue his study of science, becoming one of the greatest students of optics of all times. He stated the law of reflection more carefully than had his predecessors, specifically noting that the incident ray, the reflected ray, and the normal to the surface all lie in the same plane; and for his vigorous refutation of the Euclidean theory of visual rays and his studies of refraction, he was called by his compatriots the second Ptolemy. Certain it is that no one, from Ptolemy to Kepler, exerted a wider and deeper influence upon the development of optics than did Alhazen. His great treatise *Kitab al-manazir* or *Treasury of Optics* became the leading source book for scholars in both Arabic and Latin civilizations. He was the first physicist to make a detailed study of the human eye, and his is the first recorded use of the camera obscura to observe the shape of the sun during an eclipse. There is an important problem in optics which to this day is known as Alhazen's problem: Given a spherical reflecting surface and two fixed points not on this surface, to find the point on the surface at which light from one of the fixed points will be reflected to the other.[28] Moreover, Alhazen performed experiments with a spherical glass globe filled with water, and these might easily have been related to the particles of moisture which Avicenna postulated in his explanation of the rainbow. The stage would appear to have been all set for a major advance in the story of the rainbow. But what a rude disillusionment. It is almost as though Ibn al-Haitham had a veil over his eyes. The *Treasury of Optics* doesn't even mention the rainbow. Indirectly it is known from later comments on his other works that Alhazen supported a theory which recalls that of Anaxagoras and Aristotle. The rays of the sun are assumed to fall upon a cloud as upon a concave spherical mirror and are then, provided the cloud is in the correct form and position, reflected to the eye of the observer. In this reflection, however, one sees only color and not the form of the sun, for even though the particles of the cloud hang together, the mirror surfaces are too small.[29] Alhazen, like his predecessors, evidently was puzzled by the relationship between the cloud and its constituent elements; but he was geometer enough to appreciate Aristotle's explanation of the circularity of the rainbow.

On the matter of the colors Alhazen could do little more than suggest, in the Aristotelian manner, that they resulted from a mixture of light with darkness. The rainbow is produced by a reflection from the *inner* concave surface of the cloud, and so the varying hues correspond to differences in the extent to which the rays penetrating the medium are subjected to admixture.

Alhazen's studies on the refraction of light were a notable contribution in the history of science. He pointed out that the incident ray, the refracted ray, and the normal to the surface lie in one plane; and he showed that the angle of refraction is not proportional to the angle of incidence, despite the opinion of others since the time of Ptolemy. He gave the correct explanation of the "horizontal moon" illusion, showing that Olympiodorus was wrong in ascribing it to refraction. He even anticipated in part the explanation of refraction given long afterward by Fermat, holding that the ray is bent toward the perpendicular, on entering the denser medium, because light is transmitted less readily in the medium of greater density. Alhazen also gave tables of refraction for angles of incidence at intervals of 5°. In short, his was the most intensive study of the phenomenon since Ptolemy. One should certainly have expected of him some reference to refraction in the rainbow; but this was not to be. Evidently the refraction theories of the rainbow, to which Olympiodorus had alluded, were not known among the Arabs of his day. As a matter of fact, so thoroughly imbued was Alhazen with the idea that the rainbow results from reflection that he carried it over into the field of the halo, where, as Alexander once had reported, many others had attributed it to refraction. Here Alhazen has left behind one of the most curious theories in the history of science. Explanations of the rainbow had invoked reflections from a spherical or hemispherical surface, sometimes from the convex side, more often from the concave; but Alhazen's geometrical study of the halo was based upon a new type of reflection—a *radial* reflection of light rays with respect to a sphere! In abbreviated form this may be described as follows: Let C be the center of a circular cross-section of a spherical cloud (Fig. 14), and let O be the eye of an observer. Let R be a point on the circle which is such that if a ray of light from the sun S is reflected in the radius CR as in a mirror, it will reach the eye of an observer at O. (The angle SRQ therefore is equal to the angle ORC.) Let T be the point at which the tangent to this circle at R intersects the line COS. Alhazen went through a neat bit of calculation [30] (now handled more easily by trigonometry), to show that if $CS = a$ and $CT = b$, then $CO = x$ is a solution of the equation $\dfrac{x}{x-b} = \dfrac{a}{a-b}$. Solving this, one obtains $x = ab(2a - b)$. That is, if the positions of C, T, and S are known, O can be calculated and vice versa. How, now, can one find which other rays from the sun S will be reflected by the sphere radially to the eye of the ob-

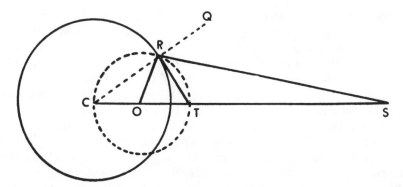

Fig. 14.—Alhazen's idea of the formation of the halo.

server at O? The answer is simple. Draw the sphere with diameter CT. (This is indicated by the dotted circle in the diagram.) The intersection of this sphere with the spherical cloud will be a circle such that rays from S which strike the circle will be reflected (according to the law of reflection) along a radius of the spherical cloud to reach the eye at O. It is the totality of these rays, Alhazen believed, which produce the halo, which is seen as the circular intersection described above. No particular quantitative significance, of course, is to be attached to this explanation of the halo inasmuch as estimates are not given for the distances a and b. The radii of the halo and rainbow were not adequately accounted for until the seventeenth century.

The direct contribution of Ibn al-Haitham to the story of the rainbow (or, for that matter, of the halo) was not of great weight, and yet indirectly his work was the most important single source of inspiration. While the main trend of thought continued along Peripatetic channels, there arose several centuries later a wave of interest in geometrical optics, spearheaded by the Treasury of Alhazen; and it was the merging of these two streams—the philosophical and the mathematical—which led on to the eventual first real success in the basic problem of the rainbow.

The age of Avicenna and Alhazen represented the high-water mark of Arabic learning. Muslim scholarship was by no means at an end, as one recognizes in the activities of the astronomer Omar Khayyam, who was born at about the time that Avicenna and Alhazen died. Omar contributed substantially to poetry and mathematics, but, like so many great figures from Archimedes to Galileo, he seems not to have been drawn to the problem of the rainbow, and no real progress was made in this connection for a couple of centuries. The status of Arabic thought on the subject can be gauged from two little manuscript works, undated and of unknown authorship, which lean heavily on Avicenna, although they, nevertheless, accept the doctrine of visual rays. These are accompanied by diagrams showing the rays from the eye reflected toward the sun at the rainbow as in a mirror.[31] (See Figure 15 for one of these.) Far more widely known was a contribution to the story of the rainbow made by Abu-l-Walid Muhammad ibn Ahmad ibn Muhammad ibn Rushd, who was born in 1126, only a couple of years after Omar had died. Ibn Rushd, better known by the Western appellation Averroës, knew no Greek, yet he wrote amazingly popular commentaries on Aristotle, making use of the versions which were at hand in Arabic. In Arabia, as in Greece, the works of Aristotle had been found difficult to read, and paraphrases on a lower level were welcomed. The comments of Averroës on the *Meteorologica* were certainly within the comprehension of all, for they cut through all of the worthwhile geometrical detail to leave a hopelessly corrupted picture of the

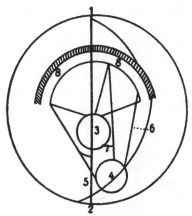

Fig. 15.—Diagram illustrating an Arabic explanation of the formation of the rainbow. (Reproduced from Wiedemann.) Rays from the sun (4) are reflected at the semicircle (8) to reach the eye of the observer (3). The semicircle (8) is to be thought of as perpendicular to the plane of the large circle (the horizon).

phenomenon. Perhaps misunderstanding Aristotle's use of the meteorological sphere, or possibly adopting a crude version of the ideas of Alhazen, Averroës accounted for the circular form of the rainbow by ascribing it to the rotundity of a hollow concave cloud, from which the rays of the sun are reflected. This amounted to a reversion to the primitive view of Pliny and the pre-Aristotelians.[32]

After repeating the Aristotelian assertion that not more than two bows are possible, on account of the weakness of the third reflection, Averroës displayed an unbecoming dogmatism on the subject of multiple bows. He held that, by the theory of perspective, it is impossible for more than two bows to be formed from one surface because the cloud is not properly disposed, with respect to distance, profundity, or reflexive properties, for the formation of a third bow. On the colors of the bows he departed somewhat from the classical doctrine, for he recognized four simple hues (red, yellow, green, and peacock blue); but he was thoroughly Aristotelian in believing in the instantaneity of light and in colors as combinations, in varying proportions, of brightness and darkness.

Averroës threw less light on the rainbow than had his eminent Muslim predecessors, but his remarks certainly were not inferior to explanations current in earlier years in Western Europe. The middle of the twelfth century, however, represented a turning-point in the cultural tides. The high point of Arabic scholarship came roughly at a time when Latin science had reached its nadir. Around the year 1000 there were few reputable works on science available to Latin scholars, for much of the scientific heritage was confined to the Greek, Arabic, and Hebrew languages. During the twelfth century, however, a wave of translation set in, with the result that by the year 1200 Europe became familiar with a large part of the scientific tradition, not only of ancient Greece, but also of the Muslim world. The so-called Arabic contributions to the rainbow were by no means restricted to the region now known as Arabia. They included works from Persia, Syria, Egypt, and many other lands. The one element they had in common was that of language, and after the twelfth century this no longer served as a barrier to the transmission of this learning to Europe. The European revival of interest in science during the thirteenth century was but one facet of a burgeoning cultural development which has led some to designate this period as "the greatest of centuries." Such an evaluation would appear to be an exaggeration; but, in the story of the rainbow, it is an exaggeration which is not difficult to pardon.

IV

European Revival

THE INTELLECTUAL LEVEL OF THE MIDDLE AGES WAS FAR FROM BEING UNI-formly high or low, either in the Arabic world or in Europe. In the Muslim world learning, beginning especially in the eighth and ninth centuries, had risen to a climax, represented by the work of Avicenna and Alhazen, in the early eleventh century. At this time Latin science was in a sorry state. The vaunted Carolingian revival had scarcely touched on scientific matters, and two centuries later the science found in the work of Gerbert (Pope Sylvester II from 999 to 1003) compared unfavorably with Arabic treatises of the same period. Before the twelfth century few of the ancient classics were available to medieval European scholars, most of whom read neither Greek nor Arabic, the tongues in which most of the story of the rainbow had appeared. Among available Latin sources one could point to Plato's *Timaeus*, Seneca's *Natural Questions* and the trivialities of Isidore, Bede and Maurus; and it is the influence of these which one recognizes, early in the twelfth century in the *De Imagine Mundi Libri Tres*, written between 1122 and 1125, of Honorious Augustodunensis, or Honoré d'Autun. The author is an obscure figure, possibly a Benedictine or perhaps a crusader, who wrote numerous treatises, mostly theological. One chapter of *De Imagine Mundi* is on the rainbow, but the account is as primitive as that of Bede almost four centuries before.

The quadricolored arc is formed in air when rays of the sun shining into a hollow cloud, are driven back toward the sun, just as when the sun shines on a water vase and the brightness is sent back onto the ceiling. From the heavens it derives the fiery color; from water, the purple; from air, the blue; and from the earth the grassy color.[1]

Not greatly different are the comments on the rainbow, toward the middle of the century, by William of Conches (c. 1080–1154). His *Dragmaticon Philosophiae* resembles the work of Bede to such an extent that it has been falsely ascribed to the scholar of Yarrow who had died about three centuries before. In this work one reads that some philosophers hold the rainbow to be substantial, while others regard it as an image of the sun reflected by so many little mirrors that the bow looks continuous. The colors in the bow are the

85

result of the mixture of light and dark; the red comes from fire, the purple from air, the blue from water, and the green from the earth (because of the plants and trees).[2] The medieval age was fascinated by the rainbow as a weather prognosticator, and here William repeated Seneca almost verbatim: a midday rainbow means heavy rain, a bow in the morning signifies light rain, and one about the time of vespers indicates clear weather.

The ideas of William of Conches, which may be taken as typical of thought current in his day, are essentially those found in Europe throughout the early medieval period; but during the later years of his life there was a fresh intellectual breeze stirring which Haskins has called "the Renaissance of the twelfth century." This period was to the Christian world what the eighth century had been to the Western European Mohammedan. For the first time it became possible for scholars to know what had been going on in the other main Mediterranean civilizations, for Greek and Arabic science was being poured into a Latin mold. Toward the beginning of the century, virtually the whole of Greek science had been translated into Arabic; and by the end of the same century, after one of the most intense periods of translation ever known, this store of learning had been converted for the benefit of those who read only Latin. European scholars, however, did not take over this new knowledge without criticism; they refashioned what they received, rejecting portions here and there, adding new ideas of their own. The works of Aristotle himself did not always fare too well. In 1210 and 1215 his metaphysical and scientific treatises were banned at Paris, although the Organon and the Ethics were recommended for study. Again in 1229 Aristotelian science was proscribed, but by 1234 there seems to have been no obstruction. (Some of the other local interdictions of Aristotelian learning, however, were more sweeping and lasted longer.) By 1255 nearly all of the Aristotelian corpus was prescribed at Paris for candidates for the masters degree. At about this time the Meteorologica was so well known that translations into the vernacular appeared. A copy of a translation into French, made between 1249 and 1270 for the son of King John of Jerusalem, is still extant.[3] Practically nothing is known about the translator, one Matthieu le Vilain, except that he was from Rouen. It is a very free translation indeed, for the mathematical treatment and the diagrams are omitted. Matthew evidently had at hand not only a Greek version of Aristotle, but also a Latin manuscript of Alexander's commentary. It was suggested tentatively by the translator that the tertiary rainbow which Alexander claimed to have seen may have been caused by the moon; but in the end Matthew rejected this explanation, realizing that the moon when full (and hence able to produce a bow) must be opposite the sun.[4] Better known than this translation were those of Aristotle's Meteors and Alexander's Commentary completed simultaneously by William of Moerbecke (c. 1215–c. 1286), Archbishop of Corinth,[5] on April 24, 1260. These translations were made directly from Greek into Latin, but there had been also, earlier in the century, a translation of Aristotle's *Meteorology* from Arabic to Latin.

The new Latin versions of the *Meteorologica* brought with them ideas on the rainbow which were ever so much more sophisticated than those which had been entertained only a century before. But it was not only Greek influences which had entered western Europe. Toward the end of the twelfth century Muslim views had penetrated as far northward as England, where Hereford had become a center of Arabic learning.[6] It is very likely that here, as well as in other parts of Europe, the philosophy of Avicenna and Averroës and the optics of Alhazen were combined with the meteorology of Aristotle to produce a new and broader matrix from which advances in the theory of the rainbow stemmed. It is a striking fact that from the year of Magna Charta in 1215 until well into the fourteenth century there appeared more new treatises and original thoughts on the rainbow than in any comparable period with the exception of the later interval from Kepler to Newton. One who would roundly condemn the medieval age as a barren period in the history of science should mark well the story of the rainbow.[7]

The manner in which Arabic science reached European centers of learning is far from clear. There are a number of transitional figures known to have been familiar with both Greek-Latin and Arabic-Latin translations during the earliest part of the thirteenth century, and among these was Alfred of Sareshel, or Alfred the Englishman (fl. 1217). He was familiar with the new Aristotle, and he was an admirer of Avicenna. It may have been during a trip to Spain that he ran across Avicenna's meteorology upon which he commented. His glosses on the meteorology, still preserved in manuscript,[8] later in the century were pillaged without acknowledgment.[9] Bacon's admiration for Avicenna, "the most important imitator of Aristotle in philosophy," very likely was a consequence of the work of Alfred. Bacon cites Alfred three times as a translator, but when using his work in meteorology he does not name him. One reason that Avicenna and Averroës were so popular is that the works of Aristotle were found obscure, and men looked for simpler paraphrases; and because the commentaries of Avicenna were tediously redundant, those by Averroës enjoyed an inordinately popular vogue. During the early thirteenth century in Europe it was customary to use glosses to explain difficult works, and Alfred supplied these for Aristotle and Avicenna. Alfred's glosses in turn were glossed later by a Master Adam [Adam de Bokefeld or Adam Rufus?], some of whose words have come down to us to furnish a clue to the works which were read at the time. Following an excerpt in which the Greeks are reported to have called the rainbow *arcus demonis* or *arcus varius*, Adam reports Alfred as saying, concerning the difficulty of the rainbow, that the truth is concealed on high, for only to God is given the knowledge of all things. "Avicenna, the greatest of all philosophers except Aristotle, admitted with regret that he did not understand this matter." [10]

The comments of Alfred certainly do not help one to understand the rainbow, but they may serve as a connecting link between Avicenna and the Scholastic philosophers from the universities who took a decisive step forward in the theory of the bow. One of the outstanding achievements of the later medieval period was the establishment of our present university system. The schools at Bologna, Paris, Oxford, and elsewhere became great centers of learning where old and new ideas were discussed—where a tradition of scholarship was built up and passed on from generation to generation. Alfred was an important figure at Paris and Oxford, and his work was referred to by scholars, such as Robert Grosseteste, Albertus Magnus, and Roger Bacon, who studied at these schools. All three of these scholars paid a great deal of attention to the rainbow, and in each case it is clear that there were both Greek and Arabic elements in their work. If it is not possible to ascribe these to Alfred directly, it is at least likely that he served as an essential intermediary. It also appears that the combination of Arabo-Aristotelian science and Neoplatonic metaphysics which figures so prominently in writers on the rainbow during the thirteenth century may well be due in no small measure to the mediation of Alfredus Anglicus.[11]

Robert Grosseteste (c. 1175–1253), also known as Robert of Lincoln, was born at the decisive moment when Greek and Arabic science became accessible in Latin versions. He was educated at Oxford (perhaps between 1199 and 1209) and Paris (about 1210), where he showed unusual interest in the study of languages and in the sources of knowledge.[12] This was especially important from the point of view of the rainbow in view of the weakness on this topic of contemporary Latin sources. Moreover, Robert became much interested in science and scientific method; and he showed particular attention to optics, which he regarded as the key to an understanding of the physical world. He was conscious of the dual approach by means of induction and deduction (resolution and composition); i.e., from empirical knowledge one proceeds to probable general principles, and from these as premises one then derives conclusions which constitute verifications or falsifications of the principles. This approach to science was not far removed from that of Aristotle, but Robert gave it a clarity and sharpness which has led at least one writer to conclude that "By relating these logical methods to scientific practice Grosseteste made the first moves towards the creation of modern experimental science"; [13] or, more boldly still, that "Grosseteste and his thirteenth- and fourteenth-century successors created modern experimental science." Certain it is that the Bishop of Lincoln placed an emphasis upon experimentation which has been regarded as characteristic only of later periods. In spite of Grosseteste's strong interest in book learning, his most famous pupil, Roger Bacon, said of him that he neglected the books of Aristotle for his own experiments, and, with the aid of other authors and sciences, treated independently scientific questions which had occupied Aristotle. Apparently it was in this spirit that Grosseteste undertook, some time between 1217 and 1235, the study of the rainbow.[14]

Robert Grosseteste's little book on the rainbow, *De Iride Seu de Iride et Speculo*,[15] opens with the assertion that speculation on the rainbow belongs both to physics, which tells "what," and to perspective, which tells "because of what." Aristotle in the *Meteorologica*, he held, had treated only the former aspect, and hence Grosseteste undertook to give the *propter quid*. He began by outlining the three divisions of perspective: the first, involving direct transmission of light in a single medium, is called "the science of vision"; the second, concerning reflection, is that part "concerning mirrors"; the third part, covering the passage of light from one medium to another, he said had remained almost untouched and unknown up to that time. The last part is much more difficult and, by nature of its profundity, much more marvellous, he reports. Distant things are made to appear near, and small things large, so that the smallest letters can be read by us at an incredible distance. (One should appreciate that this was written almost fifty years before Bacon's celebrated *Opus Majus!*) He understood, from Euclid and Ptolemy, that the size of the object seen depends on the visual angle subtended at the eye.

Grosseteste also described the well-known penny-in-a-vase experiment, a description which he may have taken from the *Catoptrics* ascribed to Euclid. He even went so far as to frame a simple (though inaccurate) law of refraction. Grosseteste appreciated keenly the part that mathematics should play in science, believing that it is impossible otherwise to understand the physical world; and he was on the lookout for geometric and numerical regularities. He sought to relate the equality of angles in the law of reflection to the principle that "the operation of nature is finite and regular." Inasmuch as the law of reflection depends upon the equality of angles, Robert tried to carry over into refraction some analogous equality. Let AB be a ray of light which strikes obliquely at B the surface of a second (more dense) medium (Fig. 16). Let AB be extended along BC, and let BD be perpendicular to the surface. Then, said Grosseteste, the refracted ray will lie along the line BE which makes equal angles with BD and BC. This law is, of course, equivalent to the statement that the angle of incidence is proportional to the angle of refraction, the ratio being two, a result which is reasonably correct only for small angles and provided that the index of refraction is two. (The index for air to water is

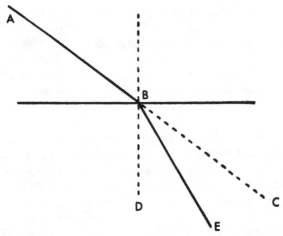

Fig. 16.—Diagram illustrating Grosseteste's law of refraction, angle CBE = angle EBD.

about 4/3.) Alhazen earlier had shown that the law of proportionality is false; but Robert Grosseteste deserves considerable credit, nevertheless, for he is perhaps the first Latin scholar to present such a law. It is, of course, quite possible that he had read Ptolemy's work on refraction, in which similar ideas had appeared, for Robert could read Greek and Hebrew. It is doubtful that he knew of Alhazen's results on refraction, for he did not read Arabic; but the *Treasury of Optics* became the *vade mecum* of workers in optics shortly after Robert's death. The question of inspiration takes on added significance in view of the fact that Robert adds the comment that the equality of angles shows that "every operation of nature is in the shortest, best ordered, briefest, and best way possible." [16] Was he, perhaps through Alfred, aware of Alhazen's adumbration of the principle of least time, or was he merely extending Aristotle's views on the orderliness of nature?

There is another important respect in which the question of Grosseteste's originality poses a problem. Continuing his discussion of refraction, he added the profound statement that "it is to this third part of perspective that the science of the rainbow belongs." [17] Since antiquity it had been realized that the direction of light rays could be altered in two ways: through reflection or by refraction. Yet for over sixteen hundred years the theory of the rainbow had belonged almost exclusively to the second part of perspective, the reflection of light. Now this one-track point of view had been broken. For some reason historians of science mistakenly have ascribed this great innovation to Witelo (born c. 1230), a physicist who wrote his work almost a score of years after Robert of Lincoln had died. There can be no question here about the assignment to Witelo of credit for the refraction theory of the rainbow, for he almost certainly knew of Grosseteste's *De Iride*. The problem is quite different. The commentary of Olympiodorus clearly implied that some sort of refraction theory had been propounded in antiquity; and traces of this would appear to have survived in Sicily until the late sixteenth century. However, it does not appear to be possible to determine whether or not Grosseteste knew of any refraction theory of the rainbow before his own. One cannot even be certain that he had access to Aristotle's *Meteorologica*, although this is quite likely. It is still more likely that Grosseteste was familiar with the work of Avicenna, and there may well be a connection between the latter's postulation of two media in the mirror producing the rainbow and the assumption by the former that several media are involved.

In presenting his novel theory, Grosseteste first refuted the older idea that the rainbow is due to reflection of the sun's rays by the surface of a cloud as from a concave or convex mirror. Were this the case, he held, the altitude of the bow would vary directly—rather than inversely, as is the case—with the altitude of the sun. Whereas reflection takes place at the *surface* of a medium, the refraction causing the rainbow takes place *within* the medium—a convex moist cloud. The details of his theory are, in the light of modern views, singular indeed. The exterior of the cloud is convex, the interior concave, "according to the nature of light and heavy." As the moisture of the cloud descends from the concavity, it forms a convex pyramid or cone in which the portion near the earth is more condensed than is the higher part. There are, therefore, four transparent media through which the sun's rays pass—the pure air surrounding the cloud, the cloud itself, the higher (hence rarer) moisture coming from the cloud, and the lower and more dense moisture in the pyramidal cone. Inasmuch as the sun's rays are refracted upon passing from one medium to the next, there must be *three* refractions in all: from air to the cloud; then from the cloud to the rarer moisture; and finally refraction within the pyramidal cone because of the varying density.[18] Refraction had indeed entered with a vengeance into the theory of the rainbow. Here, unfortunately, Grosseteste seems to have left no particular role for reflection, which also is an essential ingredient in the modern theory. And yet in another connection, in commenting on Aristotle's *Posterior Analytics*, he accepted the Peripatetic view that the echo and the rainbow are cases of repercussion, as in the reflection of light from a mirror.

But the echo is the repercussion of sound from an obstacle, just as the appearance of images is the repercussion of visual rays from the surface of a mirror and the rainbow is the repercussion or refraction of rays of the sun in a concave aqueous cloud. For, when light diffusing itself in a straight line comes to an obstacle preventing its advance, it is collected in the place of incidence on the obstacle; and because its nature is to diffuse and generate in a straight line, when it cannot generate itself by advancing directly, it can generate itself only by turning back if the obstacle is an opaque body.

This passage is equivocal with respect to the rainbow; it does not ascribe the arc to reflection and refraction, but to reflection or refraction. That the word "refraction" here is not used in the ancient sense of reflection is clear from the lines which follow:

Or, if the obstacle is a transparent body, it generates itself off the line and is not direct but penetrates the transparent object at an angle, as the ray of the sun falling on transparent water is reflected back from the surface of the water, as from a mirror, and also penetrates the water, making an angle at the surface itself, and this is properly called a refraction of the ray. Therefore, as light is reflected or refracted at the obstacle, so the rainbow is the reflection or refraction of the light of the sun in a watery cloud and the appearance of images the reflection of the visual ray at the mirror.[19]

Later on Grosseteste repeated that the echo, the rainbow, and images are caused by the reversal of light, adding that "in the rainbow it is the reversal of light because of cloud." [20] It is possible that in the interval between the composition of the commentary on the *Posterior Analytics* and the time of the *De Iride*, Grosseteste's indecision between reflection and refraction was resolved in favor of the latter. It seems more likely, however, that he intended in each case to ascribe the rainbow to the interplay of the two phenomena. The *De Iride* explains that the refractions in the descending cone of transparent media first cause the sun's rays to converge and then, on further refraction, to spread out into a conical surface expanded in a direction opposite to that of the sun. The upper half, or less, of this cone falls on the cloud opposite the sun. Grosseteste does not specifically say so, but one can presume that by this cloud the rays are reflected to the observer's eye. It is quite possible that his assumption of two distinct bodies, one transparent (the descending cone) and one opaque (the cloud), may have been an outgrowth of Avicenna's emphasis on the two components in a mirror, the transparent glass and the dark background.[21] One can reconcile the passage in the commentary on the *Posterior Analytics* with the explanation in *De Iride* by assuming that the solar rays, after undergoing the postulated multiple refractions, impinge upon a distant cloud or other background and are thereby reflected to the eye.

In a recent scholarly treatise one reads a glowing account of Grosseteste's work in optics and the rainbow.

It is in this science, in his attempt to explain the properties of mirrors and lenses and the rainbow, that his use of mathematics in conjunction with his other methodological principles is most clearly and fully seen. He was the first medieval writer to discuss these subjects systematically.[22]

One wishes that Grosseteste had indeed expounded his theory of the rainbow systematically. The account which he has left was of great importance for the suggestion that refraction is the significant agency, but there is little which could be described as a mathematical explanation or a geometrical illustration.

Grosseteste asserted that one can not understand natural philosophy without considering lines, angles, and figures;[23] but there is no diagram accompanying his De Iride. In his vigorous rejection of the notion of a simple reflection from a concave cloud, and by his introduction of refraction into the theory of the rainbow, Grosseteste performed a distinct service; but in other respects his ideas were a retrogression from those of Aristotle. His work represented a quasi-geometrical attempt to explain the bow, and yet withal it encouraged vagueness of thought and made no attempt at quantitative treatment.

The correct Aristotelian geometric explanation of the circularity of the bow was abandoned in favor of a vaguely conceived conical surface (he emphasized that it was not a solid cone) of refracted rays; two media are demanded (one of these made up of successively denser layers); and the role of discreet drops was entirely submerged in a doctrine which fastened attention on the cone as a continuous whole, and the cloud as a continuous whole. Then, too, refraction was adduced to explain the shape of the bow, not its colors. The variety of hues was explained through the admixture of light with the diaphanous media. Color was assumed to depend not only upon the purity or impurity of the medium, but also upon the clarity and obscurity of the light, and upon the multitude or paucity of the rays. These three factors, he held, generate all of the colors. The brighter colors appeared in that part of the bow where there was a greater concentration of light rays, the blue was formed where there were fewer rays. The differences in ray density were produced through the refractions which concentrated the rays in one region and dispersed them in another.

Grosseteste applied this notion of varying concentrations to one of the Aristotelian rainbow problems which was a favorite during the Middle Ages. It had been held that the larger the portion of the rainbow which was visible, the smaller was the circle of which it was a part. Grosseteste, however, held that the smallness which Aristotle intended here was smallness of illumination, due to the passage of solar rays through the multitude of vapors at sunrise and sunset.

It must be admitted that Grosseteste's explanation of the rainbow violated one of the basic criteria of a good scientific theory—economy of thought. Its importance was great, nevertheless, because of the boldness with which it broke from established tradition. Avicenna had modified the Aristotelian explanation, but Grosseteste replaced it by a totally different approach. Had the author of the new theory had leisure to think further on the problem, he might have come closer to the truth; but he suffered the fate of many another scholar attached to a university; he found himself more and more drawn into administrative posts and difficulties. In 1214 he seems to have been chancellor of the University of Oxford; in 1235 he became Bishop of Lincoln; and later he supported Simon de Montfort in his efforts to redress misgovernment. Such activities brought an end to his creative work, and one had to look to others to continue the story of the rainbow. But the works of Grosseteste exerted a powerful intellectual influence, not only at Oxford, but throughout Europe. His treatise on the rainbow undoubtedly was widely read during the thirteenth and fourteenth centuries; it is mentioned by Bacon and Duns Scotus (c. 1266–1308), and half a dozen manuscript copies of the work are extant in libraries at Madrid, Oxford, Florence, Groningen, Prague, and the Vatican.[24]

The account of the rainbow given by Robert of Lincoln stands in marked contrast, both in its brevity and originality, to that of his younger contemporary, Albertus Magnus (†1280), or Albert of Bollstadt. De Iride is a work of less than half a dozen small pages; the voluminous De Meteoris Libri III of Albert includes twenty-nine chapters (Tractatus IV) on the rainbow and halo.[25] The one presented briefly a new idea; the other is an encyclopedic history of the views of others from Hesiod to his own day. The very bulk of his writings made it inevitable that Albertus should have copied much from earlier sources, and Grosseteste was one of those who strongly influenced his science. However, his approach to problems was less mathematical than Grosseteste's had been. Metaphorical and literary allusions seem especially to have fascinated Albertus Magnus. He quoted Euripides' representation of Iris as a serpent above the room (cloud) in which Juno was confined with her children Apollo and Diana. Apollo, the sun god, escaped and shot rays like arrows, killing the enemy of his mother. The vestiges of these rays appear in the arc of the rainbow. Thus, concluded Albertus, philosophers, both natural and perspective—as well as poets—agree in this, that iris is the image of the sun on an aqueous cloud. He ascribed to the ancients, especially Posidonius and the geometer Parianus Artemidorus, the idea that the rainbow is a confusion of many images of the sun conceived on a moist and hollow cloud, and he cited Seneca's assertion that there are as many mirrors as there are drops in the cloud.[26] But Albertus did not repeat ancient ideas uncritically. He felt that the reflection could not be from the convex portion nearest the sun, for this would contradict Euclid's Optics; the reflection must be from a point diametrically opposite in the cloud. The drops in the cloud are mirrors. When they are plane and clear, a distinct image appears; when small and angular, unequal and not well cleaned, the colors are not well separated and the figure is not discernible. At this point the author makes a fresh start "so that all those things which have been said and are to be said may be more easily understood." [27] In this explanation he expected to show that his views agree with those of the Peripatetics. Hebrew scholars of the time also were expending much of their energy in reconciling Aristotelian natural philosophy with the Bible, but they showed less interest in the scientific aspects. Nahmanides, or Moses Ben Nahman Gerondi (1194–1270), a Spanish Talmudist and physician, was satisfied to report, in his commentary on the Pentateuch, "We are forced to accept the view of the Greek scientists that the rainbow is the natural result of the sun's reflection on the clouds." [28] But he himself was more interested in the new role which the bow received after the flood as a sign of God's covenant. One of the great purposes of Albertus Magnus was similar to that of his Hebrew and Arabic predecessors, the adaptation of Aristotle to the use of the Latin Christian civilizations; but his interest in the scientific explanation of the rainbow went beyond theirs.

The moist vapor or rain, he held, is like a pyramid with base on the earth and vertex in the cloud; and the heavier parts are near the base. But before the drops of rain are formed, there is beneath the concave cloud a subtle distillate in the upper part of the cone beneath the cloud. Hence there are four transparent media altogether: that in the cloud, the subtle distillate, the more coagulated moisture in the pyramid, and, finally, mixed in with all, air which is thick from coldness and damp from the moisture descending through it. The rainbow is produced when the sun shines upon the lower part of the cloud and the rays are reflected by the four media which send them. The refraction of rays is made at the extremities of the watery pyramid, just as if a globe were cut somewhat above the middle. Rays at this extremity are multiplied and generate the bow. The semicircular form is due to the fact that a cone of rays from the sun strikes the round pyramidal vapor. Only half of the circle is seen because the other rays are extinguished by the heavy material in the pyramid.[29] In this explanation one can not fail to see the influence of Robert of Lincoln, yet Albert does not mention him, even though he studied at Paris where one should have expected the studies to have included the *De Iride*. Is it possible that Albertus Magnus, a Dominican, was reluctant to admit indebtedness to a man who had been closely associated with the rival Franciscan Order?

There is trouble with Albert's use of the words *reflexio* and *refractio*. Sometimes they appear to be synonymous with reflection at a surface; at other times it is clear that they refer to a refraction of rays which pass through a medium. The context often makes clear which meaning is intended, but the confusion leaves doubt in the mind of the reader that he fully appreciated the significance of the step which Grosseteste had taken from the reflection theories to one hinging on refraction. Had Albert adequately understood this problem, it is unlikely that he would have insisted upon the agreement of his views with those of Aristotle. And yet he refers to the reflection of rays in the bow as analogous to the reflection of light which comes through a window, strikes a vase of water, and is refracted to the opposite wall. In the same way rays are refracted in the descending rain drops from the cloud, and the light is multiplied by many refractions. Were one to substitute for the vase of water a shower of many small drops, then each drop would receive the ray and reflect it to the opposite wall in an image in which the colors of the rainbow appear.

Vague as such statements are, they suggest definite advances over the views of Robert Grosseteste. In particular, Albert's reiteration of the part played by individual drops was a necessary prelude to any successful assault on the mystery of the rainbow. Seneca's microcosmic reflection theory had survived from antiquity, along with its rival, the Posidonian macrocosmic reflection theory. In a sense one can regard Grosseteste as the proponent of the first macrocosmic refraction hypothesis and Albertus as the initiator of the microcosmic refraction doctrine. One should not, of course, carry this distinction too far. Albertus, like Grosseteste, attributed the roundness of the bow to the interaction of the rays from the sun with the cone of moisture descending from a cloud, the refractions in turn producing the conical surface of colored rays the upper half of which is projected upon a second cloud in the distance.

In explaining the formation of colors, however, he again is specific about the refractions in the raindrops making up the aqueous cone. The diversity of colors is a consequence of drops in different portions of this cone. The drops near the base of the cone, being more dense, produce the green band in the rainbow; those near the top are more tenuous and cause the red circle in the bow. When droplets are sprayed into rays through a window, colors are similarly formed on the floor of the room because of the varying density of the drops in the spray.[30] The colors thus produced, as well as those in the rainbow he regarded as real colors, existing in the world of nature, even though the transparent drops themselves were colorless. Generally there are three colors in the bow: green from the heavy dampness, red from the sun, and blue from the mixture of these. If there is a fourth shade, it is due to variation in the two factors, light and darkness, producing the others. That the rainbow vanishes more quickly than the corona or halo is not an indication that the colors are apparent only, lacking reality, but simply because the bow is caused by cold vapor, the halo by warm vapor.

One finds in Albert's account of the rainbow many other details taken from Aristotle, Seneca, and Isidore of Seville—the frequency of lunar and solar bows, the reality and significance of the colors, the bow as a weather prognosticator. He even included reference to the views of Avicenna, evidencing the interchange of ideas between the Latin and Arabic worlds. Albertus said that he often had seen three or four rainbows in the same place opposite the sun, yet Aristotle had asserted that only two are possible. He believed that the third and fourth bows were caused by the appearance of a second dense reflecting cloud beyond that producing the first two bows. In this way, from two arcs four will appear, and from one, two—"as we frequently have found." He suggested also that the Aristotelian dictum against more than two bows might be interpreted to mean that there can be not more than two located on opposite sides—e.g., one in the east and one in the west.[31]

It would be superfluous to repeat all of the material in Albert's De Meteoris, for it is a confused and indecisive compilation of myth, fact, and fancy. It is the most comprehensive work on the rainbow since the Commentary of Olympiodorus, but it is not easy to find in it elements of permanent value. Apart from the stress on the individual raindrop, the most significant passages may have been those hinting that the rainbow might be brought down from the heavens into the laboratory. Several times he pointed out analogies between the cone of moisture in nature and artificial water sprays before the window in a room. The mobility of such a device was scarcely conducive to precise quantitative study; but Albert made another relevant suggestion. He reported that if a hemispherical transparent vessel were filled with black ink and placed in the sun's rays, a bright semicircular arc would appear owing to the opacity of the ink near the middle of the hemisphere.[32] The hemisphere here played the role of the Grossetestean cone, showing that even Albertus Magnus could not completely emancipate himself from macrocosmic views. Nevertheless, it may have been the suggestiveness of his indecisive comments which led his successors, about half a century later, to make laboratory studies which eventuated in a signal triumph in the story of the rainbow.

Few scholars exerted as wide an influence as did Albertus Magnus. He taught at many schools, including the University of Paris (1245–1248), where he acquired great fame and earned the title of Doctor universalis. His De Meteoris achieved considerable popularity and appeared twice in print in the fifteenth century (1488 and 1494), as well as in collected editions of his works (1518–1651). Even Roger Bacon, a member of a rival religious order whose interests differed widely from Albert's, spoke of him as "the most noted of Christian scholars."

One scholar whom Albertus Magnus clearly influenced was the Belgian Thomas of Cantimpré (c. 1200–1270). The latter had heard Albert lecture, and he undoubtedly was familiar with the material of De Meteoris, for similar ideas appear in the sections on meteorology in his *De Naturis Rerum*, a work to which Thomas devoted some fifteen years. Here on the bow one finds a collection of familiar remarks, but no systematic theory. The rainbow is a reflection of the rays of the sun from a hollow cloud, as when the sun shines on a vase full of water; the colors are due to a mixture of cloud, fire, and air; the four colors correspond to the four elements; the bow occurs less frequently around noontime; the rainbow signifies that the world will not again be destroyed by flood; the celestial arc will not be seen in the clouds for a period of forty years before the end of the world.[33]

A similar work of about the same time was the *De Proprietatibus Rerum* of Bartholomaeus Anglicus (fl. 1230–1250), a Franciscan monk. Written probably before 1260, it later was found in many languages and editions, the Latin original being printed a dozen times before 1500. The author became famous as a teacher of theology at Paris, and he cites Albertus Magnus; but his treatment of the rainbow leans heavily upon the ideas of Aristotle, Seneca, and Bede. The bow is caused by a reflection from a concave cloud as from a multiplicity of mirrors. The colors he related to the four elements; he believed only two lunar bows possible in a fifty-year interval; and he quoted Bede on the forty-year rainbowless period before the end of the world.[34] How slow is the progress of science! The Venerable Bede had lived almost half a millennium earlier, and many scholars, Christian and Muslim, had added to the story of the rainbow. Yet books which enjoyed the widest acquaintance were satisfied to ignore what had been done and said in this long interval.

The most notable scholar of the thirteenth century without doubt was St. Thomas Aquinas (1225–1274), the "Angelic Doctor," a "clear and forceful expositor . . . capable of remarkable independence of thought." [35] Albertus Magnus and his pupil Thomas Aquinus together were mainly responsible for making the Greco-Arabic Aristotelian tradition acceptable to the Christian West through the argument that there could be no real contradiction between truth as revealed by religion and truth as revealed by reason.[36] On matters of faith and morals they both followed the apostles and church fathers; on questions of science they depended chiefly on Aristotle.

Among the writings attributed to each, there are works on meteorology; that by Albertus was a compendius collection of varying views, with here and there a critical comment; St. Thomas' contribution to the story of the rainbow was a jejune commentary on Aristotle's *Meteorologica*, with trivial variations. The rainbow, like the halo, is a reflection (the words *refractio* and *reflectio* are used interchangeably) of sight to the sun; but in the case of the bow, unlike the halo, the reflecting mirrors are small, with the result that no image, but only color, is the result. The bow is tricolored, with red in the largest circle, green in the middle, and blue in the smallest; and these are the colors which painters cannot produce. Repeating the Peripatetic dictum that not more than two bows can appear, he added cautiously, in the manner of Avicenna, "except rarely." Were the third bow to appear, added a later commentator on his commentary, it would be produced in the same manner as the others.[37]

The works of Thomas of Cantimpré, Bartholomaeus Anglicus, and Thomas Aquinas were much used and copied from, even as late as the sixteenth century; and yet they can not be regarded as having added anything significant to the story of the rainbow. But then popular accounts of the history of science give the impression that there was at the time only one competent scientist, a man far ahead of his time—Roger Bacon (c. 1214–1292). He it was, we are told, who emphasized the true scientific method, based upon mathematics and experimentation. Amidst a world of ignorance and credulity, he is looked upon as a beacon light of science—*Doctor mirabilis*, he was called. Moreover, when one recalls that Bacon was especially interested in optical phenomena, and when one learns that his investigations on the rainbow form an essential part of his *scientia experimentalis*, it is natural to anticipate something worthwhile. The anticipation is heightened by the assertion of a modern historian of science and optics that, in connection with the rainbow, Bacon "represented exactly the path of the luminous ray." [38] Certainly Bacon's background for his investigations could not have been better. He was the natural successor of Robert Grosseteste, whose work he knew and admired; and he was familiar also with the writings of Albertus Magnus. A great believer in the importance of languages, he had access also to Greek and Arabic treatises. What more could one wish for than thorough knowledge of the ideas of one's predecessors, coupled with strong faith in the mathematical and experimental methods?

The results of Bacon's investigations on the rainbow form part of his celebrated *Opus Majus*, composed in 1266–1267 for Pope Clement IV. He opens in a typically critical vein, pointing out several times the falsity of the translations of Aristotle which assert that only two lunar rainbows occur in a period of fifty years. Moreover, he wrote, Aristotle, more than all other philosophical writers, has involved us in obscurities in dealing with the rainbow. "And Avicenna, the greatest authority in philosophy since Aristotle, as all insist, humbly confessed that he himself was ignorant of the nature of the rainbow. Thus it is certain regarding philosophers that no one of them has been able to gain a knowledge of the rainbow." Then follows a statement which is not calculated to heighten Bacon's scientific stature, but which shows him to be a true son of his age: "Nor is it strange," he says, "since they have not examined Scripture with the necessary diligence. For all philosophers have been ignorant of the final cause of the rainbow." [39]

The Bible explains the end for which the rainbow exists; God's bow provides against a deluge. Whenever it appears, there must be an active consumption of the aqueous moistures; and this implies the existence of something possessing the power to consume. But the rainbow is due to sun and clouds. Hence the collecting of the clouds is the material cause, and the projection of the rays is the efficient cause. That vigorous action may result, it is necessary for the rays to converge; but convergence can occur only through reflection and refraction. The rainbow therefore must be produced by infinitely many reflections or refractions in numberless drops of water, so that the truth is thus discovered regarding both its colors and its form. They are due to this multiplication of rays with respect to figures, angles, and lines, and not produced by a diversity in the matter forming the cloud, as is stated in the text of the Latins and as all believe.[40]

Regardless of the metaphysical and theological reasoning which inspired it, the last sentence above is full of significance for the story of the rainbow. Had Bacon followed this lead, he could scarcely have failed to solve the rainbow problem. But he did not pursue the clue further. Instead of applying geometry to the matter in hand, he cited Scripture. As is clear from the Book of Genesis, whenever the rainbow appears there is a resolution of the clouds into an infinite number of drops, and the aqueous vapors disappear, in the air and in the sea and over the land, since one part of the rainbow falls into the spheres of water and earth. But it is impossible for the unbelieving philosopher to reach full certainty, owing to ignorance of the Scriptures. Bacon then makes a strong plea for the usefulness of numbers for the understanding of sacred writings. Thus the fact that $1^3 = 1$ helps one to understand the blessed Trinity! After a long account of the value of experimental science (by which he apparently means inductive inference from experience), he suggests the rainbow as an example. He advises the experimenter to study the colors of hexagonal stones, observe rowers who raise drops of spray, note the dew on the grass and the colors from a glass vessel filled with water and placed in the sun's rays, and those arising from oil on surfaces. He also urges the experimenter to take "the required instrument" and look through the openings of the instrument and find the altitude of the sun and of the rainbow. This excellent advice was not without valuable result, for it led to one of the very earliest precise quantitative statements in the story of the rainbow. Bacon reported [41] that the experimenter will find that the maximum elevation of the rainbow is 42°. This figure, repeated frequently in the Opus Majus, appears to be the first estimate of the size of the bow; and it is remarkably accurate—so good, in fact, that it was not improved upon until science had reached maturity in the Age of Genius, the seventeenth century. Bacon measured also the size of the halo, with the striking conclusion that its apparent diameter is about the same as the apparent radius of the rainbow. This coincidence was to mislead later investigators who believed that the geometrical relationships between the two phenomena were closer than they really are. Bacon's estimate that the diameter of the halo is 42° is a couple of degrees shy of the modern value, 44°. Bacon did not specifically say that the radius of the rainbow is constant, but this would appear to be the implication. Influenced perhaps by Albertus Magnus, he told how the altitude of the bow varies with latitude and seasons; and he pointed out that as the observer moves, so does the bow, showing that there are at any time as many different bows as there are observers.[42]

The reader may have noted that Bacon spoke of the bow as generated by multiple reflections *and refractions*; and a question arises here, as it did with Albertus. Did the author intend the word refraction in the modern sense? That this may be answered negatively is rendered probable by further statements. The colors of a cloud, he wrote, are seen by incident or refracted rays and hence are the same for all observers, whereas in the case of the rainbow the rays are not direct but reflected. Somewhat further on he wrote, "The observer alone produces the bow, nor is there anything present except reflection"; and later he clearly says that "they" are in error who hold that the bow is due to refraction. There can be no doubt but that among those he refers to here was his teacher, Robert of Lincoln. They hold, Bacon reports, that the bow is due to the fact that rays are refracted first at the point of contact of the air and the cloud, and afterwards at the point of contact of the cloud and the higher part of the moisture. By these refractions the rays converge in the lower and denser part of the moisture; and there the rays, refracted as from the vertex of a cone, spread out into a figure conformed to the curve on the surface of a cone. Here one sees clearly the triple-refraction theory of Grosseteste; but Bacon rejects it categorically. The rainbow shape is seen in water sprays where there cannot be three refractions.[43]

Earlier Bacon had held that the truth with respect to the colors of the rainbow is to be found in terms of "figures, angles, and lines." The only clear-cut explanation along such lines had been given by Aristotle, whom Bacon rather ungraciously rejected; yet he did not himself follow out his own promising lead. He confessed that the problem of the semicircular shape is difficult, but he explained that it is due to the fact that "all parts must have the same position with respect to the solar ray and the eye. But such a position of identity cannot exist except in a circular figure because of the equal declination of the parts." [44]

Bacon's explanation of the circular form of the bow is far from clear, and his theory of the colors is no less devious. All authorities, he says, agree that the colors are due to the varying density of the texture of the moist cloud; but Bacon objects to this on the grounds that when light passes through crystal, the density is uniform, yet the colors vary. He agrees with Posidonius, as over against Albertus Magnus, that in the case of the rainbow there is the appearance only of color, and the phenomenon is not a reality. The number of colors he takes to be five, in contrast to the seven which he attributes to Aristotle—white, blue, red, green, and black. For those who look upon Bacon as an isolated spirit born before his time, it is of interest to note the reasons given for choosing the number of colors in the bow:

> For the number five is better than all other numbers, as Aristotle says in the Book of Secrets. . . . Because the number five distinguishes things more definitely and better, nature for this reason rather intends that there shall be five colors. Therefore, these five colors are in the rainbow, rather than other colors, in accordance with the general arrangement of nature, which carries into effect and purposes that which is better.[45]

In confirmation of this idea, he noted that there are five bodies in the eye: three humors and two coatings (the uvea and the cornea); and the five colors of the bow appear according to the properties of these five parts. (Alhazen had recognized three fluids and four membranes.)

It must be confessed that Roger Bacon had failed utterly to clear up the problem of the rainbow. His views are in general those of Seneca, together with an admixture of theology and numerology. However, through other work in optics he contributed indirectly to the story of the rainbow. He and Alhazen, for example, were among the few who refused to follow Aristotle in believing that the speed of light is infinite. Bacon correctly conjectured that as also in the case of sound, the speed depends on the medium.[46] Then, too, Bacon wrote on the use of lenses as magnifying glasses (following the lead of Grosseteste) and burning spheres, tracing the paths of rays of light through spheres of glass; and he wrote on the structure of the eye and on the mechanism of binary vision. Nor should one overlook the fact that Bacon was the first person to make a quantitative contribution to the rainbow problem. If to measure is to know, then Bacon's estimate of the apparent radius is sufficient to entitle him to an honorable place in the story of the rainbow. And Bacon was prophetic in at least one verdict: "Many experiments are needed to determine the nature of the rainbow, both in regard to its color and its shape."[4]

Grosseteste had introduced the important phenomenon of refraction into the theory of the rainbow; but a prophet is not without honor save in his own country and among his own people. If Bacon failed to appreciate his work, the use of refraction did find acceptance in far-away Poland, where the physicist and philosopher Witelo (born c. 1230) effectively established it in optics and the rainbow. Witelo, who called himself "filius Thuringorum et Polonorum," probably was brought up in the neighborhood of Cracow, but he had been educated at Paris, as well as at Padua and Viterbo, and hence he may well have been acquainted with the work of Robert Grosseteste.[48] But the book which most influenced him, was Alhazen's *Treasury of optics*. He wrote, sometime between 1270 and 1278, a treatise on *Optics*, dedicated to William of Moerbeke, which is derived so largely from the work of Ibn al-Haitham as to earn for Witelo the soubriquet, "Alhazen's ape." It will be recalled that the *Treasury* failed to include a section on the rainbow, but Witelo's *Optics* (sometimes known as *Perspectiva*) closes with a tediously long account of the bow, a score of folio pages, in which the author shows distinct independence of thought.[49]

Witelo, like Robert of Lincoln, first refuted the older view of the rainbow as resulting only from reflection. Were this the case, he held, the bow should not move from side to side as the observer moves laterally. Instead, the bow is caused by an aggregation of reflected and refracted rays.[50] One should not read too much into these words, for Witelo apparently did not have in mind the modern idea of rays which are refracted, then reflected, and once more refracted before reaching the eye. In his theory some rays were reflected directly from the convex surfaces of drops, others were refracted through drops before being reflected at the outer surfaces of other drops lying further within the medium. Clouds are a mixture of dry and moist vapors. Light does not penetrate the dry vapors, and hence some rays are reflected from the surface of the cloud; other rays penetrate the moist vapors and are refracted at the surfaces of the dense portions of the cloud. It is a peculiar combination of these rays, he thought, which causes the bow. Refraction served primarily to condense the light; the drops served as spherical lenses, causing the light to make a stronger impression upon the eye. Mistakenly he believed that the reflections, as well as the refractions, participated in the formation of the colors.

Ever since ancient Greek and Roman days there had been conflicting and confused ideas as to the relationship between the raindrops and the cloud. Aristotle had regarded the dewy cloud as a multitude of tiny drops which reflected the solar rays; but the droplets were too small to serve as mirrors in which images are seen, and so he had based his geometrical study upon the aggregation of drops. One of the moot questions since his day had been the way in which the individual drops fitted into the scheme as a whole. Grosseteste had disregarded the problem of atomicity in a rain-cloud, treating his diaphanous media as continua; but Albertus Magnus had speculated on the importance of the drops. Witelo raised the old question once more, although he gave no satisfactory answer. Between the continuous aqueous vapor and the discrete drops of rain water there is a deep transition stage in which the rarer parts of the vapor are beginning to be round, to condense, and to take

VITELLONIS THV·
RINGOPOLONI OPTI·
CAE LIBRI DECEM.

Inftaurati,figuris nonis illuftrati atque aucti:infinitisq; erroribus,
quibus antea fcatebant,expurgati.

A'

FEDERICO RISNERO.

BASILEAE.

Fig. 17.—Title page of Witelo's Optics.

on a downward motion. These dewy particles are somewhat like little mirrors (in the sense of Aristotle) in which color, but not form, is presented; and, as Bacon had asserted, only those rays are reflected which make the proper angle. The colors of the bow result from the weakening of light by the mixture of dry and moist vapors. Evidently the role of refraction in the rainbow was quite a minor one, although to refraction within the moist vapor Witelo ascribed the putative constancy in the obliquity of the plane of the rainbow to the horizontal surface. And yet an historian of physics has asserted that "Vitello came marvelously close to the correct thing." [51] It is well worth noting, incidentally, that, whereas Aristotle and Alhazen had explained the halo in terms of reflection, Witelo held that it is due to the refraction of rays by humid vapor.

Witelo's studies on refraction are particularly worthy of notice because the phenomenon was so little understood at the time. Following the description of Alhazen, he gave detailed instructions for making an instrument, provided with a graduated scale, for measuring angles of refraction. Ptolemy and Alhazen had given tables of refraction from air to water or glass, but Witelo furnished also tables in the other direction—from water or glass to air—as well as for refractions between water and glass. Some of his values were derived ultimately from Ptolemy, others were calculated, not always correctly, from the reciprocal law. This principle—that the path of a refracted ray between a point A in one medium and a point B in another is independent of the sense in which the path is traversed, whether from A to B or from B to A—may have been first used by Witelo.[52] That his table was not entirely based directly on observation is apparent from the fact that it includes angles of refraction from water to air purporting to correspond to angles of incidence of 50°, 60°, 70°, and 80°, when actually there can be no refraction in these cases, but only total reflection, a phenomenon of which he was unaware. Witelo tried, unsuccessfully, to find general mathematical relations between the angles of incidence and refraction; and he applied his ideas on refraction to the study of lenses. A hasty reading of Witelo's account of refraction can leave the impression that he anticipated Newton in the discovery of dispersion.[53] Witelo believed that the refraction of different rays through different angles produced the various colors; but there is no evidence that he was aware that parallel rays falling on a uniform medium at the same angle of incidence are variously refracted according to the color. In fact, one can not avoid the impression that Witelo, despite his continuation of Alhazen's studies on refraction, had no clear idea of the application of the phenomenon to either the halo or the rainbow. His diagram for the halo, for example, resembles closely the traditional illustrations accompanying Aristotelian reflection theories of the corona.

It indicates rays of light passing through a sphere with a single, instead of a double, refraction.

Witelo compared the colors of the bow with those seen when a round glass vase full of water is exposed to the sun's light as the rays pass from air to glass, then glass to water, then water to glass, and finally from glass to air again. But he added cautiously, "Yet these prismatic colors are not truly like the colors of the rainbow, for the former are seen directly while the latter are seen by reflection." Moreover, he held that the number of colors is not the same in the two cases.[54] One can not help wondering whether these ideas were not suggested by Grosseteste's multiple refractions and Bacon's distinction between prismatic and rainbow colors. In any case, Witelo's words seem to have a more modern ring to them, even though one is disappointed by the vagueness of his over-all picture of the rainbow.

That Witelo was familiar with Bacon's work seems to be borne out in his statement that, on the basis of certain experiences, it has been assumed that the altitude of the sun added to the altitude of the bow is 42°. Did he verify this for himself or did he merely borrow it without giving credit? This figure for the size of the bow runs like a golden thread through treatises of the sixteenth century, and it is likely that it was derived in many cases from Witelo's *Optics*. Witelo himself, however, believed that there were variations in the radius of the bow resulting from fluctuations in the density of the atmosphere, but he added that possibly the difference is not perceptible inasmuch as "even Aristotle wrote nothing on this." [55]

The secondary rainbow had not come in for much attention since Aristotle, but Witelo tried his hand at it. The colors of the exterior bow, he thought, are weaker because they are further from the perpendicular and more remote from the eye. The reversal in the order of the colors bothered him but he suggested that this may be due to the fact that color depends not only upon the angles involved but also upon the darkness of the cloud. He reports Aristotle's belief that not more than two bows are possible, but avers that while at Padua he himself had seen four at one time. As far as the order of colors is concerned, however, he admits that only two *kinds* of rainbows are possible. Inasmuch as there are only three colors in the rainbow, there are only two ways of disposing the colors with respect to each other; for, since the middle color, green, is the generative cause of the other colors, it is possible only to interchange the extremes, red and blue. Hence, from one surface there can be not more than two bows.[56] The multiple bows he had seen he therefore attributed to different media.

Witelo's followers admired him greatly, not realizing how heavily his *Optics* was indebted to the Arabs. Throughout the next three centuries no writer on optics and the rainbow, with the exception of Aristotle, was cited with greater regularity. This admiration served to give his ideas wide circulation, with the result that his refraction-reflection theory of the rainbow became a serious rival to the conventional Aristotelian doctrines. But views less sophisticated than those of Witelo and Aristotle also were not hard to find in succeeding centuries.

Among the treatises of the century which appeared under the title *Summa Philosophiae* was one formerly thought to be by Grosseteste but now attributed to a member of the English school of about 1265–1275. This shows the persistence of Grossetestean refraction ideas on the rainbow, for the author held that the four colors are produced by the mixture of the darkness of the medium with variously refracted light. Although suggesting that the shape of the bow may be attributed in part to the concave shape of the cloud in which it appears, or to the rotundity of the incident rays, the author finally came much nearer to the truth. The refraction of rays in the individual drops gives the appearance of equality and continuity and uniformity of position; and this, either in part or in entirety, is the cause of the circular figure.[57] Such a quasi-modern statement could at the time carry little weight because there was no way of testing it. The typical catch-all explanations of the time show how desperate was the need for crucial experiments in the story of the rainbow.

The most popular optical treatise of the thirteenth century, apart from Witelo's *Optics*, was the *Perspectiva Communis* of John Peckham (†1292), Archbishop of Canterbury.[58] This is a much smaller book than the *Optics*, and hence correspondingly less space is devoted to the rainbow. The work, which also is heavily indebted to Alhazen, is divided into three "books," of which the first includes extensive discussion on the properties of refraction, the structure of the eye, and the requirements for vision. Book II is on the properties of mirrors, and this leans heavily upon the geometrical treatment by Euclid and Alhazen. The last book is on "common perspective," and propositions 18-21 of this are on the rainbow. In the formation of the bow, wrote Peckham, all three types of rays enter—direct rays, because the iris is generated opposite the sun; reflected rays, because the spherical drops are like mirrors; refracted rays, because light from the sun enters into the depths of the water. Fundamentally, then, the rainbow is generated by the reflection of rays in spherical drops which serve as mirrors. However, part of the cause lies in the rays which penetrate the water in the dewy vapor, converge to a point (as in refraction), then diverge again into a pyramid the middle of which falls on a cloud. The semicircular impression thus formed is then reflected to the observer. Hence the circularity of the bow is due principally to the cloud rather than the rays, a statement reminiscent of the *Summa* of Pseudo-Grosseteste. "When the cloud is regularly suspended, equidistant from the earth, it is certain that the moisture descends regularly; and this suffices for circularity, for water suspended irregularly would not have a regular impression." Such circularity in reasoning does not add conviction to an already weak argument! Perhaps for this reason the author added another justification. Rays of the sun falling on a hexagonal prism give a figure not like the rays (orbicular) but like the prism, which is columnar. (Here Peckham seems to be so close to the crucial observation which long afterward led Newton to the discovery of the dispersion of light, but he failed to make the correct inference.) This indicates that figure depends not upon the rays (as some say) but upon the shape of the medium; and hence the circularity of the bow is in the cloud. (In later centuries the most significant portions of Peckham's explanation were forgotten, and he was recalled, along with Posidonius, Pliny, and Averroës, as one who attributed the form of the bow to the roundness of the cloud.) The diversity of colors arises partly from the cloud and partly from variations in the light rays. The rain descends to a center, forming a round cone with gradually increasing density; and hence the nobler colors are along the higher or exterior part of the bow. The concourse of rays reflected from the cloud with direct rays brings about an attenuation of vapors; and hence the formation of the rainbow, which accompanies the consumption of the substance of rain, precludes a cataclysm. The noble colors, which artists are unable to imitate, are not admitted by a dense cloud; and hence the bow indicates an insufficiency of moisture which would preclude the possibility of a deluge.[59]

It is curious that throughout the century the secondary rainbow had exerted so little influence on scientific theory. Some scientists had disregarded it altogether, others had alluded to it perfunctorily; yet the outer bow shortly was to play a far more prominent role. In this respect poetry appears almost to be an intermediary, for Dante's *Divina Commedia*, the action of which is set in 1300, twice pointedly called attention to the double form of the arc. Assuming that the second bow is a reflection of the first, the poet compared it with an echo:

Two bows parallel and of like colors are turned across a thin cloud when Juno gives the order to her handmaid (the outer one born of that within after the manner of the speech of that wandering one whom love consumed as the sun does vapors).[60]

And again at the close of the *Paradiso*, speaking of three circles of colors appearing in the Light Eternal and vaguely depicting the mystery of the Trinity, Dante says of them, "and one appeared reflected by the other, as Iris by Iris." [61] Poets and scientists of the time evidently were not in agreement on the number of colors in the rainbow and the halo. Three, or sometimes four colors were all that science would grant, but Dante, in the *Purgatorio*, wrote of

> . . . seven stripes all in those colors
> whereof the sun makes his bow,
> and Delia her girdle.[62]

Peckham's explanation of the bow is, in a way, a summary of the most significant contributions of the century—those by Grosseteste, Albertus, Bacon, and Witelo. If, as is commonly believed, the medieval period was an age of excessive reliance upon the authority of Aristotle, then the study of the rainbow in the thirteenth century must be regarded as quite exceptional. From Grosseteste on there was an earnest criticism of earlier writers, Aristotle not excepted; and the search for new and improved explanations was carried on with remarkable enthusiasm. One must admit that the efforts of the century had failed to effect a satisfactory solution of the rainbow problem; but this failure should not obscure the fact that three important contributions had been made: (1) the introduction of the essential idea that refraction is necessary for the explanation of the rainbow; (2) the measurement of the radius of the bow; and (3) revival of interest in the function of the individual raindrops. What one misses most in the explanations of the time is a clear-cut geometrization of dioptrics comparable to that of optics and catoptrics. Lack of a mathematically precise law of refraction may have discouraged attempts in this direction, but, as the next century was to prove, a correct theory nevertheless lay entirely within their power. Grosseteste had, in fact, proposed a crude law of refraction but neither he nor Albertus Magnus nor John Peckham associated this with a quantitative geometrical study of the rainbow. Bacon and Witelo, on the other hand, measured the bow; but they failed to study it with the precision afforded by Euclidean geometry. Every scholar of the century had missed the key to the solution, the discovery of which consequently fell to the lot of a less illustrious century.

V

Medieval Triumph and Decline

THE THIRTEENTH CENTURY REPRESENTED THE HIGH WATER MARK OF SCHOLASTIC thought and medieval culture. Contributions of the time to literature, philosophy, science, and the pure and applied arts rose to the highest levels since antiquity. One may be disappointed in the outcome of the determined efforts to solve the problems of the rainbow; but it can not be denied that the century was responsible for promising new approaches to these. But what advance could be anticipated during the fourteenth century, during which the decline in learning was accelerated by war and pestilence? Yet, strangely enough, the century seems to have been responsible for more striking additions to science and technology than was its predecessor.

Gunpowder, the compass, mechanical clocks and spectacles may have been adumbrated before 1300, but it was the fourteenth century which saw their effective introduction into European civilization. In mathematics the graphical representation of functions, and in physics the revival of the concept of impetus or inertia and the suggestion of new laws for the motion of a freely falling body, are among the outstanding examples of the development of scientific thought of the time. Yet even these are less startling than one which may be taken as the greatest contribution of the medieval age to physical science—an explanation of the rainbow which anticipated, in all but its quantitative aspect, the correct geometrical theory given more than three hundred years later.

Some time shortly after 1304 Theodoric (or Dietrich or, in French, Thierry) of Freiberg (†c. 1310), a Teutonic member of the Order of Preachers who spent his later years in France, was encouraged by the master of his order to put into writing his novel ideas on the rainbow. He had been a professor of theology in Germany, and the author of at least thirty works on metaphysics and optics, of which about a dozen are now lost. From 1285 to 1303 he had occupied important administrative offices in his order in Germany, becoming embroiled in disputes with the Franciscans. He had earned a degree in theology at Paris in 1297. In 1304 he was sent as German elector to the General Chapter of the Order; and in this year at Toulouse he undertook, at the instance of the Master-General, Aymeric de Plaisance, his work on the rainbow. In 1310 he went to Paris as Master in Theology. After 1311 there is no further information about him, from which one may conclude that this is approximately the year of his death.[1]

Theodoric's book on the rainbow, *De Iride et Radialibus Impressionibus*, is a lengthy work, running to a couple of hundred printed pages. The very length of the treatise is unusual, for Grosseteste's *De Iride* had been equivalent to less than half a dozen pages. Albertus Magnus had written at considerable length on the bow, but then the bulk of his uninspired work on meteorology was devoted to recapitulation of the views of others. This is in marked contrast to the exposition of Theodoric who makes virtually no reference to earlier theories of the rainbow. One almost wishes that he had not departed so thoroughly from the Scholastic custom of encyclopedic commentary, for it is almost impossible to reconstruct the historical development of his ideas. It appears nevertheless that he was certainly familiar with the *vade mecum* of all optical investigators of the time—Alhazen's *Treasury of Optics*—as well as with some Greek sources (including Aristotle) and the work of Avicenna. It may be presumed from circumstantial evidence that Theodoric was acquainted also with the chief Latin treatises of the preceding century, for he had been sent to Paris to study; yet he makes no mention of any of his predecessors on the rainbow with the exception of Albertus. Dietrich's *De Iride* is not a work of exegesis but a report of the author's experimental studies and novel conclusions.

Experimentation was far from one of the pronounced characteristics of the Middle Ages; but then neither was it the sudden discovery of modern times, as the cases of Petrus Peregrinus in the thirteenth century and Theodoric of Freiberg in the fourteenth show. Galen in ancient times had distinguished between the *via experimenti* and the *via rationis*, and varying combinations of these had been used, more or less consciously, from the earliest days of science. Grosseteste had seen the need for balance between the two, but he evidently had seen no way of setting up experiments on the rainbow. The story of the rainbow shows that mere observation, no matter how accurate nor how long continued, could not carry man far toward an understanding of the phenomenon. The bow is too elusive, remote, and ephemeral for concentrated study. What was urgently needed was some way of making the bow more tangible—of bringing it into the laboratory, where it could be studied at close range whenever one felt the urge to do so. Just so long as men thought in terms of clouds as the agency producing the bow, no practicable laboratory study seemed at hand. Frequently explanations of the rainbow had referred to raindrops, but it generally was the totality of drops which seemed to be important. When the colors of the bow had been compared with those in the spectrum produced by a globe of water, the globe was equated to the cloud or to an aggregation of drops. Alhazen had made many experimental observations on globes of water, but he was fettered by preconceptions: he thought of the sphere either as a magnifying glass or as a reflector. Never once did the thought seem to have occurred to him, or to anyone (with the possible exception of Albertus Magnus) before the fourteenth century, that a globe of water can be thought of, not as a diminutive spherical cloud, but as a magnified raindrop. This was the brilliantly simple idea which came to Theodoric; and with it he coupled an equally simple postulate—that the rainbow is but the aggregate of the effects produced by each individual raindrop, without reference to the properties of the cone or sphere or other figure which as a totality they might resemble. The deviousness of the psychology of invention is apparent when one realizes that it is quite likely that the idea came to Theodoric through his study of Alhazen's *Thesaurus Opticae*, coupled possibly with hints derived from the remarks of Albertus Magnus or other writers of the thirteenth century.

Theodoric had pierced the veil which seemed to have been drawn over the eyes of Alhazen; but this flash of insight was not in itself sufficient. As an ingredient in scientific success, inspiration undoubtedly constitutes more than one per cent, but it must be supplemented by a large proportion of the proverbial 99 per cent of perspiration; and Dietrich of Freiberg would appear to have possessed experimental skill and persistence as well as theoretical imagination.[2] Then, also, he was thoroughly grounded in the optical learning available at the time. He was a Neoplatonist who had been motivated in his studies by the light metaphysics which was patent in the work of Grosseteste, and had mathematical training. Thus he was admirably equipped to exploit to the full his key to the rainbow, for, as he put it, "It is necessary for optical and philosophical reasoning to be used together in this matter."[3] In reading his treatise, however, one must not expect to find an approach which is totally modern. It has been said that Dietrich, in his work on the rainbow, was three hundred years ahead of his time.[4] In a sense this is true; but in many respects he was also a product of his time and, especially, of the treatises he had read. Passages showing medieval preconceptions will be found side by side with others of striking modernity.

Theodoric's *De Iride* is made up of four parts: first a general treatment of the theory of optics; then the theory of the primary rainbow; next the explanation of the secondary bow; and, finally, a consideration of the other types of ray-induced impressions. His opening remarks remind one strongly of those of Grosseteste: it is the function of physics to tell what the bow is (e.g., a concentration of rays, in damp air or a cloud extended in a colored arc above the horizon), and for optics to determine the *propter quid*, showing the manner in which this sort of concentration can be produced by rays directed to a specific place in the cloud and there, by determined refractions and reflections, redirected to the eye.[5] And a little further on he notes the usual tripartite division of light paths: direct, reflected, and refracted.[6] His enumeration of "radial impressions," including the primary and secondary rainbows, red and white bows, white and colored lunar and solar halos, and parhelia, he believed to be exhaustive; and all of these he said depended upon five basic types of radial phenomena: (1) a single reflection, (2) a single refraction, (3) two refractions with an intermediate internal reflection, (4) two refractions with two intermediate internal reflections, and (5) a total reflection at the boundary of two transparent media.[7] These five types of ray-phenomena he believed could take place either in continuous vapors and clouds, or else in clouds, rain, and mist made up of discrete particles.

The second book opens with sections on the nature of light and colors, topics more fully treated in his separate works, *De Luce* and *De Coloribus*. He admitted that the authority of Aristotle was worthy of respect, but he recognized the higher authority of the evidence of the senses:

We say that one should teach that which the Philosopher said, for the authority of his philosophic doctrine and for the respect it deserves; and each one should interpret that which is said according to his knowledge and ability. But it is to be understood, according to the same Philosopher, that one never should depart from that which is evident from the senses.[8]

Consequently he wrote that "it can be most certainly asserted that the said colours which appear in the rainbow are in truth really four." [9] Theodoric was, of course, unwarrantedly categorical; but at least he supported his argument, not from numerology, as had Roger Bacon, but from experience. (The idea that three is more appropriate for the bow because of the importance of the Trinity he rejected with the observation that man does not have three eyes or three teeth.) His collection of instances would have pleased even Francis Bacon, the reputed founder, some three centuries later, of the inductive method. He cited colors in spider webs, in the rainbow arcs seen in water sprays from mill wheels, in drops of dew on the grass (if one applies his eye close to them), in hexagonal crystals held in the sunlight, in water drops sprinkled into the sunlight and observed by someone standing in the shade. In each and every case the colors and their arrangement are the same: first red, then yellow, next green, and finally purple.[10] His explanation of color formation was a modification of the Aristotelian versions of his contemporaries, the fundamental basis being the combination of light and shade. He believed that there are two formal principles (clear and dark) and two material principles (the bounded and the unbounded); and the four colors corresponded to the various combinations of these, just as in medieval chemistry the four elements were characterized by combinations of two pairs of opposite qualities. Red is clear and unbounded; yellow is clear and bounded; green is dark and unbounded; blue is dark and bounded. In the refraction of light, colors are produced by the weakening which results from the resistance due to the medium. Light falling perpendicularly to the surface, he argued, retained its original strength, and hence rays in this case were untinged. When it falls obliquely, those colors which are less weakened (red and yellow) appear on the side nearer to the line which the incident ray would have followed if undeviated; the colors more weakened are farther from this line. And the colors always follow "in the same inviolable order." [11]

Various *ad hoc* assumptions were necessary before Theodoric could reconcile his theory of color with the facts of experience, but these are of much less interest and value than are his geometric explanations of the formation of the rainbow arcs. Here his experiments with the spherical flask of water, his magnified raindrop, led him to an observation which was crucial in his rainbow theory. His predecessors in optics had studied reflections of light rays by concave and convex mirrors, and they had passed sunlight through transparent spheres. They had noted that some of the light, in the case of transmission through transparent spheres, is reflected at the outer convex surface; but, as is seen most clearly in the case of Witelo, they had overlooked the reflection of light rays by the *inner* concave surface. It is quite possible that Theodoric's discovery of internal reflection in spheres of water was a consequence of his prior discovery of the phenomenon of total reflection which had vitiated part of Witelo's table of refractions.

In the case of light rays which traverse raindrops, the internal reflection is not total; some of the light passes through the rear surface, as Theodoric well knew. But the reflection of light rays at the inner surface is of sufficient intensity to make an impression at the eye of an observer, he found, and this he held to be the explanation of the rainbow. In his words:

Let the radiation enter the oft-mentioned transparent body and pass through it to the opposite surface and from that be reflected internally back to the first surface by which it originally entered, and then after passing out let it go to the eye; such radiation, I say, inasmuch as it is produced by a transparent spherical body, serves to explain the production of the rainbow.[12]

Theodoric went on to explain that rays of the sun strike the upper portion of the globe of water, are refracted into the sphere, are reflected backward at the inner concave surface at the rear of the globe or drop, and then are refracted once more on leaving the lower front portion of the spherical surface to travel toward the eye of the observer (Fig. 18).

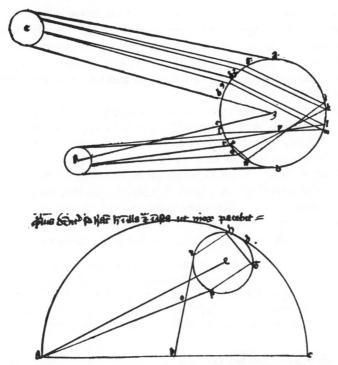

Fɪɢ. 18.—Diagrams accompanying Theodoric's explanation of the rainbow. (Reproduced from Crombie, *Robert Grosseteste.*)

Here, in a clear-cut and unambiguous explanation, one suddenly finds the correct qualitative explanation of the rainbow. The reflection and refraction of rays is not blended vaguely, as in Witelo; their interplay is vividly indicated in diagrams which would not be out of place in modern textbooks. The circularity of the bow had been correctly understood by Aristotle, but his argument had been associated with so devious (and incorrect) a geometrical theory that most of his successors had failed to appreciate it. Bacon, too, had correctly said that circularity was due to the fact that "all parts must have the same position with respect to the solar ray and the eye"; but such an equivocal assertion could take on meaning only when supplemented by precise geometric diagrams indicating the relationships between the sun, the eye, and the raindrops. Obviously, not all rays which enter and leave the drops after an internal reflection are returned in the proper path to reach the eye. In experimenting with his globe of water, Theodoric found that there are only certain determined positions of the drop, with respect to the sun and the eye, at which the rays become visible. This, he understood, was why the rainbow had a circular form. The angle between the path of the rays from the sun to the drop and the path of the rays from the drop to the eye must always be the same, for this is determined by the laws of reflection and refraction. This is the angle which determines the radius of the rainbow arc—that is, the maximum altitude of the bow.[13]

This portion of his work was not, of course, new, for a correct explanation of the circularity had been known to Aristotle and many of his successors. Nor was there anything novel in Theodoric's iteration of the fact that when an observer changes his position, he sees a new rainbow in the sense that a wholly different set of drops is required for its formation. A discovery that was in a sense new and surprising was Theodoric's observation that for any one position of his globe, rays of but a single color could be seen. This meant, of course, that although there appears to be continuity in the colors of the rainbow, in reality each drop is responsible for but one color in the bow.

The colors do not all come simultaneously to the eye when it is in one and the same position with respect to the drop, but different colors come to the eye according to the different positions in which it is put with respect to a particular drop. Consequently, if all of the colors are seen at the same time, as happens in the rainbow, this must necessarily result from different

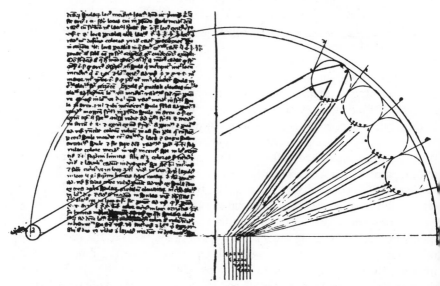

Fig. 19.—Diagram from a manuscript copy of Theodoric's De iride. Note that each of the four drops sends to the eye a different color. (Reproduced from Crombie, Robert Grosseteste.)

drops which have different positions with respect to the eye and the eye to them.[14]

Theodoric's explanation is an eminently clear and essentially correct description of the mechanism producing the rainbow. Had he stopped here, his work would yet have been vastly superior to that of any one of the eminent scholars before him who had sought unsuccessfully to explain the bow. But Theodoric was not through. In Book III he went on to the formation of the secondary arc. Just as he had discovered, through careful observations with his oversized raindrop, that the primary bow belonged in his third category of "radial impressions" (involving two refractions and an intermediate reflection), so also did he find that the secondary arc belongs to type four (caused by two refractions separated by two internal reflections). In the latter case, rays from the sun impinge upon the *lower* portion of the drop, are refracted into it, are twice reflected at the interior surface, and finally leave, with another refraction, from the *upper* portion of the drop to reach the eye of the observer (Fig. 20). In this case, too, the positions of the drops for any one color and any one observer are characteristically determined so that the angle between the rays from the sun and those returned to the eye is constant. Hence the outer bow also is a portion of a circle, and the two rainbow arcs are concentric.[15] And, as in the case of the inner bow, any one drop participating in the formation of the outer bow sends to the eye rays of a single color; and the drops producing the outer arc for one observer are different from those producing it for another.

The story of the rainbow up to the time of Theodoric had centered largely about the primary bow. This could scarcely have been otherwise; until men felt that they correctly understood this, there was little point in attempting an explanation of the secondary. Scientists generally had taken refuge in one or the other of two interpretations—either they held vaguely that it was generated in a manner analogous to that of the primary (in which case the reversal of colors embarrassed them), or they took refuge in a plausible belief that the secondary is a mirror-image of the primary (in which case the concentricity of the bows was a stumbling block). There appears to have been no systematic theory for the secondary comparable to Aristotle's explanation of the primary. Theodoric, however, showed clearly that the explanation of the one is scarcely more involved than is that of the other. One additional internal reflection within each drop was all that was needed. And this simple explanation should have accounted easily for the relative paleness of the exterior bow and for the fact that often it fails to appear when the inner bow is clearly visible. The double internal reflection weakens the light, some of which is lost at each reflection.[16]

Theodoric believed, however, that there are various other factors making for the paleness of the secondary bow. Among these he pointed to the supposed greater distance of the exterior arc from the observer, betraying that he could not completely resist the blandishments of some of the familiar Aris-

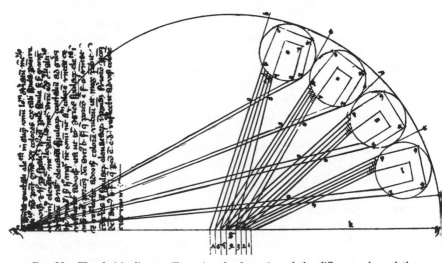

Fig. 20.—Theodoric's diagram illustrating the formation of the different colors of the secondary rainbow. (Reproduced from Crombie, *Robert Grosseteste.*)

totelian ideas. Then, too, the rays causing the higher bow return more obliquely from the drops, and this tends toward greater debility—another Aristotelian argument.

Theodoric went to considerable pains to explain the inversion in the orders of the colors in the two bows. He had noticed, in his refraction experiments, that red invariably appeared on the side of the emergent ray nearest to the path of the incident ray (extended into the medium), blue on the side farthest from the prolonged path of the incident ray. Hence a reversal in the sense of rotation should be expected to cause a reversal in the order of the colors. Theodoric may have had this in mind when he referred to the sunlight as turned "by its incidences and reflections round the opposite concave surface of the drop"; but his reasoning is based on the assumption that, as the rays traverse the drop, those nearest to the outermost portion of the tiny drop are red. (For large spheres, on the contrary, he believed rays near the center are red.) He held that for the primary bow the outermost portion emerges above the blue, for the secondary arc it emerges below the blue. "Therefore it is plain that the colors of this upper rainbow are in reverse order to that seen in the lower rainbow." [17] It must be confessed, however, that this is not nearly so "plain" as are other portions of his reasoning on the rainbow.

Theodoric's work on the rainbow is so obviously superior to anything else which had appeared on the subject that one might get the impression that he had found the explanation of the rainbow. This is, in fact, what can be read in many an otherwise reliable history of science; but it is far from the truth. It is perhaps ungracious to suggest that Theodoric failed on the most important point of all, for he was unable to account either for the size or the shape of the bow. Rays following paths such as those he had indicated should send light to the eye from many directions, and the problem of why only two narrow arcs are brightened, rather than the whole heavenly vault, remained unsolved. Then, too, there was the Aphrodisian paradox to be accounted for. Theodoric, like Alexander, realized that not all rays reach the eye with force sufficient to make a visual impression, and he was one of the first to appreciate how incredibly small a fraction of the solar rays are of consequence in the formation of the rainbow. But why are the rays effective at a particular angle only? His first answer was uninformative: "It is thus ordained by nature." Not really satisfied with this, Theodoric sought to reconcile his own correct qualitative explanation of the rainbow with the incorrect Aristotelian quantitative theory. He invoked the familiar meteorological sphere with the eye of the observer at (or near) the center, with the sun at one end of a diameter, and with the drops of rain arrayed along the inside of the sphere (Figs. 19 and 20). (Theodoric seems not to have appreciated any more keenly than his predecessors the incongruity of placing the sun and the drops at equal distances from the observer!)

Modern though much of his thought was, Theodoric could not break with the Aristotelian notion that the phenomena of the rainbow, including Alexander's dark band, are to be explained in terms of the positions of the raindrops on the beguiling meteorological hemisphere.[18] In the dozen or so diagrams which he used to illustrate the formation of the two rainbows, the drops invariably are drawn along the vertical great semicircle (of this hemisphere) at one end of which the sun is located. The prominence given by Theodoric to this illusory "circle of altitude" (Figs. 19 and 20) indicates how close in quantitative theory he was to Aristotle. He accepted without question the misdirected theory of the effective ratio—the Peripatetic idea that the efficacy of the reflection hinges upon the ratio which the distance between the sun and the drops bears to the distance between the drops and the observer—and he devoted pages of calculations to problems concerning this ratio.

If the ratio were known, the angular radius of the rainbow could be determined; but Theodoric, like Aristotle, was not ready to justify a priori any particular value of the ratio. Instead, he solved the converse problem; he computed, from the known radius of the bow, the effective ratio, and from this he found (in relation to the radius of the meteorological sphere) the height of the bow and the distance from the eye to the center of the bow. His computations are more laborious than those which a student of elementary trigonometry would carry out today, for he used sexagesimal fractions and the ancient Greek geometry of chords. But worse than this is the fact that all of his calculations are practically vitiated by an egregious error in the radius of the rainbow. Whereas Bacon and Witelo had set the apparent angle of the bow at 42°, Theodoric placed the radius at only 22°, not much more than half the correct value. And that this numerical value is not due to the carelessness of those who copied the manuscripts is clear from the further context. Theodoric wrote that the radius of the bow is twice as great as that of the halo of 11° (the *ratio* is essentially correct, but the radius of the halo should itself be twice as large, as Olympiodorus and Bacon had known), and that the radius of the exterior (secondary) rainbow is 33°, or 11° larger than the primary (here the *difference* is approximately correct, but the secondary should be about 20° larger). Could it be that numerological considerations had influenced him in his selection of the values 11°, 22°, and 33°, an arithmetic progression? There is, however, an alternative explanation which, although only speculative, would appear plausible.[19] Theodoric found the Aristotelian ratio in the tables of Ptolemy by reading off the chord of the supplement of the angle DCB (Fig. 21). (The diagram of Theodoric has been slightly modified for purposes of exposition.) That the angle DCB is about 42° was well known from the thirteenth century on, and the supplement of this would be 138°. Theodoric wrote 158° instead of 138°. Perhaps this was a slip of the pen, perhaps it was the outcome of grossly faulty observation. (He repeated that a radius of 22° would be found by one who measures the angle of the rainbow with an astrolabe!) In either case, all of his quantitative work[20] hinges on the single erroneous figure 158°. The chord of this, angle ACD,

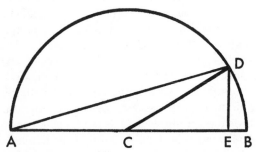

Fɪɢ. 21.—Diagram illustrating the method Theodoric used in finding relative distances associated with the rainbow.

Theodoric found from the tables to be 117 parts and 48 minutes if the radius CB is taken as 60 (as generally was done in ancient calculations). That is, the crucial ratio AD:DC, expressed in fractional form, was found to be $117\frac{48}{60}$: 60. Expressed decimally this would be about 1.96; but decimal fractions seldom were used in mathematics before 1585. (Using the correct angle, the ratio is about 1.87.) The height DE of the bow is of course half of the chord of twice the angle DCB. Again from the tables of Ptolemy, Theodoric found DE:CB to be $44\frac{57}{60}$: 60, equivalent to about .75. To find the distance CE from the observer to the center of the bow, Theodoric took half the chord of twice the complement of the angle DCE (equivalent to finding the cosine of angle DCE). This he found to be 55 parts and 46 minutes, if the radius CB is taken to contain 60 parts, equivalent to saying that CE/CD is about .93. Theodoric's sexagesimal calculations, typical of his time, are clumsier than the modern decimal equivalent; but one who would criticize the medieval age in this respect should note first that vestiges of this ancient Babylonian system of numeration still survive today, in measures of time and angles, in our generally decimal civilization. Although Theodoric evidently placed considerable emphasis on these calculations, they have in reality little permanent value. The ratio AD/DC in particular is utterly meaningless in terms even of ancient and medieval estimates of the solar distance; and in the light of modern astronomy and atmospheric physics this ratio is millions of times that given by Theodoric. On the other hand, it should be noted that his estimate for the secondary rainbow seems to be the first to appear in the literature of science. If upon Bacon's more accurate radius for the primary bow one were to superimpose Theodoric's estimate for the interval between the two rainbow arcs, one would obtain for the outer bow a radius of 53°—a very respectable approximation indeed.

Despite his patient calculations related to the effective ratio, Theodoric must have felt uneasy about the Aristotelian theory, for he linked the orthodox macrocosmic geometric explanation to a new microcosmic consideration of the geometry of the raindrop. In so doing, he proposed the only quasi-quantitative theory of the rainbow to appear in the long interval from Aristotle to Copernicus. Inasmuch as this represents the first effort to associate the size of the bow with the measures of angles relative to the raindrop itself, his theory is of particular interest, although historians seem generally to have overlooked it. Theodoric noted, in connection with his diagrams illustrating the formation of the primary rainbow, that the higher the position of the drop on the circle of altitude, the larger is the arc on the surface of the drop (the arc nz in Fig. 19) between the points at which the incident and emergent rays pierce the drop. Consequently he believed that it is the magnitude of this arc—which we shall call the "Theodorican arc"—which governs the effectiveness of the rays reflected to the eye. The outermost part of the primary bow is red, he argued, because the Theodorican arc in this case is longer, and hence the reflected rays are stronger, than for rays striking other drops at lower positions on the circle of altitude. The lower rays are weaker because of the smaller arc on the drops, and hence the colors are less bright, tending toward blue. For drops at still lower altitude the Theodorican arc is still smaller, and the reflected rays produce only a weak albescent light such as is seen within the primary rainbow arc. Above the red band, on the other hand, the arc is too great, he thought, for rays to be reflected to the eye, for reflection comes only at the proper altitude; and hence the region just outside the primary bow is quite dark.[21]

For the secondary rainbow, Theodoric thought, the situation was just the opposite, for the arc on the drop between the points of incidence and emergence of the rays (the arc px in Fig. 20) seemed to vary inversely with the altitude of the drop on the circle of altitude. Hence the lowest portion of the secondary bow is red (the strongest color) because the arc is longest, and the uppermost part is blue (the weakest color) because the arc is shortest. The region beyond the blue is illuminated by a weak light which tends to white toward the zenith, for the critical arc becomes smaller; and within the red band, he held, no rays at all are transmitted to the eye, following the double internal reflection, because the Theodorican arc is too great. He concluded, therefore, that for the secondary bow, as for the primary, it is the magnitude of the arc on the raindrop between the points of incidence and emergence of the rays which determines the efficacy of the reflection, and hence the position of the rainbow in the heavens.[22]

Theodoric's ingenious theory, accounting apparently for the properties of the two rainbow arcs, and the intervening dark band of Alexander, possesses a high degree of plausibility; but it has two drawbacks. In the first place, it is incomplete, for no reason was adduced to show why the Theodorican arc should determine the quality of the reflection—why color begins to appear only when the arc has attained a particular magnitude, and why, when the arc has reached a certain critical length, all reflection suddenly ceases. In other words, his theory failed to account for the radius of the rainbow. In the second place, as will be seen in connection with the work of Descartes, the Theodorican theory is wrong. To correct the theory, two changes in particular must be made: (1) one must discard the Aristotelian macrocosmic circle of altitude; and (2) one must replace the Theodorican microcosmic raindrop arc (between the points of incidence and emergence) by the geometric angle between the incident and emergent rays. Theodoric, like other ancient and medieval scholars, strangely failed to observe that the drops producing the bow can be either at arm's length from the observer or miles away, providing only that the appropriate angle is formed between the incident and emergent rays. Not until the spell of the meteorological hemisphere was broken did mankind take the decisive step in the theory of the rainbow; and this was not a medieval achievement.

One of the favorite topics of commentators on Aristotle's *Meteorologica* was the non-appearance of a tertiary rainbow. One should have expected that Theodoric, who had done far more on the secondary bow than had anyone before him, would show keen interest in the problem of multiple arcs. Moreover, in the light of hindsight it would appear that the solution of the question of the tertiary bow was at hand. The relation between the first two bows virtually points the way to bows of higher order; but, obvious though it now seems, it escaped Theodoric completely. He had not hesitated to question the authority of Aristotle on other points on the rainbow, but here he found himself in agreement. Impressed perhaps by the fact that there are only two senses for a deviation, clockwise or counterclockwise, he concluded, with respect to the primary and secondary bows, that "except for these two modes of radiation and reflection in the generation of the rainbow, it is not possible to imagine another." [23] Even the potentiality of a tertiary bow, which Peripatetic theory had not denied, is here precluded. And yet Theodoric himself had noted the rainbow-colored arcs which sometimes appear inside, and contiguous with, the primary bow, dismissing them with the inadequate remark that they are generated in the same manner as the primary. He cogently recognized that his theory implied the possibility of analogous arcs on the convex side of the secondary, although he confessed that he had never seen these, perhaps because of the weakness of the radiation. Long afterward these supernumerary arcs were to play an important part in the story of the rainbow, as was also the white fog-bow, to which Theodoric was one of the first to call attention.

Although he correctly held that the white bow is generated in the same way, except perhaps for color, as is the ordinary primary rainbow, Theodoric, nevertheless, included it not with the proper rainbows, but in Book IV, along with halos, parhelia, and similar appearances. Aristotle and his successors had devoted about as much space to the halo as to the rainbow, but for Theodoric the rainbow was the important phenomenon. Nonetheless, he was one of the first to write on the larger halo, and this he correctly ascribed to a double refraction in a medium between the observer and the sun. (Only in the eighteenth century was the precise nature of these refractions determined.) For the smaller halo (the one more frequently noticed and for which he had assumed a radius of $11°$) he adopted a highly novel (but incorrect) explanation.

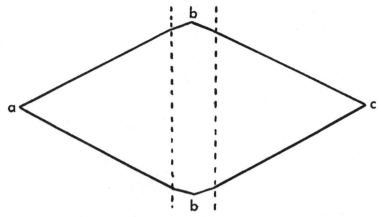

FIG. 22.—Diagram illustrating Theodoric's idea of the formation of the halo.

Light from the sun a strikes the surface of a medium between the sun and the observer c, is refracted into the medium (Fig. 22), then undergoes some sort of internal reflection at points b, then is refracted as it leaves the extremity of the medium nearer the observer to travel to the eye. (His diagram does not show the refractions, but his account makes it clear that he made allowance for these.) The uniformity of the angles of reflection and refraction account for the circularity of the halo.[24] Evidently his success in applying internal reflections in the rainbow led him to carry over the idea to phenomena where it really has no place. Nevertheless, his use of refraction for the halo, when Aristotelians continued to cling to reflections alone, marks his explanation as closer to modern views.

The summary which has been given above of Theodoric's De Iride covers the more salient points. The treatise is long, and it contains other good and bad features. On the credit side one might include the arguments by which he showed the falsity of rival theories. On the debit side one should note that his diagrams picture rays from the sun as non-parallel, those emerging from a given drop being drawn parallel, whereas this is just the reverse of the true state of affairs. But any candid examination of the work can lead only to the conclusion that it represents one of the greatest scientific triumphs of the Middle Ages.[25] It is difficult to overestimate the significance of his step in immobilizing the raindrop. For the first time the rainbow in the laboratory could be studied more thoroughly than the astronomer in the observatory scans the heavens.

The clarity and precision of Theodoric's work is in marked contrast to the equivocal and confused ideas of most of his predecessors, and one may well be amazed at the suddenness with which views of such modernity emerged. But the situation becomes almost incredible when one learns that similar ideas were presented at almost precisely the same time in quite another part of the civilized world. Between the years 1302 and 1311 the Persian scholar Kamal al-Din al-Farisi (†c. 1320) presented a rainbow theory very much like that of Theodoric of Freiberg. Kamal says that he was greatly assisted by a helper who opened the door for him, referring apparently to his teacher Qutb al-Din [or Kittab al Din or Kotb ed Din] al-Shirazi (1236–1311), one of the greatest Persian scientists. Hence the discovery of the theory presumably should be ascribed to the latter, its elaboration to the former.[26] One's immediate reaction is to ask which discovery of the correct qualitative theory of the rainbow came first, the Latin or the Arabic, and then to try to find how the knowledge passed so quickly from the one civilization to the other. The truth of the matter is that we seem here to be faced by two completely independent and virtually simultaneous discoveries. Priority of discovery presumably belongs to Qutb al-Din, inasmuch as Kamal al-Din commented on this work at the same time that Theodoric was writing (during the first decade of the fourteenth century); but there is no evidence of any link between what was being written in Persia and the analogous explanation in Europe. There is every indication that the only connections between these Muslim and Christian explanations of the rainbow were through past ages rather than contemporary intermediaries. The basis for the amazing case of simultaneous discovery lay in the common intellectual heritage available to them. Qutb al-Din and Theodoric of Freiberg both derived their inspiration from two great treatises, the *Meteorologica* of Aristotle and the *Kitab al-Manazir* (*Treasury of Optics*) of Alhazen. But just as there were transition figures in the Latin world of the thirteenth century, so also there were scholars in the Arabic domain between Ibn al-Haitham and Qutb al-Din.

During the twelfth century the flow of learning between the Muslim and Christian worlds was almost wholly in one direction—from the Arabic world toward Western Europe. By the following century the Latin world had made such strides toward intellectual recovery that the two worlds were no longer on grossly disparate levels. Moreover, despite periodic hostilities occasioned by the crusades, there was relatively easy intercommunication between occidental and oriental scholars. Early in the thirteenth century, for example, a series of questions was sent by the scientifically curious emperor, Frederick II (1194–1250), to King Kamil (fl. 1218–1238). These questions, which became famous in the Arabic world as the "Sicilian Questions," served as the basis for much scientific discussion.

In particular, the questions of Frederick II led Al Qarafi (†c. 1285), an Egyptian theologian, to write a scientific work on optics, investigating fifty questions, at least some of which were among the Emperor's questions. In this book, entitled (in Arabic) *The Revelation of What the Eyes May Perceive,*" Al Qarafi took up the following question:

What are the causes of the colors and of the circular shape of the rainbow why does the rainbow appear at certain definite times especially, and why is it sometimes small, sometimes large? [27]

The explanation gives no hint of the revolutionary changes which were to be introduced a generation later. It is, in fact, less original than that given at about the same time in Europe by Witelo, for Al Qarafi had no Grosseteste to work from. His chief sources were Aristotle and Avicenna, and his account is heavily indebted to them. The rainbow is produced by the reflection of the sun in the vapors in the air; but the reflecting mirrors are too small to reproduce the shape of the object. Nevertheless he asserted that the circular shape of the bow is caused by the fact that the disc of the sun is round. That he borrowed from Avicenna is betrayed in the passage in which he explains that to be effective, the little raindrop mirrors must have behind them a hill or a dark cloud, just as a piece of glass will not act as a mirror unless it is backed with some dark material.

The theory of color given by Al Qarafi depends less heavily on Aristotle and Avicenna; it definitely betrays an original touch here and there. He recognized four colors in the rainbow, but of these, one (yellow) is really caused by the mixture or contrast of red and black. The rainbow colors are produced by layers of moist vapor with different properties, the vapors farthest from the earth being "stiffened into stones because of the cold existing in these high regions," the lowest vapors being warmed "through the effect of the heat in the earth." [28] Here one finds a rare instance in which ice crystals are adduced to account in part for the rainbow. But Al Qarafi did not give a mathematical treatment of the bow. In fact, he seems to have made but little use of the geometrical properties found in the great *Treasury of Optics* of Alhazen Neoplatonic light metaphysics did not, in Arabic works, play the same role that it had in thirteenth-century Europe, and so the Muslim study of optics consisted largely in the examination of isolated problems and questions. But then the rainbow generally had occupied a peripheral position with respect to optical theory, for it was more likely to form a portion of a more philosophical meteorology. In the case of Al Qarafi, optics, meteorology, and the rainbow all were far from the focus of his attention, for he was at heart a theologian. As in Europe, so also in the Arabic world it was to be the mathematically minded scholars who were to reap the rewards in the period of medieval triumph.

The Persian scholar, Nasir al-din al-Tusi (1201–1274), was practically contemporary of Al Qarafi, but his main interests were quite different. He was one of the greatest mathematicians and scientists of Islam,[29] known particularly for contributions to the problem of the Euclidean parallel postulate, the development of trigonometry, and the compilation of the Ilkhanite astronomical tables for the Mongol conqueror, Hulagu Khan. Nasir al-din (or Eddin) was especially fascinated by the mathematical, astronomical, and optical works of classical antiquity; but this, unfortunately, was not the kind of background that would lead one to the rainbow. It will be recalled that it had been the philosophers, rather than the mathematical scientists, who had appropriated the study of the bow in classical times. Moreover, the ancients had known far less about refraction than had Nasir al-din's own countryman, Alhazen, whose *Treasury of Optics* he did not properly value. Even if Nasir al-din had turned to a study of the rainbow, he might have done little for it because of his Greek--derived propensity for confusing the phenomena of reflection and refraction.[30] He is, therefore, a link in the story of the rainbow not so much for what might have been, but because he was the teacher of Qutb al-din, the astronomer and physician who did for the rainbow in Islamic science what Theodoric had done for it in Europe. How Qutb al-din discovered the secret of the rainbow is not definitely known, but the presumption is strong that it was his discovery of the worth of Alhazen's *Treasury of Optics* that set him on the right track.[31] He, like Theodoric, had the happy thought of using Alhazen's globe of water as a glorified raindrop and studying the passage through it of rays of light. He too found that the primary rainbow is caused by rays of light twice refracted and once internally reflected in the drop, the secondary bow being due to two refractions and two internal reflections.

Qutb al-Din, unlike Theodoric, did not write a treatise on the rainbow. His explanation appeared in the *Nihayat*, an astronomical work which includes questions on geometrical optics.[32] Further details on his correct theory are found in a commentary by his student, Kamal al-Din, on the *Optics* of Alhazen. Having been shown the hint by his teacher, Kamal made a systematic laboratory study. It is perhaps significant that commentaries, whether Greek, Latin, or Arabic, on the *philosophical* works of Aristotle had not made much progress in unraveling the mysteries of the rainbow, whereas commentaries on the *geometrical* treatise of Alhazen led to triumph over the intractable phenomenon; but it probably was more the new experimental approach, rather than the mathematics itself, which accounted for the success.

Kamal al-Din opened his work on the rainbow by saying that, with the help of God, he would give his ideas as a supplement to the *Treasury of Optics*, taking the best from his predecessors. His mathematics he said he would take from Alhazen and his philosophy from Avicenna. This modest assertion notwithstanding, there is little resemblance between his theory and those of either Alhazen or Avicenna. He cites their ideas with great respect, only to go on with the words, "But I say . . ." Inasmuch as the halo and the rainbow are due to thickened air, Kamal held that they can be treated physically; and since they are circular, they can be handled mathematically. But whereas Alhazen had assumed the thickened air to be a cloud, Kamal postulated a "damp" medium; and where the former had depended upon reflection, the latter asserted that "only through refraction and a bending toward the perpendicular will one come on the correct explanation." [33] Consequently Kamal made a careful study of the paths of rays of light through simulated raindrops. In a dark room he placed a glass sphere full of water in such a position that it would be struck by rays of the sun which entered the room through a hole, and he studied the colored bow which was formed. Like Theodoric, he was systematic in his experimental study. Anticipating a practice adopted by Descartes long afterward, he interposed an opaque screen between the light source and one-half of the spherical surface while he studied the bow formed by light on the other half. He traced correctly the paths of numerous rays (from a point source of light) through the sphere of water for various cases—two refractions and no reflections, two refractions and one reflection, and two refractions and two reflections (Fig. 23). This work was a systematic extension of ideas of Alhazen; but Qutb al-Din and Kamal al-Din were the first Arabic scientists to see in it the key to the rainbow.[34] The last-mentioned also came close to answering the problem of the tertiary rainbow. He closed the report of his experimental work with the statement that there is a ray thrice reflected in the spherical globe of water. "One can observe this, but it is difficult." [35] But when speaking of the rainbow he seems to forget all about this, for he categorically rejects the possibility of the appearance of three bows. As a matter of fact, he cites the non-existence of the third bow as an "indestructible contradiction" of an earlier Arabic dissertation. Scharaf al-Din al-Masudi, he says, had held the second bow to be caused by the rising up of a second portion of the cloud. Kamal argued that in this case, if a third portion were to obtrude itself, a third bow should result, whereas at most two bows are seen, never three.[36]

As was true of Theodoric's work, so also in the case of the Arabic co-discoverers, the explanation of color formation was weak. Kamal al-din said vaguely that the colors blue, green, yellow, red, black, and white were due to differences in the strength of the refracted and reflected light, rays touching a sphere tangentially being weaker than those incident near the axis. Moreover the work of Qutb al-Din and Kamal al-Din lacked the quintessence of physical theory—a *quantitative* explanation. These two men were no more able than was Theodoric to explain why the size of the bow is what it is. Worst of all

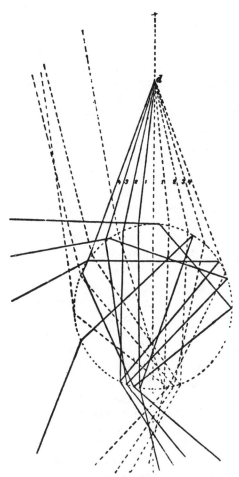

Fig. 23.—The passage of light rays, according to Kamal al-Din, from a point source d through sphere in which they undergo 0 or 1 or 2 internal reflections. (Reproduced from Wiedemann.)

however, was the fact that in the Islamic world there were no successors to carry on the work they had so well initiated. In the story of the rainbow the Arabs share in the brilliant medieval triumph, but then they fade from the scene. The future belonged to the Latin civilization; and there Alhazen and Avicenna continued as wells of inspiration, but the ideas of Qutb al-Din and Kamal al-Din remained, so far as we are aware, quite unknown.

The opening decade of the fourteenth century was a memorable one in the theory of the rainbow, both in the Islamic world and in Christendom; but the growth of science is far from uniform. As in biological evolution, so also in human culture mutations must be expected. Some of these, such as the idea of Qutb al-Din and Kamal al-Din, lack survival value. Was Theodoric's theory to share this fate or was it viable? Certainly the immediate reception of his work was far from auspicious. Interest in the rainbow, both in Europe and elsewhere, must have fallen off sharply, if the scarcity of surviving material for the first half of the century is accepted as an indication of this. Toward the middle of the century, however, there are at least half a dozen works in which one again picks up the thread of the story. One of these was the *Buch der Natur*, written in 1350 by Konrad of Mengenberg (1309–1374), a book which was especially influential for well over a century. The author was a Frank who had studied at Erfurt and Paris and then settled at Regensburg to become canon of the cathedral there and to write works popular in the Bavarian-Austrian world. His *Book of Nature* includes a moderately long chapter on the "Regenbogen" which is not entirely original, being based on a manuscript by Thomas of Cantipré (although Konrad believed it to be by Albertus Magnus). The bow is caused by the reflection of the sun's rays in small droplets behind which there are dark clouds, just as the maker of mirrors covers the back of the glass with lead or pitch. Inasmuch as large mirrors reflect the sun more easily than small ones, the drops causing the green color are smaller than those producing the yellow and larger than those yielding the red. The rainbow can not appear near noon in summertime because the sun is then too high over our head. For a bow to be visible the sun must be on one side, the rays must be reflected from the other, and the observer must in between. When a rainbow does appear around midday, this foretells a heavy rain, for watery clouds are in the air; when it appears near sunset, this signifies good weather. There evidently was no conflict in Konrad's mind in the use of the bow as both a natural and a divine sign—a harbinger of weather and a seal of God's promise.[37]

What a far cry are Konrad's mid-century musings from the sophisticated geometrical explanations which had appeared only a generation before in one of the very cities in which he had been educated! The very word refraction had disappeared, carrying the story of the rainbow back to pre-Grossetestean ideas. But, fortunately, Konrad was not the only writer of his time who penned thoughts on the bow. There was at the University of Paris during the time that Konrad flourished a quartet of brilliant scientists, two of whom are sometimes referred to as initiators of the modern age in science. One of them was Jean Buridan (†c. 1358), whose opposition to Aristotelian doctrines made him one of the founders of modern dynamics. One eminent historian of science, Pierre Duhem, would place the precise line separating ancient from modern science at the time when Buridan applied the theory of impetus to the heavens, thus breaking with the ancient distinction between terrestrial and celestial motions.[38] It has been claimed that from Buridan, Galileo borrowed the idea of momentum, Descartes the principle of the quantity of motion, and Leibniz the doctrine of *vis viva* (kinetic energy).[39] While such claims must be properly qualified,[40] and although the notion of inertia had roots in antiquity, nevertheless Buridan was by any standard an outstanding figure in the history of science.

A second member of the quartet, a close associate of Buridan, was Nicole Oresme (c. 1323–1382), Bishop of Lisieux and advisor to Charles V of France. He has been hailed as a genius "whose contributions to economics, physics, astronomy, and mathematics marked the beginning of the modern period in experimental science." [41] The holder of more precursorship claims, advanced by modern admirers, than any other medieval scientist, he has variously been regarded as a forerunner of Copernicus, Galileo, and Descartes.[42] His appreciation of the relativity of motion led him to advance arguments in favor of the diurnal rotation of the earth; but more important, probably, were his contributions to mathematics. His introduction of generalized powers in algebra, his anticipation of the graphical representation of functions in analytic geometry, and his study of instantaneous rates of change, all show the originality and power of his "hunches." [43] In particular, his observation that the rate of change of a variable quantity is least at its maximum value later played a key role in the calculus and the theory of the rainbow.

Albert of Saxony (1316?–1390), a third member of the scientific quartet at Paris and transmitter of this science to Vienna, wrote on logic, physics, and geology. His work on the void, on centers of gravity, and on terrestrial erosion may well have been used later by Leonardo da Vinci. Albert's associate, Themon Judaei (or Judaeus) or Themo Judoci or Thimon le Juif (fl. 1349–1361), author of commentaries on physics, astronomy, and meteorology which also probably were used by Leonardo, formed the fourth member of the closely-knit Parisian group. So intimately related was the work of the four men that it is virtually impossible to disentangle the literary threads to determine the original authorship in cases in which all of them wrote similar works on the same topic. All four of them were, of course, strongly influenced by Aristotle; but during the thirteenth and fourteenth century a reaction had set in against servile acceptance of Peripatetic teaching. This reaction probably had been a factor in the condemnation at Paris in 1277 of Averroism, for, as we have seen in his work on the rainbow, Averroës had given to Aristotelian ideas a dogmatic expression which was not in keeping with the intentions of the original author. The more critical form of Scholastic Aristotelianism formed a hospitable background for novel views in physics, as is quite apparent in the temper of the fourteenth century. Then, too, there was an increasing awareness of the need for a mathematical treatment of physics. The Aristotelian four causes were not necessarily abandoned, but there was a definite inclination to regard quantitative formulations as adequate explanations of natural phenomena. Emphasis on the metaphysical "why" was being shifted, in scientific discussion, to a mathematical "how." This attitude boded well for advances in the theory of the rainbow. Moreover, the four Parisian scientists mentioned above were not only leaders in the mathematization of science; they all were interested in the rainbow to the extent that each composed a work on the subject. In each case the title was, with minor variations, *Quaestiones in Libros Meteororum* (i.e., questions on Aristotle's *Meteorologica*). Elucidations of Aristotle in general were appearing at the time in the new genre, the series of *Questions*, replacing the older commentaries and glosses. They seem to have been to that day what textbooks, or the professor's notes, are in ours. (Themo was a prominent teacher at the University of Paris, being thrice elected proctor of the English nation.) [44] For this reason the *Quaestiones in Meteora* are important to the story of the rainbow for the portrayal of the nature of university instruction in the subject. It will not be necessary to analyze each of the four versions of the *Quaestiones*—by Buridan, Oresme, Albert of Saxony, and Themo—for the content of these is close to identical.[45] In some cases the questions are listed in the same order, they generally are stated in much the same language, and the answers are similar. Were the four versions reports on original research, one would charge plagiarism; but in this case the use of similar lecture material presumably means that the four authors all were associated in some way with the same course given at the University of Paris. One would like to know which version came first, but no definite decision can be made. (Extant manuscripts cannot be relied upon, for the *explicit* date, 1366, of a copy of Buridan's *Quaestiones* is posterior to the death of the author.) Buridan being the oldest of the four, we shall refer to the group as the "Buridan school"; but this does not necessarily mean that the ideas all are to be ascribed to him. If survival value is used as a yardstick of intrinsic worth, the *Quaestiones* of Themo was superior to the other versions. Again and again he is cited by medieval and early modern writers on the rainbow, and his *Quaestiones in Meteora* was published several times between 1480 and 1518. The versions by Buridan, Oresme, and Albert of Saxony, on the other hand, rarely are referred to in later works, and they seem never to have been printed.

Themo's *Quaestiones* was not a systematic exposition of the theory of the rainbow; it was to be used simply as a companion volume to Aristotle's *Meteorologica*. The two works, at least from the pedagogical point of view, turn out to be strange bedfellows, for Themo seems to be determined to deny whatever Aristotle affirms and to affirm what Aristotle denies. (This is scarcely what one has been taught to expect of the medieval period, with its presumed respect for authority.) And one's sympathy must go out to the students at the

University of Paris around 1350, for it is difficult indeed to make out just what Themo really believed about the rainbow. Following a question, he sometimes gives several interpretations and objections, and objections to the objections, until the reader (and the author also, if one may judge by the apparent contradictions) is lost. One finds himself longing for the clarity of Theodoric—or even for the unambiguity of Aristotle! And yet in the *Quaestiones* one is in touch with the foremost teachers of the age.

The opening question on Book III (the book in which one finds the material on the rainbow) is relatively direct and free from difficulty, but even here one gets a touch of the writer's captious attitude. It inquires whether rays of light always are refracted on passing from one medium into another of differing density; and it is argued that this is not the case inasmuch as a ray perpendicular to the surface is not refracted, contrary to the authority of Aristotle, who had not made this exception.[46]

The second question is practically a repetition of the first. The next two questions are on reflection from the surfaces of transparent and non-transparent media, and here Themo makes the point that wherever there is refraction, there is reflection, but not conversely. The following five questions concern the halo; and here the author is uncertain. At first he argued that, contrary to Aristotle, it is caused by refraction; but later he recognized two possibilities, either reflection at an obtuse angle or direct refraction. On the position of the vapor producing the halo he again held Aristotle to be in error, saying that it does not lie between the eye and the source of light. Theodoric had proposed a halo theory involving refractions and a reflection in a medium with plane surfaces. Themo says nothing about the internal reflection, but he does argue against two refractions through such a medium, showing that the emergent rays diverge. He proposed instead a refraction through many spherical transparent bodies, a theory not too far from that now accepted. His accompanying diagrams show a generally satisfactory understanding of the refraction of rays through transparent spheres and flat plates.[47] In question 9, "Whether the halo is the periphery of a circle," he again raises a quibble against Aristotle. The answer is "No," because, unlike the pure circle of mathematics, the halo arc has width. In question 10, "whether the visual ray can be reflected by uniform non-condensed air," the affirmative reply is supported by references to Grosseteste and Alhazen.

Themo raised fifteen questions on the rainbow itself. The first of these, question 11, is without doubt the favorite of medieval times: "Are the colors in the rainbow real, and are they where they appear to be?" At first he adopts the affirmative view, citing Albertus Magnus, and arguing that the colors appear from the diversity of refractions "according to the diversity and disposition of the body upon which the light is incident." When rays of the sun are refracted through a prism and are reflected by a mirror, the colors seen are real. In the same way two globes of water can be so arranged that rays are refracted in one and reflected by the other to produce the rainbow colors. This arrangement evidently was suggested by Grosseteste's theory. Nevertheless, Themo objects to the idea that the colors are a real thing projected on the cloud. In this case they could be seen from the rear of a diaphanous cloud, which is contrary to experience; nor are they visible from the side, as is clear from artificial rainbows. Themo, therefore, believed such a theory of the rainbow to be false.

The twelfth question, whether the bow is a real shape impressed by the cloud or an illusory image of the sun, is argued pro and con without a definite conclusion. Question 13, whether the rainbow is caused by reflection or refraction, would appear to be a crucial one; yet here too Themo is indecisive. He argues first that the bow is caused by refraction; were it caused by reflection in a concave mirror, the rays would have a focus, which is not the case. In the end it is argued that it can be produced by reflection and refraction, and here Themo seems to have in mind the ideas of Witelo. The indecision continues into question 14, which is hypothetical: If the iris is caused by reflection, is it from a cloud or from drops? Here, for a change, the reply is definite; the reflection is not from a continuous surface, but from many drops. In question 15, whether the iris is tricolored, Themo breaks again with Aristotle. The diversity of color is due to the intensity of light in reflection and refraction, with red being refracted more intensely than the others through a denser medium. Inasmuch as many more than three types of reflection and refraction are possible, there are infinitely many potential colors. The weakening of rays can be of many degrees in intensity, but Themo does not make clear how this comes about. At one point he says that it is not due to diversity in the media but rather to the nature of the rays. This statement is vaguely suggestive of what technically is known as the dispersion of light later discovered by Newton; but one can not identify Themo's diffuse remark with any clear-cut discovery.[48]

The explanations of Themo, at least up to this point, may not compare favorably with those of Theodoric or Kamal al-din; but it must be admitted that they show, presumably on the part of the Buridan school, a high degree of originality and independence of judgment—as well as of captiousness. The answer to question 16 brings this out strikingly. In reply to the familiar query on the possibility of more than two rainbows, Themo's first reply is characteristically anti-Aristotelian: There are as many bows as there are eyes. This, of course, was no new observation; but he went on to say that just as there are many different kinds of artificial bows, so also in nature there can be bows from different materials, some subtle and others grosser. Moreover, natural bows can be reflected to give others in different places, so that from two original bows one gets four in the end. Citing Witelo's observations, Themo concluded that there can be infinitely many bows beneath the primary arc or above the secondary. In a sense this was a remarkably prophetic assertion; but the lack of anything like a systematic approach to the theory of the rainbow in Themo's work meant that it played no role in the ultimate solution of the question of multiple rainbows.

The most important statement by far in the whole of Themo's *Quaestiones* is one which is made almost casually in connection with question 16. The author is comparing the order of colors in the rainbow and those seen in a spectrum produced by a globe of water. He was puzzled by the fact that the spectrum from the globe showed red at the bottom and blue at the top, whereas in the primary rainbow the opposite is the case. (He was not aware of Theodoric's discovery that, for a given observer, the globes (drops) which produce the red of the bow are not those which cause the blue.) He therefore continued as follows:

From this I imagine that not only the convex surface of a sphere of moisture reflects rays to the eye, but also that after it falls obliquely on the convex surface and is refracted and penetrates farther to the other surface on the opposite concave part with respect to the center of the sphere, it is reflected in this so that the angle of incidence is equal to the angle of reflection. And because the eye is beyond the focus of the rays from the concave mirror, therefore the image appears reversed.[49]

Note that here Themo, like Theodoric, appears to be in possession of the secret of the rainbow; but he failed to see its significance. For him it was not *the* explanation of the rainbow; it was simply a device which was necessary to reconcile his ideas on color formation with the observed facts. He did not even feel it necessary to call attention to the refraction of the rays on leaving the drop, although an earlier diagram on refractions through spheres (without an intermediate reflection) makes it appear likely that he was aware of the need for this. It certainly is too much to say, as has Pierre Duhem,[50] that "the system of Thierry of Freiberg, at least that part relating to the primary rainbow, was reproduced about 1360 by Themon, 'Son of the Jew.'" Examination of the two works will show how great was the difference. Theodoric's explanation is clearly set forth and is bolstered with beautifully precise diagrams. Printed editions of Themo's questions include some suggestive figures for the halo, but there are no illustrations accompanying his casual remark on the internal reflection which takes place in the case of the rainbow. Nor is there a satisfactory extension of his key passage to cover the secondary bow. For the lower bow he had assumed that the red rays were produced by those

rays which penetrate less of the drop (reminding one of Theodoric's statements), the green by those which pass nearer the middle and hence traverse
more of the medium. For the upper bow Themo again postulated an inverted
impression from the reflection by the concave surface; but he asserted that
in this case the drops are higher, and hence the rays which for the lower bow
pass through the thickest part of the medium are those which for the upper
bow traverse the lesser portion of the medium, and conversely.

Thermo's lack of precision in the statement of a problem or its solution
is particularly apparent in question 17, whether the colors of the second rainbow arc always are ordered in a manner opposite to that in the primary. Citing artificial bows, in which he believed the colors are not necessarily con-
traposed, and pointing to Witelo's observations, he incorrectly answered in
the negative. No question, perhaps, better illustrates the retrogression from
the work of Theodoric to that of the Buridan school than does Themo's question 18, whether it is necessary for the colors in the upper rainbow to be
paler than those in the principal bow. Theodoric had given a beautifully clear
explanation of the double rainbow arc, but Themo's account of the generation of the secondary bow is far more vague and diffuse even than that for
the primary. Aristotle had argued that the greater paleness of the secondary
is due to the greater height and distance from the eye; but this Themo easily
disproved by pointing out that at times the altitude of the secondary arc is
as low or lower than the inner bow is at other times, yet even then it is not
so bright as a primary bow of comparable height would be. After denying
that the difference in brightness is a result of the diversity of higher and
lower humors, or that the exterior bow is a reflection of the inner arc, he suggests that it is due to the fact that the rays producing the outer bow are
farther from the "axis," or perhaps that the refraction takes place at the edge
of the cloud. "Many causes combine in the weakness of the outer bow," he
said, "and Aristotle had not uncovered them." One is prone to add, "And
neither did Themo."

There is nothing in the remaining questions which is likely to salvage the
reputation of the vaunted Buridan school in its handling of the theory of the
rainbow. Question 19 raises the banal query, whether the bow is the periphery of a circle, if there is no impediment; and the reply is a quibble—"No,
because it is semicircular." Themo's iconoclasm leads him to deny even that
the sun, the eye, and the center of the bow are collinear (question 20), or
that the diameter of the iris is always double that of the halo (question 21).
He realized that others have claimed that the diameter of the halo and the
radius of the rainbow both are 42°, but he reiterated his position, taken presumably from Witelo, that the radius of the rainbow is subject to variation

n the medium producing it. (In this he was technically correct, although for all practical purposes he was wrong.)

Question 22, "Can the rainbow be seen at all hours on all days?" is first answered by a characteristic bit of sophistry: "Yes, if one includes all people and all artificial rainbows." But then the author admitted the restrictions on place and season, depending on a maximum altitude of 42° for the bow, which long had been a part of Aristotelian commentaries. In question 23 one finds the old chestnut, "Can lunar bows appear more than twice in fifty years?" necessarily answered in the affirmative; and in answer to question 24, inquiring whether the radius of the bow is larger when the visible portion is smaller, Themo answers no, harking back to his view that the radius depends on the density of the vapor in which it is refracted. He is less certain, in question 25, about the reality of the circular arcs seen about candles, and in question 26, the last one concerning the rainbow, he could not be sure that the causes of parhelia are the same as those of the halo and the rainbow.

The account of Themo's views on the rainbow which has been given above is a considerably abbreviated summary of what he actually wrote. In the printed edition this material covers over fifty pages of a good-sized book, indicating that the author regarded his thoughts as more than mere *obiter dicta*. The significance of the work, moreover, is heightened by the fact that it is in a real sense a group project, representing the best thought then current among the outstanding scientists of the day. The notable weakness on the quantitative side cannot be attributed to any mathematical ineptitude on the part of the author, for examination of the *Questions* of Oresme, probably the foremost mathematician of his day, shows the very same deficiency. This work is, in fact, virtually identical with that of Themo, making it certain that they are products of the same school of thought.[51]

Oresme once had boasted that everything measurable can be represented geometrically; and in other connections the Buridan school tried to give quantitative treatment of things they could not even measure, such as intensity of heat and light. This school was primarily responsible for the triumph of the mathematical over the philosophical approach to physical problems,[52] and yet their handling of the rainbow looks almost like a revolt against the careful geometrical approach presented by Theodoric about half a century earlier. It is, of course, possible that Theodoric's *De Iride* was not known to the Buridan school; but this would appear unlikely. In the first place, Theodoric had been associated with the University of Paris, and this was the very center of the Buridan group. At least three manuscript copies of the *De Iride* survive,[53] showing that someone had found the work worth reproducing. His pupil, Berthold of Mosburg, in 1318 quoted his ideas in a commentary on the *Meteorologica*; and at the University of Erfurt at least portions of his explanation were taught and published in the sixteenth century.[54] How could his work escape the attention of the alert and progressive Paris group headed by Buridan? And yet Themo, who cited so many other writers on the rainbow, never once mentioned Theodoric. Oresme envisioned the possibility of worlds beyond our own, but paid no attention to a sophisticated theory of the rainbow which had been proposed at his own university. Here and there in the above summary of Themo's work, minor points of similarity between Theodoric's De Iride and Themo's Quaestiones have been pointed out, but these can be accounted for by making the very reasonable assumption that the two men made use of common sources. After all, the experiments on globes of water carried out by Kamal al-din were inspired by Alhazen. Is there then any reason why one should look to Theodoric, whom Themo does not mention, rather than to Alhazen, whom he does name, for the source of Themo's analogous experiments? Three independent discoveries from different sources would constitute a strange coincidence; three stemming from a single source should occasion no great surprise. Then, too, the alternative to admitting that the Buridan associates were not familiar with Theodoric's explanation is the less credible assumption that they knowingly disregarded it because they failed to appreciate it. The superiority of De Iride over Quaestiones is, at least in the light of modern hindsight, so clearly apparent that one is inclined to reject impatiently any such suggestion. One must constantly bear in mind, however, that the portions of past scientific works which strike a reader of today as of particular significance are not necessarily those upon which contemporaries placed the highest value. Theodoric had not been able to explain the size of the bow, and his theory of color was far from convincing. It may be that his immediate successors, unlike ourselves, found that their questions on the rainbow were not materially simplified by his explanations. From our vantage point many of the medieval criticisms of Aristotle appear to be trivial; but to those brought up within the Peripatetic system they loomed larger and had greater significance.

The neglect of Theodoric's explanation is difficult to account for, but it is easy to appreciate the continuing influence of Themo's work. His Quaestiones formed part of a recognized course of study given at one of the greatest universities by the foremost teachers. It is consequently only natural that the Questions of Buridan, Oresme, Albert of Saxony, and Themo should find their way to other centers of learning. Paris and Oxford in particular were at the time closely related by scholarly ties, and Buridanian doctrine flourished in England as well as France. There, at some time between 132 and 1369, was composed a commentary, formerly ascribed to Duns Scotus but now believed to be by Simon Tunsted († 1369), on the Meteorologica which shows the influence of the Paris physicists.[55]

One reads that light from the sun is colored on going through drops of water, just as in passing through a spherical globe of water. The light then is reflected from drops behind, as in little mirrors. (Note that Themo's unobtrusive reference to an inner reflection has been overlooked.) The light is returned in a direction determined by the law of reflection. The colors are due to the fact that some of the light traverses the more tenuous portion of the drop and is therefore clearer, some passes through the more dense portion near the center and emerges as purple (an obscured red). "And what is said of one drop can be said of all." Some color, nevertheless, is produced by the mixture of light reflected directly with that which is reflected after having traversed the drops. The colors appear at a position determined by the angles of incidence and reflection.

Well-worn questions—whether the colors of the iris are real or illusory, if the bow always is circular, whether or not more than a semicircle can be visible, whether reflection from small mirrors results in color rather than image—are argued with a diffuseness reminiscent of the ancient commentaries and the Paris questions. A geometrical argument, however, obtrudes itself incongruously into the discussion when the pseudo-Scotus deals with multiple rainbows. Here he argued, with ingenuous geometrical naïveté, against Themo's belief in the possibility of indefinitely many rainbows. He drew a semicircular bow of radius 42°, and inside this he constructed another semicircle of radius 39°; then inside the latter he drew a third semicircular arc of radius 36°. (The figure 42° was, of course, the traditionally accepted radius of the rainbow, but there seems to be no particular significance in the other values, other than the equidistance.) From the obvious fact that only two equidistant arcs can exist at a fixed angular distance (in this case 3°) from the 39° circle, the author concluded that a tertiary bow is an impossibility! [56]

Simon Tunsted did not mention either Themo or his suggestion of an internal reflection in raindrops. Nor does one find this idea in the *Questiones super Perspectivam* ascribed to a slightly later writer, Henry of Hesse († 1397). This is a commentary on Peckham's *Perspectiva Communis*, in which, of course, the influence of Grosseteste and Bacon are obvious.[57]

The fourteenth century, which had opened with so auspicious a chapter in the story of the rainbow, was showing signs of decline. The precipitousness of this becomes apparent in the *Tractatus Super Libros Meteororum* composed, probably, between 1372 and 1395, by Cardinal Pierre d'Ailly (1350–1420). The author, Bishop of Cambray, achieved scientific eminence through his *Imago Mundi*, a book which Columbus perused with care and which may, therefore have played a part in the discovery of America.[58] If Pierre d'Ailly had not lived, the history of the New World might have been differ-

ent, but certainly the story of the rainbow would have been much the same
His *Tractatus* is a brief commentary on Aristotle which, for primitiveness, re
minds one of Latin works from the twelfth century. So far as Pierre d'Ailly'
account of the rainbow is concerned, the brilliant contributions of the pre
ceding two hundred years were non-existent. He mentions no one but Aris
totle, and his own explanations show little trace of other influence. How
could a writer on meteorology, more especially one who had been educated
at the University of Paris and who flourished there, ignore the thoughts no
only of Grosseteste, Albertus Magnus, Witelo, and Theodoric, but ever
of the Buridan School?

In 1384 Pierre d'Ailly became the director of the celebrated Collège d
Navarre, where Nicole Oresme had taught and translated Aristotle but a few
short years before; and yet one finds no thread linking the ideas on the rain
bow entertained by these two distinguished scholars, the "best mathematician
and astronomers of the age." [59] That Pierre d'Ailly was active in administra
tive capacities (playing a prominent role in the Council of Constance which
healed the Great Schism in the Roman Church) scarcely explains his dis
regard of the rainbow literature of the two centuries before him. Perhaps he
placed too great a value on his own scientific powers, for his explanations ar
characterized by a marked degree of ingenuous novelty. The order in his dis
cussion of the "impressions of the air" is that of Aristotle—first the halo, ther
the rainbow, and finally parhelia and virgae—and the aspects he touches or
are largely those of the *Meteorologica*; but every now and then Pierre
d'Ailly proposed alternative notions—or let slip a peculiar *faux pas*. He ac
cepted a tricolored arc; but he repeatedly denoted the highest "periphery" a
purple, the innermost as red, with green between them, reversing the orde
seen in nature.

His explanation for the order he gave is that a higher arc is formed in
colder region, and the higher rays therefore are weaker. The color in the oute
band, tending toward darkness, consequently is purple; that of the inner o
lower (warmer) region is clearer, tending toward white, and thus is red. Th
middle band, in a mean thermal position, is green. [60] He confirmed his order
ing of the colors through another egregious rationale which he admits was no
given by Aristotle. The bow, Pierre d'Ailly held, was caused by reflection from
a cloud, and he suggested that differences in color resulted from diversity i
the density of the cloud. This was not, of course, a new idea; but the manne
in which he interpreted it was indeed novel. When the watery cloud is abou
to be resolved into rain, he believed that the lower part is more subtle thar
the upper because of "a certain dilatation which is necessary in the separatior
of the inferior part from the superior." Hence the lower portion of the bov
(that is, the part nearest the place of separation) is caused by the rarer me

dium, and this periphery is red; that part of the bow which is higher, and farther from the place of separation, is produced by the denser medium, and this arc is purple. Green occupies the middle or mean position. This extraordinary interpretation served the author as a ready device in accounting for the inverted ordering of the colors in the secondary rainbow. This inversion, he noted, commonly produces great admiration; but it is easily explained in terms of the disposition of the cloud. In the case of the secondary rainbow, the "place of separation" is toward the upper portion of the cloud, and hence the outer periphery of the outer bow is of the same color as the inner periphery of the inner bow.[61]

Pierre d'Ailly's theological bent is apparent in his argument that before the days of Noah there were no rainbows seen, despite the admitted presence of moist clouds and bright solar rays. He held that at that time there was an insufficiency of cloud mass in direct opposition to the sun, for God wished to reserve the appearance of the bow until the day when he should use it as a sign of the covenant that the world was not to perish in a flood. As a work of science perhaps nothing better illustrates the futility of Pierre d'Ailly's account than does his "explanation" of the roundness of the bow. "It is to be noted that the cause of the circularity of the iris is a disposition of the cloud toward a certain circular disposition, partly from the circular figure and partly from the incidence of the rays in the form of a circle." [62] When the greatest cosmographer of the day could prefer such a fatuous statement to the correct geometrical reasoning of the author upon whom he was commenting, should one expect more of his contemporaries?

The scientific literature that has come down from the fifteenth century is considerably smaller than that from its predecessor, but so low is the quality that the recovery of more of this material seems scarcely calculated to alter the history of the theory of the rainbow during this unhappy and retrogressive period. The work of Theodoric had not actually been lost, as is known from the fact that Regiomontanus (1436–1476), the greatest astronomer of the century, at one time planned to print the work. He had set up his own printing establishment to facilitate the publication of scientific works, but he died prematurely in 1476. This is particularly regrettable in the story of the rainbow because Theodoric's ideas were not generally known or appreciated at the time. In the year of Regiomontanus' death there appeared a commentary by Gaetano di Thiene (1387–1465) on the *Meteorologica* of Aristotle, and this shows no familiarity with internal reflections in raindrops. It leans heavily on the ideas of Albertus Magnus, who is frequently cited. The author, a philosopher of Padua, is on the whole a good Peripatetic, repeating Aristotelian arguments with general approval; but on one point he disagrees:

Aristotle doesn't seem to have distinguished between reflection and refraction. . . . But there is a big difference.

At first the author held that the halo is caused by refraction, the rainbow by reflection; later he attributed both to refraction. This sounds promising, but closer examination shows that he had no clear idea of the operation of this phenomenon. The diagram accompanying the account of the halo is identical with those inherited from the Aristotelian reflection tradition. Diagrams for the rainbow also are in the traditional Aristotelian manner.[63] There is little point in following Gaetan de Thiene (or, for that matter, others in the fifteenth century), over the familiar battleground of Peripatetic controversy, for there is nothing in his work which was not earlier in that of Albertus Magnus a couple of hundred years before. One finds, for example, the familiarly vague remark that some say that there can be only two bows from one cloud, but that others may be possible from other causes. All of the ground which had been won by Theodoric, and even the modest advance by Themo, had been lost.

The development of science is unlike the smooth and monotonically progressive path which one would hope for. Gains are so easily followed by loss; and what one generation builds, another tears down. And yet all is not lost. Man still by nature desires to know; and if he will but appreciate that the preservation of knowledge is just as important as its acquisition, the spiral course of science may yet be directed always upward. The level of science in the medieval world was not nearly so abysmal as is our ignorance of its contributions. May the successor to our age be less ungrateful.

Comments on Rainbow Photographs
by Robert Greenler

IN THE ORIGINAL EDITION of this book, a selection of rainbow photographs was included but, I'm sure to the author's disappointment, they were reproduced in black and white. Following eight paintings depicting rainbows in various cultures, we present a series of eight recent photographs, each of which illustrates a special feature of rainbows. Most of these features are discussed in the text. The last photograph, however, shows something that Boyer did not discuss—and it is reproduced in black and white.

9. This wide-angle view of the rainbow shows that, as predicted by Descartes, the primary and secondary bows lie on circles that are centered on the antisolar point. Here the antisolar point is marked by the shadow of the photographer's head (actually his camera) on the ground. The brighter sky inside the primary bow is easily observed. On the right, the brighter sky inside the primary bow and outside the secondary bow shows Alexander's dark band between the two bows.

10. A sun, higher in the sky, produces a rainbow lower in the sky, so that only the upper part of the arc is seen. The primary bow in this photo is accompanied by a beautiful set of supernumerary bows. Several of these interference bows can be seen just below the main arc of the primary bow.

11. Boyer mentions the influence of the index of refraction of the raindrops on the angular size of the bow. This unusual photograph shows the effect in a dramatic way. The upper part of the primary bow is the usual arc resulting from raindrops in the sky. The lower part results from the saltwater drops of the ocean spray. The higher refractive index of saltwater, compared with freshwater, gives a bow of noticeably smaller radius. The saltwater bow lies about one degree closer to the antisolar point than the freshwater bow.

12. A rainbow appears reflected in the smooth surface of a pond. The upper bow extends slightly below the horizon. If the lower, reflected bow were extended it would intersect the upper bow just at the horizon. Although the reflected bow appears to be an ordinary reflection of the primary rainbow, it is actually formed by a different set of water drops from those producing the primary bow.

13. Two strange rainbow arcs intersect the primary and secondary rainbows at the horizon. Sunlight, reflected from a smooth water surface behind the observer, produces those arcs. The reflected rays have an antisolar point above the horizon that lies at the center of the reflected-light rainbows.

14. For the rainbow arc to exhibit intense colors, it must be produced by relatively large water droplets (1 to 3 millimeters in diameter). For smaller droplets the width of each color band increases and overlaps neighboring colors. In the white rainbow, or fogbow, shown here, the droplets have a diameter of about 0.01 millimeter. Note the faint reddish tinge to the outer side of the bow. As the droplets get smaller, the supernumerary bows move farther from the primary. One such widely separated supernumerary can be seen in the photo.

15. The red rainbow was seen near sunset time. It appears to be rather broad and under other circumstances would probably have been seen as a white rainbow. However, light from the setting sun traveled through a long atmospheric path on its way to the water droplets, and atmospheric scattering selectively removed the blue end of its spectrum. The result is the red rainbow.

16. This photograph was taken with a filter that is opaque to visible light and transmits only invisible, infrared rays. The film, which is sensitive to these rays, recorded the infrared rainbow. The tree leaves are seen to reflect infrared rays from the sun. Both the primary and secondary bows and several supernumerary bows are seen, recorded by infrared light that is invisible to the human eye. (See a further description in *Rainbows, Halos, and Glories* by Robert Greenler, Cambridge University Press, 1980).

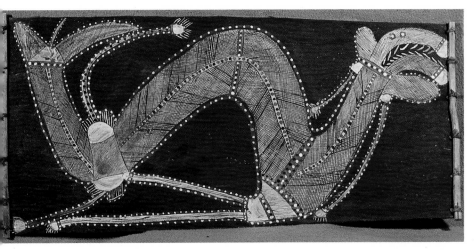

Namirrgi (1924-73), Dangbon Group, Arnhem Land Plateau, Australia. Bark painting:
ᴀɪɴʙᴏᴡ Sᴇʀᴘᴇɴᴛ (*Reproduced by permission of Professor Edward L. Ruhe, University of Kansas*)

ʜirteenth-century mosaic, Cathedral of San Marco, Venice: Nᴏᴀʜ, Hɪs Fᴀᴍɪʟʏ, ᴀɴᴅ ᴛʜᴇ Aɴɪᴍᴀʟs
ᴇᴀᴠɪɴɢ ᴛʜᴇ Aʀᴋ (*Reproduced by permission of Dumbarton Oaks*)

Jacob van Ruisdael: THE CEMETERY (*The Detroit Institute of Arts*)

Frederic Church: RAINY SEASON IN THE TROPICS
(*Fine Arts Museum of San Francisco*)

douard Manet: FISHING IN SAINT-OUEN, NEAR PARIS (*The Metropolitan Museum of Art, Purchase, Mr. and Mrs. Richard J. Bernard Gift*)

ince Eugen of Sweden: THE RAINBOW
he Metropolitan Museum of Art, Gift of L. W. Wilkens, 1954)

Henry Mosler: ABOVE THE RAINBOW (*Reproduced by permission of Schweitzer Gallery, New Yor*

Awa-tsireh, Hopi: THE RAINBOW (*School of American Research, Santa Fe, New Mexico*)

Wide-angle rainbow, photographed in Wisconsin by Robert Greenler.

ıbow with supernumerary arcs, photographed in British Columbia. Photo © Alistair B. Fraser.

Saltwater bow. Photograph provided by G. P. Können, taken by J. Dijkema in the Pacific Ocean, 500 miles southeast of Japan.

Reflected rainbow seen in a pond in Massachusetts, next to the Atlantic Ocean. Photo by David G. Stork.

ted-light rainbows, photographed in the evening near the California Coast by Allen L. Laws.

rainbow or fogbow with a supernumerary bow, photographed in a layer of fog
tario by Robert Greenler.

Red rainbow photographed in Washington state. Photo © Alistair B. Fraser.

An infrared rainbow, photographed by light that is invisible to the human eye. Photo by Robert Greenler.

VI

The Sixteenth Century

THERE IS NO CATACLYSMIC EVENT TO MARK THE SEPARATION OF MEDIEVAL FROM modern times. Politically, the fall of Constantinople in 1453 serves as a convenient dividing line, although artistically the separation might well be advanced by at least a century. In science the so-called renaissance, or the events which may be said to usher in the modern world, came almost a century afterward, and centered around the date 1543. This was the year of publication of two great scientific classics: the *De Revolutionibus Oribum Coelestium* of Copernicus and the *De Humani Corporis Fabrica* of Vesalius. One could cite also for this year Tartaglia's edition of some of the physical science of Archimedes, and the attack by Ramus on the logic of Aristotle. In mathematics the date 1545 is somewhat comparable, for it marks the publication of the *Ars Magna* of Cardan, the treatise in which was first printed the solution of the cubic equation, the earliest great algebraic discovery of modern times. These works have one element in common—they represent a break with the authority of medieval tradition. An exaggerated form of this iconoclastic attitude was presented by Ramus who defended the audacious thesis that everything Aristotle had said was false. In some cases what the writer had in mind to attack was not so much the ancient Aristotle as the medieval Peripatetic tradition which had become embedded in Scholasticism. Throughout Europe the prestige of medieval Latin and Arabic works declined; and this tendency was accelerated by the rising tide of Humanism. The leading figures in the literary renaissance were scarcely interested in the growth of science, and hence they found little of value in the newer treatises in medicine, astronomy, and physics. Their attention was held especially by Greek literature, manuscripts of which became increasingly easy to procure. As ancient treatises were being rediscovered and printed in Latin, medieval works faded more and more into the background. The availability of the works of Archimedes was a boon to the study of statics, but it is regrettable that the popularity of these so heavily overshadowed medieval steps toward dynamics. The return to antiquity was not an unmixed blessing, and this is particularly evident in studies on the rainbow. There can be no doubt about it; the only significant extant ancient theory of the rainbow, that of Aristotle, was not to be compared with

143

the medieval triumph of Theodoric. And yet commentaries on the work of the former appeared every few years for the next two centuries, whereas the ideas of Theodoric were timidly referred to in but one obscure treatise throughout this period. In 1514 at Erfurt the Summa in Totam Physicam of Jodocus Trutfetter of Eisenach contained a brief account of the main features of Theodoric's theory.[1] Presumably, the theory was taught there in the university, but it must have made no real impression. In 1517, the year Luther posted his theses on the door of the castle church at Wittenberg, Theodoric's ideas appeared once more in another edition of Trutfetter's Summa; but after this they disappeared for almost three hundred years. When rediscovered in 1814, the theory of the rainbow had developed to a point where his De Iride could no longer be an inspiration.

One important element in Theodoric's theory, that of an internal reflection in raindrops, had appeared again in the fourteenth century in the work of Themo. This survived at least one year beyond 1517, for Themo's Quaestiones, thrice published before, was printed again at Paris in 1518. This work was not entirely forgotten thereafter, for Themo's name appeared every now and again in later treatises on the rainbow. However, Themo's internal reflection idea was buried in such a confusing welter of arguments and counter arguments that there is no indication that it was noticed by later writers. Those who cite him made no use of an internal reflection, and not one of those who adopted reflections inside raindrops admitted any indebtedness to Themo. It appears that the brilliant discovery of the fourteenth century may have had no part in the later theoretical developments. Although ideas of the thirteenth and fourteenth centuries are perceptible here and there, the story of the rainbow in the sixteenth century began largely where Aristotle had left it—tied to a vague sort of reflection from a moist cloud. In the case of the rainbow the modern period opens, not with something new or some significant departure from tradition, as in so many other aspects of science, but with a decisive return to Peripateticism. The philosophical climate of the fifteenth century had not been conducive to a successful study of the rainbow, and Neoplatonism had nothing to contribute in this respect. The vehement anti-scholasticism of the humanists served to aggravate the situation, and early sixteenth-century writers on the rainbow appear to have labored under an inferiority complex. Recognizing that the ancients, notably Aristotle, knew more about the phenomenon than they, and faced with a garbled text on the subject, commentators were discouraged from thinking up plausible explanations, and they tackled instead the more difficult task of justifying the text. As far as the theory of the rainbow is concerned, the vaunted renaissance brought about a retrogression, for man began again, like Sisyphus, where he had been many centuries before.

No century in the history of mankind was more concerned about the rainbow, if one can judge by the multiplicity of books on the subject, than was the sixteenth. Only in part was this due to the then recent invention of printing, for of the more than 9,000,000 books, representing over 30,000 titles, which had appeared before 1500, remarkably few concerned the rainbow. The growing use of movable type made possible the mass production of scholarly works, and nothing acts so effectively as a leaven to new publication as does reading about the ideas of others. But printing also is a two-edged sword, for the aims of a printer are not necessarily those of a scholar. The immediate effect of printing on science was not especially favorable, for those works printed most often were old-fashioned.[2] A publisher is more keenly aware of what men wish to read than of the obligation to supply them with what is best for them to read. The story of the rainbow almost tempts one into accepting a sort of Gresham's law for sixteenth-century publications. The bad explanations seem all too often to have driven out the good ones. In 1503, for example, there appeared the *Margarita Philosophica* of Gregor Reisch (†1525), prior at Freiburg in Breisgau. This served somewhat as a textbook for a typical arts course in the universities of the day;[3] and it enjoyed an amazing vogue for over a century and a half. Reisch's *Margarita* went through one edition after another.[4] Yet the pages on the rainbow in the *Margarita* give but a crude explanation, quite inferior to those of the thirteenth and fourteenth centuries. The bow is assumed to be produced by reflection when the sun shines upon a dewy and concave cloud which appears opposite to it. The drops are so numerous as to give the impression of continuity, yet so small that only colors appear, as when rays of the sun shine upon a glass full of water and are reflected on a wall. (The author here probably had refraction in mind, his lack of clarity indicating a revival of the old confusion between the phenomena of reflection and refraction.) The diversity of color arises from the varying density of the cloud: where it is dense, the reflection is strong and the color is yellow; where it is less dense, the reflection is weak and the color is green. The bow is round chiefly because it is an image from a round mirror. This assertion is supported by the remark that the iris from the hexagonal crystal is not round but hexagonal, even though the rays come from the spherical body of the sun.[5] Among the ancient errors that reappeared was the doctrine that the weakness and inverted order of the colors in the secondary arc are due to the fact that it is a reflection of the primary bow. Foreshadowing the future conflict between science and religion, the question of the existence of rainbows before the time of the flood was raised. Reisch answered the question scientifically enough, believing that the rainbow had indeed appeared earlier, but that it had not then been a sign of a covenant, but only a natural occurrence.[6]

The explanation of the bow given by Gregor Reisch is of a piece with

Meteorologia Ariſtotelis. Eleganti

Iacobi Fabri Stapulenſis Paraphraſi explanata.
Cómentariocṗ Ioannis Coclæi Norici declarata
ad fœlices in philoſophiæ ſtudiis ſucceſſus
Calcographiæ iamprimū demandata.

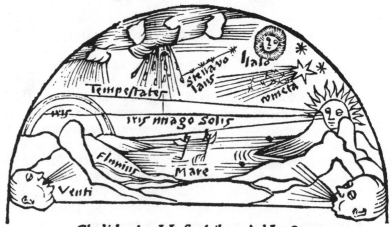

Chelidonius Muſophilus. Ad Lectorem.

Eſt citra decimæ quicquid curuamina ſphæræ
Vna non facie ſiue tenore manet.
Sedibus ipſa ſuis mutantur ſidera & orbes
Induit & vultus Luna ſubinde nouos
Plus etiam phœbes ſub fornice.quattuor inter
Corpora/iuris habet lis & amicicia.
Acris hinc vaſticṗ maris.telluris & ignis
Apparent miris plæracṗ monſtra modis
Quæ tibi præſenti Lector cernenda libello
Producit Cocles eruta mille locis.

Fɪɢ. 24.—Title page of a commentary on a paraphrase of Aristotle's *Meteorology* (1512).

those found in other popular books of the day. Pierre d'Ailly's *Tractatus* appeared in 1504 and 1506; Thiene's *Commentary* was printed again in 1507. Nor were newer commentaries on the *Meteorologica* an improvement over the old. In 1501, 1512, and 1540, for example, a paraphrase by Jacques Lefèvre d'Étaples expressed views not far removed from those of the twelfth century. The commentator does indeed quote Albertus Magnus on the generation of the rainbow through the cone of rays from the sun which strike the surface of the round pyramid in which the moisture descends from the cloud and are projected onto a thick cloud behind it; but there is no appreciation of the role of refraction. The circularity of the bow comes from the roundness of the pyramid of descending moisture and from the fact that rays go out pyramidically from the sun. The reality of the colors is justified by reference to four reasons ascribed to St. Thomas Aquinas: (1) because they affect vision; (2) because sight judges them different colors; (3) because sight judges them true colors; and (4) because the chief colors seen in material things are those which are seen also in the bow.

The primary rainbow is understood to be a reflection of the sun's rays by a cloud, and the secondary arc is an image of the primary bow. It is argued that the third is so weak as scarcely to be visible, for it arises from the already weakened secondary by a reflection (refractio) and has colors arranged in the same order as in the primary.[7] So confident were men of the time in the operation of the reflection principle that the editions of 1501 and 1540, issued by Josse Clichtove or Jodocus Clichtovaeus (†1543), assured the reader that the colors of the quaternary bow, when it appears (as a reflection of the third bow), are arranged in the same order as those in the secondary.[8] How far from understanding the primary bow men were, and yet with what confidence they spoke of the bows of third and fourth orders! The iterated reflection doctrine, although really of ancient origin, came to be associated, during the sixteenth century, with the name of the editor of Lefèvre d'Étaples' *Paraphrase*; and many a later writer used Clichtove as a whipping-boy in connection with this theory of multiple rainbows. In the very first year of the sixteenth century there had appeared at Paris the *Philosophiae Naturalis Paraphrasis* of Clichtove, a mathematician and churchman who received a doctorate at the Sorbonne and became deacon at Chartes. Clichtove died in 1543, the year of Copernicus' death; but where Copernicus achieved fame through a revolution in astronomical thought, Clichtove became relatively widely known—apart from his opposition to Luther—for a mistaken view on the rainbow. In his *Philosophiae Naturalis* he had developed the idea that the secondary rainbow is a mirror image of the primary; and throughout the century this doctrine continued to be ascribed to him—a dubious honor.

The year 1540 is important in the story of the rainbow, not because of the

edition of Lefèvre d'Étaples, but because in that year the Commentary of
Alexander of Aphrodisius became available in the Latin translation by Alessandro Piccolomini (1508–1578), Cardinal at Siena and scion of a distinguished family, two members of which became popes (Pius II and III). Recovery of the verbose Aphrodisian work was of considerable significance.
Coming as it did when the published works on the rainbow were derived
mostly from the weaker medieval writers, it could not but serve to enhance
the reputation of ancient authors. Aristotle's thoughts on the rainbow were
headed pretty much in the wrong direction, because of his disregard of refraction; but his *Meteorologica*, as also Alexander's *Commentary*, was far more
sophisticated than the sickly paraphrases which often masqueraded under
Aristotle's name. The result was that Piccolomini's translation met with a very
favorable response and went through half a dozen editions within a quarter of
a century. From the short-term point of view this was all to the good, for it
raised appreciably the level of popular views on the rainbow. It included the
quantitative portions of Aristotle's theory which, because of their relative difficulty, had been blithely disregarded by the great majority of medieval and
early modern writers. In another respect, however, the success of the work
was unfortunate in that it gave the theory of the rainbow renewed impetus in
the wrong direction. Piccolomini was familiar with medieval departures from
orthodox views, for he cited Witelo, Albertus Magnus, and Themo. But he
remained faithful to the main tenets of the Aristotelian theory. Moreover, he
raised the same elusive and time-worn questions which had been argued without conclusion for so many centuries. What was needed was a new idea, not
endless refinement of the old sophisms; but this Piccolomini was not able to
provide. Not that he didn't try. To the translation of the Aphrodisian *Commentary* he appended a little tract entitled *De Iride* in which he boasted that
one who reads it carefully will no longer remain in doubt about the rainbow.
But the boast is an idle one. There is no real geometrical treatment, in spite
of the warning that "to understand this treatise one must know the principles
of Euclid's *Elements*, Book I." After calling attention to the fact that where
Aristotle—and the mathematicians of his day—spoke of visual rays, one should
substitute the passage of rays from the source to the eye, and after pointing
out that he will restrict himself to *celestial* rainbows, Piccolomini outlined his
intention as follows:

(1) To give the material cause, formal cause, and effective cause (which
in nature coincides with the final cause)
(2) To show that the rainbow has real form and color, not apparent only
(3) To show why there are only three colors
(4) To show why the colors in the secondary are reversed, and why no
color exists between the bows

(5) To show why the bow is always circular

(6) To show why the bow is semi-circular when the sun is rising or setting

(7) To show why the rainbow is always less than a semi-circle when the sun is not on the horizon

(8) To show why the greater the portion of the circle which makes up the rainbow, the smaller is the radius of the circle

(9) To show when the rainbow is formed

The argument opens auspiciously with some elements of geometry, mostly from Euclid; but it soon lapses into typical dialectical commentary. It is cast along thoroughly Aristotelian lines, with reference to other authors from Seneca to Gaetano di Thiene. The material cause is the drops of rain; the effective cause is the light of the sun which is reflected by the drops. The drops are too small to reproduce form; but there must be continuity, and so the sun is seen colored in the rainbow. Piccolomini admits that there is great dissension among the Latins as to whether the form and color of the bow are real or not. Some say the bow is not real because it changes position with the observer, but Piccolomini timidly avoids a commitment on this. The Peripatetic commentators, Alexander Aphrodisias and Olympiodorus, agree that the form is apparent, but they differ on the colors, the latter criticizing the former for holding that they have substantial existence. Piccolomini is inclined to agree with Alexander, although he admits the existence is not permanent, for Aristotle spoke of "colors," not "apparent colors." The author agrees with Aristotle also that the number of colors is three; and he believes he can explain why the order of the colors is reversed in the secondary bow, a "problem which perplexed the ancients." He thinks this is due to the density of the cloud and to contrast (as black looks blacker near white). In the primary the green is due to the lesser penetration and admixture of light; but in the case of the secondary it results from the greater elongation. On the question of the existence of a third or fourth bow he does not commit himself, for he wrote his tract simply "to make clear what Aristotle had said." [9]

With respect to the size of the rainbow Piccolomini was on firmer ground. He reported that it had been observed that the bow can be as high as 42°, and he himself had observed one at Bologna with an altitude of 38°, measured by the astrolabe. If the sun is higher than 42°, no rainbow is visible, for the higher the sun, the lower is the bow, That the iris can be seen only if the elevation of the sun is less than 42° is confirmed "by many observations." The most sophisticated portions of Piccolomini's *De Iride* are his constant references to quantitative aspects of the bow. Witelo and Theodoric had furnished numerical values for the Aristotelian ratio, but since their time little attention had been given to the relative distances entering into the Peripatetic theory. Following Witelo, Piccolomini reported that the height of the clouds on which the rainbow is projected does not exceed 5,000 German feet; yet he continued to use the conventional type of diagram in which the cloud and the sun are equidistant from the observer.

The abiding influence of Aristotle is clearly evident in Piccolomini's work. It is hard to see what Duhem meant when he wrote that at least that part of Theodoric's system which related to the primary rainbow "reappeared in the writings of Alessandro Piccolomini." [10] There is in Piccolomini's *De Iride* of 1540 virtually no hint of the novelties in Theodoric's *De Iride*. There is no trace of the internal reflections in individual drops; there is not even a clear idea of the role of refraction. One reads that the red rays are strongly refracted, the green less strongly, and similarly for the others. [11] This sounds almost like a knowledge of dispersion, but one should not make too close a comparison. The phrase "strongly refracted" does not mean through a greater angle; if it did, Piccolomini would have to reverse the order of the colors. One cannot even be certain that he did not have reflection in mind when he wrote this, but his reference here to Albertus Magnus leads to the presumption that he was using the word "refraction" in the modern sense rather than the Aristotelian. Yet so vague was his understanding of the phenomenon that he saw no conflict between this and the Aristotelian reflection theory. It is not without justice that history has judged Piccolomini's sonnets, essays, and comedies above his scientific works.

Panegyrics on the Renaissance are prone to make much of the fact that there was a revolt against authority. In the second half of the sixteenth century, especially, there were vitriolic attacks on the doctrines of Aristotle, although they perhaps did not outnumber the authors who rallied to the defense of the Philosopher. [12] In the story of the rainbow, however, this controversy was not of particular significance. The weight of authority had perhaps held back the advance of science in certain fields, notably astronomy, partly because science and theology had become inextricably interwoven, and an attack on Aristotle's cosmology appeared to be tantamount to an assault on Christianity. It was feared that an attack on one essential detail might threaten the collapse of the whole magnificent structure. The theory of the rainbow, on the other hand, was peripheral in Aristotelian doctrine, for only that part which related to color had a significant relationship to other portions of the master's teachings; and Biblical passages on the bow are relatively few and unimportant. Yet ancient ideas on the rainbow were questioned more tardily than were more fundamental doctrines. One gets the impression that any reluctance there may have been to break with older views on the bow was induced more by the inherent difficulty of the topic than by an exalted respect for authority. The greatest mathematician of antiquity had not attempted an explanation, and Renaissance scholars with mathematical training were far from successful in their half-hearted efforts.

Today the darling of Renaissance figures in science undoubtedly is Leonardo da Vinci (1452–1519), one of the "intellectual giants" of the age. Had he not said, "He who scorns the certainty of mathematics will not be able to silence sophistical theories which end only in a war of words"? [13] Leonardo seems to have known of the work of Themo, and he might have turned the sophistical theory into mathematical form; but the restless mind of Leonardo did not dwell long enough on the rainbow to reach any worthwhile conclusions. His random jottings on the subject remind one of pre-Aristotelian ideas:

The mirror does not take any images except those of visible bodies, and the images are not produced without these bodies; therefore if this arch is seen in the mirror, and the images converge there which have their origin in this rainbow, it follows that this arch is producd by the sun and by the cloud.[14]

He compared the colors of the rainbow with those formed when rays of the sun pass through a globe of water, but he concluded from this merely that the eye has no share in the creation of the colors of the rainbow.[15]

Mathematicians of the sixteenth century seem to have been as unsuccessful as philosophers in attacking the rainbow problem. The most celebrated mathematician of the age undoubtedly was the eccentric genius, Jerome Cardan (1501–1576). As a physician his reputation was such that he was called to Scotland to cure the king; and he is acknowledged to have been the greatest algebraist of his day. Yet, in spite of his enormous success as a scholar, his life was beset by difficulties. He was an inveterate gambler and he was addicted to astrology, and both of these weaknesses led to his imprisonment—the first for debts and the second for impiety in casting the horoscope of Jesus. Then, too, his two sons were scoundrels; one was executed for poisoning his own wife, and the father himself punished the other by cropping his ears. Yet in spite of all his troubles Cardan wrote prodigiously—on medicine, mathematics, physics, gambling, and a wide variety of other topics. The work of greatest permanent value was the *Ars Magna*; but the most ambitious, and the one which enjoyed the greatest success at the time, was the *De Subtilitate Rerum* of 1550, a sort of encyclopedia of natural history. Cardan had a penchant for completeness, and hence one is not surprised to find some half dozen pages of the *De Subtilitate* devoted to the rainbow. Was his treatment in keeping with his reputation?

The rainbow, Cardan said, is an image of the sun due to rays reflected perpendicularly from a cloud. Reflection from the cloud is not unlike that from drops of water. The sun is infinitely far away, and the sun, the eye, and the center of the bow are on a straight line. (Aristotle, Leonardo, and many others had, of course, made this last observation.) The maximum elevation of the bow in his locality he said was 42°; and by as much as the center of the bow is below the horizon, by this amount is the upper semicircle concealed. The rainbow is not seen on the meridian between the equinoxes [in summer] because then the altitude of the sun is too great. On the other hand, if one ascends a mountain, more than a semicircle can be seen. The colors are due to varying intensity in the illumination: the interior is more obscured, and so is blue; the middle is more luminous and hence is green; the outer is brightest, and is therefore red. Purple and violet are not primary colors but result from combinations of others. The colors of the rainbow are not real in the sense that those from a hexagonal prism or a vase full of water are. Hence, Cardan does not believe that the bow imparts a sweet scent to trees.

Cardan's views of the rainbow are mostly true to the Aristotelian tradition —without the most valuable portions: the explanation of the circularity and the attempt to justify the size. On the secondary bow he adopted the facile idea that it arises from the primary by reflection; being much weaker than the primary, it is impossible for a third to be formed from the second. Cardan was on safer ground in his laborious calculations on the distance and actual size of the bow; but these have no particular significance in the story of the rainbow, for he gave no new numerical data. (He cited Albertus Magnus for the estimate that the vapors producing the bow are not more than fifteen stades—roughly a mile and a half—high.) They are simply trigonometric exercises, similar to those given earlier by Peripatetic scholars, in which ratios of distances are determined. One is hard put to it to find a single new idea in Cardan's verbose account.[16]

Cardan was personally cantankerous and intellectually contentious, and so he had made numerous enemies. One of these was Julius Caesar Scaliger (1481–1558), a scientist who seized every opportunity to needle the hypersensitive Cardan. Scaliger had no need to express the well-known desire, "Oh that mine adversary would write a book," for his antagonist always had rushed headlong into print. The De Subtilitate in particular afforded fruitful ground for controversy, and Scaliger promptly published an attack in Exotericarum Exercitationum lib. XV De Subtilitate ad Hieronymum Cardanum. On the rainbow Scaliger contradicted virtually everything that Cardan had written. Cardan had said that the reflecting cloud is round and opaque, so Scaliger held it to be irregular and pellucid; Cardan had ascribed the colors, which he thought an optical illusion, to a weakening of light, so Scaliger insisted on their reality and said that they resulted from the variety of material in the cloud (blue from earth, green from water, yellow from air, red from fire); Cardan had believed the secondary bow to be an image of the primary, so

Scaliger pointed out that this contradicts Aristotle. Cardan had reiterated the popular idea that the smaller the visible portion of the rainbow, the greater is the radius of the circle of which it is an arc, attributing this to the fact that the eye is deceived by a comparison with terrestrial distances; Scaliger accepted the variability of the radius, but insisted that the smaller bow is less curved because it is nearer to its chord. Scaliger even attacked two of the most invulnerable parts of Cardan's explanation: the collinearity of the sun, the eye, and the center of the bow; and the fact that more than a semicircle may be seen from the top of a mountain.[17]

Cardan replied in a book entitled *Quaedam Opuscula*, but he added nothing essentially new. He admitted that the rainbow presented grave difficulties, but he felt that what he had written "saved the phenomena"—that is, was consistent with the results of observation. The bitter controversy did not result in any positive achievement; it pointed up, rather, the hopeless confusion of ideas which continued to prevail. The phrase, "saving the phenomena," had been a familiar one in the exact science of astronomy, where precise mechanisms had been devised which accounted quantitatively for the regular motions of the heavenly bodies; but with respect to the rainbow there had not yet appeared a single scheme which could account for the regularity in the size of the arc.

The sixteenth century is noteworthy, apart from the revival of classical mathematical science, for an increasing interplay between philosophical learning and the technology of practitioners. The medieval period has bequeathed to us an imposing array of mechanical innovations—paper, gunpowder, compass, printing, spectacles, mechanical clocks, horseshoes, to name a few [18]— and the "scientific revolution" sometimes is attributed to the marriage of medieval manual skill with the rediscovery of classical erudition. Under such a thesis the theory of the rainbow must be regarded as a stepchild. The utilitarian drive in the study of the bow is insignificant; and the deformed youngster, cherished by its philosophical mother, was neglected by its technocratic father. About the only use to which the gamin could be put was in the hazardous game of weather forecasting. In England in 1555 the mathematical practitioner, Leonard Digges (who died in 1558, the year of the death of Scaliger and of Mary Tudor), made much of this aspect of the rainbow in his *Prognostication of Right Good Effect, Fructfully Augmented Contayninge Plane, Briefe, Pleasant, Chosen Rules, to Judge the Wether for Ever*. This was a remarkably popular almanac and practical nautical guide which went through more than half a dozen editions within less than half a century; but the explanation of the rainbow is a far cry from those given centuries earlier by his countrymen.

The Raynbowe is the shynyng, and rebounding of beamys of light, that tourne to the contrarie vapour agayne, in the cloude. It declareth sometyme rayne, and many tymes fayre wether.[19]

Earlier the author had given the "general prognostication" as follows:

If in the mornyng the raynebow appere, it signifieth moysture, onlesse great drouthe of ayer woorke the contrarie. If in the evening it shew itself, fayr weather ensueth: so that aboundaunt moyste ayer take not awaye the effect. Or thus. The rayne bowe appering, if it be fayr, it betokeneth fowle weather: if fowle, loke for fair weather. The grener, the more raine: redder, wynde.[20]

If mathematicians could do no better than this with the explanation of the rainbow, it is scarcely to be wondered at that men continued to look for help to the far more learned and sophisticated accounts in the omnipresent Peripatetic paraphrases of the *Meteorologica*. Can one blame the age for not lightly throwing off the mantle of Aristotle before a better was available? The availability in Latin of the commentaries of Alexander (after 1540) and Olympiodorus (after 1551) gave to the ancient doctrine an aura of scholastic respectability which served to conceal its deficiencies. Instead of looking for new approaches, men spun out the old arguments into ever finer sophistic detail.

This was particularly true in the case of Francesco Vimercati, or Franciscus Vicomercatus (†1570) of Milan, whose *Commentary* of 1556 (and 1565) was widely cited. This work really is a commentary on commentaries, for it makes frequent use not only of Aristotle, but also of Alexander, Olympiodorus, and medieval authors. For each passage of Aristotle the Greek is given first, then the Latin translation, and then a tediously discursive Latin paraphrase. The author, physician to Queen Eleanor of France, cited the same old questions and answers, invariably coming to the conclusion that Aristotle was right. Even on the foolish old corruption that held only two lunar bows in fifty years to be possible, which Albertus had denied, Vimercati takes the banal position that Aristotle may have meant only two *clear* bows in fifty years. He was aware, through Alexander and Olympiodorus, that many ancient writers had argued that the halo is a refraction phenomenon, and he must have known, through Albertus Magnus, that a medieval refraction theory of the rainbow had been proposed; but the literal-minded Vimercati rejected such heresies. Both phenomena are assumed to be produced in the same way, by reflection, differing only in the distance, position, and nature of the medium in which the reflections take place.

One of the points on which Vimercati felt strongly was that of the cause of multiple rainbows. He attributed to Ammonius and Olympiodorus the notion that the secondary arc is merely the reflection of the interior bow on an exterior cloud, an idea which, like Alexander, he unhesitatingly rejected. He explained at great length that the question of a tertiary rainbow does not hinge on three successive reflections, but rather, according to Aristotle, on a third type of reflection, higher and at a greater distance than those producing the other bows. At first he denied the possibility of three or more rainbows, but later he wrote that three arcs may be possible in rare cases when the cloud is very high, but that when this happens, the third bow is so pale that colors are scarcely apparent. The multiple arcs seen by Witelo he ruled out as not being rainbows in the strict sense, but he accepted the actuality of multiple artificial bows in water sprays, believing that colors in these are arranged in the same order.[21]

The importance of the Aristotelian meteorological sphere had not abated in traditional studies on the rainbow, and Vimercati's lengthy exposition is accompanied by half a dozen diagrams in the usual manner, following Olympiodorus. The argument hinges upon the Aristotelian little-mirror reflection law in which the ratio SN:NC was regarded as the critical factor in the formation of the bow (Fig. 25). Vimercati, like his predecessors, proved in detail that for a given constant K [between 0 and 2] there is one and only one

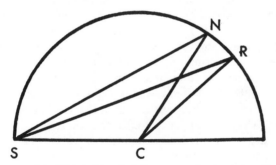

Fig. 25.—Diagram used by Vimercati in calculations concerning the Aristotelian rainbow ratio.

point N on the semicircular arc for which SN:NC = K. (The assumption that SN:NC = SR:RC is easily demonstrated to contradict the *Elements of Euclid.*) To the objection that this Aristotelian explanation of the rainbow violates the equality of angles in the familiar law of reflection, Vimercati replied, as Olympiodorus had earlier, that two answers are possible. (1) The equality in this case is perhaps to be interpreted not to mean that the angles of incidence and reflection with respect to the semicircle SNR are equal, but simply that the relations between these angles is the same for all points along the arc of the rainbow, just as in the case of the circle of the halo. (2) The point C at which the eye is placed may be only an apparent center, rather than the real center, of the circle.[22]

Vimercati was fascinated also by another of the favorite Aristotelian-Scholastic rainbow theses—the presumed variation in the radius of the rainbow with the altitude of the bow. "I have not perceived this inequality," he admitted honestly; but this inability in no way obstructed his speculations on the reasons for the phenomenon. He reported that there were some who held the putative variation to be an optical illusion, to be accounted for in the same way that Alhazen had explained the "horizontal moon." Vimercati rejected this simple explanation for others of a more involved nature: As the sun gets higher the rays are stronger and can be reflected from a greater distance, thus producing a bow with a larger radius; or perhaps the chord of the visible bow is of constant length, so that as the rainbow sinks with the rising sun, the circle of which the bow is a part necessarily becomes larger.[23] One of the disheartening reflections on the story of the rainbow is the alacrity with which scholars discussed phenomena which did not exist. Careful observations with the astrolabe, plus a few simple trigonometric calculations, would have sufficed to show that there is no appreciable change in the radius of the rainbow corresponding to variations in the altitude of the sun. There are indeed changes in the size of the rainbow, as the eighteenth century was to show, but they are quite unrelated to the position of the sun or the bow. No astronomer would have persisted in such blind speculations, unaided by the available evidence of the senses; but then the theory of the rainbow had been fostered by philosophy, whereas astronomy always had been akin to mathematics.

Perusal of Vimercati's commentary makes more understandable the impatience of later scientists with Peripatetic thought. It was not so much that the ideas of Aristotle were at fault; it was the unbridled passion of his successors to heap commentary upon commentary until the whole stratified structure fell under its own weight. It was not enough that Vimercati had commented on Olympiodorus who had commented on Alexander who had commented on Aristotle. A sixteenth-century possessor of the copy of Vimercati's *Commentarij* now at the University of Michigan carried the process one step further; he supplemented it by an equal volume of manuscript commentary, dated 1566, on the already ponderous *Commentarij*! It is clear that little progress in the story of the rainbow could be expected from Peripatetic writers. What was needed was a fresh start from a whole new set of ideas. One such start was indeed made, during the very years that Vimercati wrote, by the Sicilian mathematician Franciscus Maurolycus (1494–1575) of Messina.

Maurolycus, Abbot of Castronuovo and probably the greatest geometer of the century, was not unfamiliar with at least some of the voluminous material on the rainbow, but he had read deeply in another source which Aristotelian of his day were all too prone to disregard. Alhazen's theory of the rainbow had not been well taken, but his *Treasury of Optics* had twice, and possibly three times, served, nevertheless, to inspire a correct theory of the bow during the fourteenth century. Alhazen's *Treasury* was published in Latin in 1572, but substantial portions of the *Treasury* had appeared earlier in the printed editions of Witelo's *Optics* (1533) and Peckham's *Perspective* (1542). It probably was through his own geometrical study and through perusing those sections of optical treatises which concerned reflections in spherical mirrors that Maurolycus came to the conclusion that the key to the rainbow lies in the sphericity of the raindrops. Did he know of the earlier work, similarly inspired by Alhazen, of Theodoric or Themo? It is not impossible; but nowhere does Maurolycus name either one. His candor in admitting sources, the absence of appreciable similarities, and the intrusion of striking novelties would indicate that his work was independent of that of the fourteenth century.[24]

Maurolycus long had been interested in optics, having completed by 1521 the first part of a treatise which appeared under varying titles: *De Lumine et Umbra*, or *Theoremate de Lumine*, or *Diaphaneon*, or the playful *Photismi de Lumine*. The opening portions of this cover reflection in plane and convex mirrors. The second part, written years later, is on the passage of light through various media; and it comprises three books: one on the refraction of light; a second on the rainbow; and the last on the structure of the eye. The section on the rainbow is dated February 12, 1553; but to the volume the author later added a section on "Questions pertaining to optics and the rainbow," and this closes with the date October 12, 1567. The first publication was in 1611, long after the author had died.[25]

The most original portion of the *Diaphaneon* or *Photismi* is unquestionably the book on the rainbow, for it presents one of the most intriguing theories ever devised. Here one finds the first clear-cut break with the Aristotle tradition. It seems to have been less obvious to meteorologists that reform was needed in the theory of the rainbow than to cosmologists that the system of Ptolemy was inadequate. Then, too, Maurolycus' reform of the rainbow in 1553 was in a sense more revolutionary in attitude than was the reform in astronomy by Copernicus a decade before. Copernicus had displaced the earth, but he had retained the ancient and medieval Aristotelian-Ptolemaic concept of uniform circular motion. He greatly simplified the mechanism, but he did not modify the over-all pattern. Maurolycus, on the other hand, took a step which even Theodoric had not dared to take. Theodoric had given a new and essentially correct theory, but he had forced it into a Peripatetic framework by retaining the meteorological sphere. Maurolycus abandoned this sphere completely. Moreover, he focused attention for the first time on the basic question to which Aristotelian writers had given only desultory consideration: How can one account for the size of the rainbow? *Hic opus, hic labor est.* As Maurolycus well put it, nearly the whole of any demonstration of the rainbow is to show why the apparent size of the bow is what it is.[26] The proximity of the angle of the bow to half a right angle attracted his attention, and, being influenced by Pythagorean ideas of mathematical harmonies in nature, he looked for a simple geometrical explanation of this coincidence. Rays of the sun strike a given drop at many different points and are reflected at varying

Fig. 26.—Maurolycus' conception of the way in which the rainbow is formed by rays of the sun reflected in raindrops.

angles from the convex surface. Why, he asked, should only that ray be effectively reflected for which the angle between the incident and reflected rays is about 45°? He realized that only a portion of the light from any incident ray is reflected by the exterior surface, for part of it penetrates the drop. In the case of the effective rainbow ray, he believed, part of the light which has entered the drop must, upon leaving the drop after reflection on the inner concave surface, reinforce the portion which had been reflected by the convex exterior surface. Half of a right angle, reasoned Maurolycus, is an eighth of a circle, and so he divided the circle of the raindrop into eight equal parts, constructing within it an octagonal star polygon (Fig. 26).

Maurolycus thus became the first scientist of the modern period to base the explanation of the rainbow on internal reflections within the raindrops. And it is of interest to note that whereas those of the fourteenth century who had anticipated him seem to have been led to the idea by observations on globes of water, Maurolycus seems to have been guided only by the geometry of spheres, with little if any experimental assistance. Perhaps this was the reason that he went badly wrong on the details of an otherwise basically correct approach to the problem. Theodoric knew from experience that there was a particular path through the drop designated by nature as appropriate for the production of the primary bow, and he successfully traced this even though he could not explain it in terms of number or measure. Maurolycus the mathematician seems to have felt that the geometrical basis was discoverable without recourse to experimental observation. He noted that if a ray enters the drop along one of the lines of this polygon, and if it is reflected within the drop seven times at vertices of the eight-pointed star, then at the eighth vertex the ray will leave the drop along the very same path as that traversed by the portion of the incident ray which is reflected at the external surface. Moreover, at the point of emergence, that portion of the ray which, instead of leaving the drop, is again reflected internally will retrace the star path once more. At the eighth vertex another fraction of the light will leave the drop to further reinforce the effective ray. At angles which "do not agree with the octagon" the rays are not thus sufficiently reinforced to reach the eye with the necessary effect. Here, he thought, was the secret of the rainbow's 45° radius. It is the angle at which externally reflected rays are reinforced by those internally reflected seven times or more; and, *mirabilu dictu*, seven is precisely the number of colors in the rainbow! Could anyone ask for a more beautiful confirmation?

But Maurolycus did give a further argument, more closely related to Aristotelian views, in favor of his theory. He argues that the reflection of a ray is most effective when the incident and reflected rays make an angle of 45°. When the angle is smaller, he believed, the incident and reflected rays, being closer together, tend to interfere with each other; and when it is larger, the greater obliquity entails a weakening of the reflection. He realized, probably through Witelo, that the angle of the bow is slightly smaller than 45° and suggested a clever explanation:

But how does it happen, you ask, that the altitude of the rainbow is not exactly forty-five degrees, but a little less as ascertained by observation? I do not know how to answer this or what reason I may offer, unless it be that the falling drops are somewhat elongated or somewhat flattened, and thus, varying from the spherical form, change the angle of reflection and hence also the straightness of the ray which in the case of a perfect sphere comes back at an angle of forty-five degrees.[27]

The rainbow theory of Maurolycus is geometrically fascinating but physically impossible. Its author must have been aware of thirteenth-century references to the role of refraction in the bow, and yet he himself made no allowance for this. This situation is not easy to understand, for his *Photismi* marks the opening of the modern study of refraction. Grosseteste's bisection refraction law had been published in 1503 in a treatise entitled *De Phisicis Lineis Angulis et Figuris per Quas Omnes Acciones Naturales Complentur*, but a greater impetus was given to the subject by Witelo's *Optics*. Maurolycus was not noted as an experimentalist, but he gave a systematic geometrical exposition of the known facts of refraction. He was among the first to make use of the modern "angle of incidence," but he retained the traditional *angulus fractionis*—the angle between the refracted ray and the prolongation of the incident ray (now called the "angle of deviation"). Maurolycus, like Ptolemy, believed that his angle of refraction is proportional to the angle of incidence. Although this is not far from correct for small angles, it fails badly in most cases. Nevertheless, even this crude law should have made clear to him the utter untenability of his reflection hypothesis of the rainbow. Here is a striking case of what Huxley later described as the tragedy of science—the slaying of a beautiful theory by an ugly fact. Here and there Maurolycus had used the word "refraction" in connection with the rainbow, but from the context he betrays that it is in the older sense of a reflection.[28] How does it happen that he overlooked Witelo's suggestion on refraction in the theory of the bow? Was the authority of other available writers so strong that Maurolycus rejected Witelo's idea? Or was it because Witelo "multiplied his lines and makes work for his readers, but never proves anything?" [29]

Did Maurolycus perhaps think that the drops were so small that refraction could be disregarded? After all, he had made a study of the refraction of light in glass spheres, going so far as to note what later was called spherical aberration. As it stands, the explanation of Maurolycus is physically absurd, for a ray can not enter or leave a spherical drop obliquely without undergoing a refraction. It is odd that he should have overlooked (or suppressed) this fact and attributed the rainbow to reflection alone; but his work nevertheless is superior to all earlier reflection theories. More insistently and consistently even than Theodoric, he focused his study on the geometrical optics of a single spherical drop. On the physics of the rainbow Maurolycus may have been pre-medieval, but in his quantitative geometrical approach he was quite iconoclastically modern. On the width of the rainbow, too, he gave half of the correct answer when he assumed that the width of the bow derives from the apparent diameter of the sun. This breadth is, in reality, several times as great as the angular diameter of the sun or the moon, which had been known in antiquity to be just about half a degree. In amplifying his half-truth Maurolycus betrayed the *ad hoc* nature of parts of his work, for he had recourse to a reflection from a macrocosmic sphere (Fig. 27). This sphere cannot be a

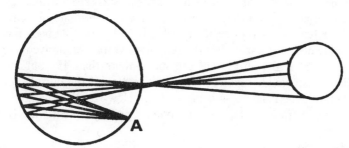

FIG. 27.—Diagram used by Maurolycus in explaining the width of the rainbow band.

raindrop, for the eye of the observer, A, is on its circumference; nor can it be the Aristotelian meteorological sphere inasmuch as the sun DE does not lie on the periphery. It presumably is but part of one of the vague geometrical schemata to which the author resorts when in trouble.

Maurolycus considered carefully the hypothesis that the secondary rainbow is an image of the primary. He cited three factors supporting it: (1) the reversal in the order of the colors, (2) the fact that the second is never seen unless the first is visible, and (3) that the colors of the secondary are weaker than those in the primary. But he concluded that this hypothesis is "a very great error"; and he rejected the theory in favor of another star-polygon explanation. Evidently he was hypnotized by the approximation of the radius of the secondary bow to a simple numerological relationship. It may be recalled that Theodoric had taken the difference in the radii of the two rainbows to be 11°. Maurolycus could scarcely have known of this work, which presumably disappeared while he was a young man; and yet he estimated the difference as 11¼°. This gave to the secondary bow a radius of 56¼°, the first reasonably accurate value. Now, reasoned the geometer, 56¼° is an eighth of a right angle more than half of a right angle. He believed, therefore, that the secondary bow was generated in the same manner as the primary except that in this case the inscribed polygonal figure has four times as many vertices. Inasmuch as the angle of incidence for which the externally reflected ray is reinforced by the ray multiply reflected at the inner surface is in this case more than 45°, the optimum for reflection, and because the light is further weakened by the larger number of internal reflections (thirty-one are implied), the secondary bow is less bright than the primary. Inasmuch as Maurolycus categorically asserted that no other angle can produce any sort of a rainbow except 45° and 56¼°, one may infer that he regarded a tertiary bow as an impossibility.[30]

A modern reader with little feeling for the geometrical harmonies of nature may be pardoned a tinge of bemusement at such speculations. He may be inclined to admire Maurolycus instead for a more prosaic but eminently defensible statement on the secondary bow. It had invariably been held that the secondary never appears unless the primary is visible. Maurolycus correctly pointed out that if the altitude of the sun is more than 45° but less than 56¼°, then it is possible that a certain small portion of the outside bow may be seen without the inside bow being visible. "How rare this phenomenon is can be imagined from the fact that no one has yet ever mentioned it or observed it." [31]

The formation of colors was ever a stumbling block in the story of the rainbow; and Maurolycus could do no better than his predecessors. He believed in a doctrine of four principal colors: orange, green, blue, and purple. But the mixture of these colors with light and moisture leads to changes in hue. If darkness is mixed with yellow, purple is produced; and when shade is added to green, the result is blue. And so Maurolycus presumed that one may fairly speak of seven colors in the rainbow—one between each of the four primary hues. These are due to the quality of sunlight poured into the raindrops and to the rarity and density in the distribution of the drops. The colors of the secondary are produced in the same manner as for the primary. The reason that the colors appear in inverted order is far from clear, the argument being based on a tortuous discussion concerning the density of the drops and the obliquity of the rays. The geometrical background again is a reflection in the concave surface of a large sphere, much as in the argument on the width of the bow.

The Diaphaneon of Maurolycus closes with a series of twenty-four more speculative problems or queries on perspective and the rainbow. Among these one finds many of the favorite ancient and medieval questions, but the twenty-fourth problem in particular deserves attention. The author had noted the "zones of color" which appear just inside the primary bow. He gave no adequate explanation of these, suggesting merely that they are due to "manifold, reverberating, and oft-repeated reflections of the rays"; but the query is significant as an indication that the problem of the rainbow is more complicated than most people realize. Maurolycus was not far wrong when he reported that, "If among the problems of optics the most obscure is the one which is concerned with transparent bodies, the most difficult one is that of the rainbow." [32] He confessed that he had spent almost thirty years on his books on optics, and this is not hard to believe. As he said, perhaps Euclid and Ptolemy "acted wisely in not exhibiting any great eagerness concerning the theory of the rainbow." [33]

Maurolycus was familiar with earlier literature on his subject, citing many of the names which had figured in the story of the rainbow to 1553—Aristotle, Pliny, Averroës, Peckham, Witelo. He seems to have been on the lookout for sources, for he closed his work on the rainbow with the words, "I hear that certain pamphlets have been discovered in Germany, as we have learned through an inscription on some old copies of Andreas Stibonius (i.e., Stibonius), a Viennese Canon, in which a discussion of this subject is presented; but I have not seen it." [34] Could these pamphlets by any chance have contained the views of Theodoric? This is not an impossibility, but the probability is high that they were but further instances of the typically Peripatetic views which dominated the age. Nevertheless, while the Photismi was being completed, there was at Breslau an obscure Lutheran clergyman and physicist who was laboring on a thick and somewhat unconventional volume devoted entirely to the rainbow. This man, Johann Fleischer or Fletcher (1539–1593), had worked on the problem of the rainbow for half a dozen years before he published his De Iridibus Doctrina Aristotelis et Vitellionis in 1571 (the title page incorrectly carries the date 1579), and his effort was rewarded with partial success. Contemporaries had investigated the rainbow primarily as an aspect of natural philosophy, but Fleischer, like Maurolycus, coupled it with the precise study of optics. The chief inspiration for the philosophers continued to be the Meteorologica of Aristotle, while the most widely available source for the physicists was the Optics of Witelo; and it was the conjunction of these two streams which led to Fleischer's theory. Aristotle, said Fleischer, had explained the rainbow intricately, but he hadn't adduced all the causes; Witelo, on the other hand, had given the causes, but he had explained them obscurely. Many others had written on the bow, but Fleischer felt that none of these had presented a complete and verified theory such as he proposed. Basically his ideas are not new, but he thought them through with a clarity quite unusual for the age. He gave a lengthy explanation of the Aristotelian theory, applying trigonometry in connection with the exposition. [35] This in itself was significant, for most men were satisfied with only the vaguest idea of what Aristotle had had in mind. But, as Fleischer wrote, Aristotle had overlooked an essential cause of the bow—the refraction of the rays. Fleischer cited Witelo as authority for the assertion that if the rays did not penetrate the vapor and were not refracted, the rainbow would not be colored. [36] But Fleischer was not satisfied with Witelo's fuzzy ideas on the interplay between the reflected and refracted rays, and on the relation between these and the individual drops. He, therefore, proposed an ingenious and not implausible theory. Rays from the sun, he believed, traverse spherical raindrops, being refracted both upon entering a drop and upon leaving it. Some of these rays subsequently are reflected at the convex exterior surfaces of other raindrops and enter the eye of the observer. On the way to the eye, reflected rays may indeed undergo further refractions through drops lying in their path. In fact, rays may pass through many a corpuscle in penetrating the body of the vapor until at length a spherical body or drop presents itself which, on account of the density acquired by the action of the cooling agent, does not permit the light to pass through, but reflects it. In the case of the reflected rays there necessarily may be further refractions through individual drops until at length the light is carried to the eye. The book contains numerous diagrams illustrating exactly how Fleischer visualized the operation of his theory (Fig. 28); and one must admit that, at least at first glance, the explanation looks convincing. There is nothing fundamentally incorrect in the way that the phenomena of reflection and refraction are portrayed. Yet there is, of course, one big question. Amidst the confusion of the myriad of drops, how is it that rays, after following the paths indicated, reach the eye only from certain particular directions? Fleischer's ideas are attractive from a superficial qualitative point of view; but they lack a sound quantitative basis. Why is the rainbow always seen at about the same angle? Fleischer himself was aware of this uniformity. He cited the 42° figure of Witelo, although he, too, thought that the radius was subject to slight variations, and he seems to have preferred 42° 30′.

The generation of the secondary rainbow is assumed to take place in somewhat the same manner. Fleischer says only that it results from new refraction and reflections different from those for the primary. For the reversal of color he proposed an ingenious geometrical explanation. He believed the colors of the bows to result from a weakening of light as it passes through the medium —and this weakening depends somehow upon the distance traversed and the angle of incidence. Now the radius of the primary bow is somewhat smaller that of the secondary somewhat larger, than 45°. Fleischer, therefore, assumed the eye of the observer to be at the lower end-point O of a diameter OD of a circle, with the diameter inclined at an angle of 45° to a horizontal line OH (Fig. 29). Let HOP and HOS be the angles under which the middle of the primary and secondary bows, respectively, are seen, the former being less than the latter greater than, 45°. Now it is clear that chords from O to the lower portion of the primary bow are shorter than those to points which are higher that is, as one goes downward from D toward H along the circle, the chords decrease in length, and the angle between the circle and the chord become

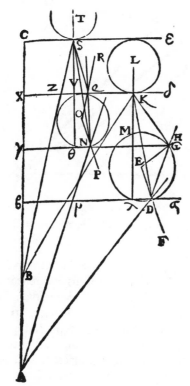

Fɪɢ. 28.—Fleischer's idea of the formation of the rainbow. Rays from the sun at A are refracted through one drop of rain and are then reflected by another raindrop to reach the eye of the observer at B.

maller. But for the secondary bow, the chords become shorter, and the angle between the circle and the chords become smaller, as one goes upward from ⊃ along the circle. This geometrical diagram the author felt somehow explained the inverted order of the rainbow colors, although even he accepted it with some hesitation. In fact, Fleischer showed a commendable suspension of judgment with respect to other moot points on the rainbow. After citing the familiar rules on the rainbow as a weather prognosticator, he added that these do not always follow, as he had found. On the matter of multiple rainbows, too, he was cautious. Aristotle had limited the number of possible bows to two, whereas Witelo claimed to have seen four. Fleischer believed that for any one medium only two bows appear; but he advanced the possibility that there may (upon rare occasions) be two different media, one nearer than the other, each of which produces two bows according to theory already outlined.

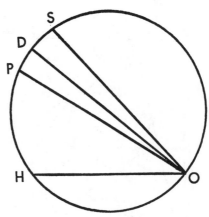

Fɪɢ. 29.—Fleischer's suggested explanation for the reversed ordering of the colors in the sec ondary rainbow.

There are, of course, physical difficulties inherent in Fleischer's theory. Fo: one thing, the rays, following the refractions and reflections indicated, woulc be too divergent to stimulate the eye of the observer. Then, too, the rays o the sun are pictured as diverging, instead of being nearly parallel; and they d(not make a constant angle with the rays reflected to the eye. Nevertheless, th(clarity of his thought represented a welcome advance over the ubiquitou, obscure works of the century. In the first place, it reintroduced refraction as : sine qua non; and secondly, it diverted attention from the cloudy continuun to the optical properties of discrete spherical raindrops. The importance o these two steps should not be underestimated; and yet one must bear in minc that Fleischer had not attempted the most important problem of the rain bow—to show why the bow has a very nearly invariable apparent size.

The thorough and original treatments of the rainbow by Maurolycus an(Fleischer pointed in the right direction—the application of geometrical optic to the study of the passage of rays of light through spherical raindrops. Wer one to combine the refractions of Fleischer with one or two of the interna reflections of Maurolycus, the result would be a good approximation to th modern theory; but this was not the way the story of the rainbow unfoldec Fleischer's De Iridibus of 1571 quickly became a rare book, and Maurolycus thoughts on the rainbow, although composed earlier, did not appear until th following century. There was, nevertheless, a bright spot in the picture. I 1572, the year after Fleischer's book appeared, the Optics of Alhazen an(Witelo was published in an Editio Princeps, by Risner. This excellent wor: further increased the already pronounced influence of Witelo upon earl modern optical investigations. In the ideas of Fleischer and Maurolycu one notes a merger of philosophical and optical treatments of the rainbow; and it was through such a synthesis that the riddle of the rainbow ultimately was solved. But this confluence of speculative and mathematical streams was atypical of the sixteenth century. Peripatetic complacency was not easily dislodged.

Unruffled Peripateticism characterized most of the published treatises on meteorology and the rainbow during the sixteenth century. Among these it is worth noting one of the first vernacular commentaries on Aristotle's *Meteorologica*, the *Trattato* of Francesco de Vieri, published in 1573 (and again in 1582). This treatise, dedicated to Francesco de Medici, Grand Duke of Tuscany, is derived in large measure from the four most-quoted sources: Aristotle, Alexander, Olympiodorus, and Witelo. The level of the exposition can be judged from one explanation, the inverted order of the colors in the secondary rainbow. De Vieri drew a square C to represent a cloud directly opposite the observer's eye E, and he drew lines from C to E to represent rays of the sun reflected by the cloud to the eye to form the rainbow (Fig. 30). In

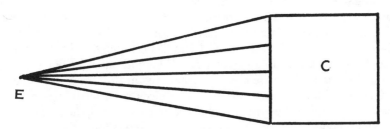

Fig. 30.—De Vieri's idea on the reversed ordering of the colors in the secondary rainbow.

this diagram it is obvious that the middle ray is the shortest; whether one goes upward or downward from this position, the rays get longer. In an analogous way the rays get longer as one proceeds upward or downward from a position midway between the two rainbow arcs. But increasing distance means successively weaker rays, and hence the red, or strongest, rays are nearest to the middle position, the blue, or weakest, are furthest from the middle position.[37] This type of argument, a more refined form of which had been given by Fleischer, shows the futility of early modern explanations of color formation.

In 1576, the very year after the publication of Maurolycus' *Opera*, there appeared at Paris the *Iridis Coelestis* of Jean Demerlier, a royal professor. This is again thoroughly in the tradition of natural philosophy rather than optics, with no quantitative or geometrical treatment. The rainbow once more is described as a confused and distorted image of the sun on a moist spherical concave cloud lying opposite. The author at one point makes a feeble effort to follow Witelo in ascribing it in part to the inflection (refraction?) and in part to the reflection of light. The outer bow is an image of the inner. Because of the smallness of the drops and because of inequalities in rarity and density, and of form and position, a variety of colors is produced. There are, however, three primary colors: yellow, green, and purple. Color depends on the extent of penetration of the cloud by the rays of the sun; those at the greatest angle penetrate least, and these rays are most brightly colored. Since rays of the sun can not penetrate a dense cloudy concretion, a bow signifies a lesser accumulation of moisture. Consequently, if a rainbow appears in the east, it forecasts fair weather; if in the west, this is a sure sign of rain.[38] The author cited Aristotle and Witelo frequently, but his interpretation of their ideas falls far short of that given by Fleischer.

The treatises of Aristotle had been paraphrased many times throughout the sixteenth century, but the commentaries which enjoyed the greatest reputation were those which were dictated to students during the early part of the last decade by the Jesuit professors at the University of Coimbra. Because these notes or commentaries came to be fraudulently printed, to Father Peter Fonseca was entrusted the task of revising them for publication. The results of this large-scale "masterpiece" were so admired that Father Fonseca became known as the "Aristotle of Portugal," and the commentaries often were referred to under the simple designation "Conimbricenses," from the Latin form of Coimbra.[39] The *Meteorologica* appeared in this series in 1592 under the full title *Commentarii Collegii Conimbricensis Societatis Jesu in libros meteorum Aristotelis Stagyritae*, and this formed a part of the curriculum in the Jesuit schools of Europe. This may be taken to symbolize the fact that the authority of Aristotle on the rainbow, occasionally challenged throughout the sixteenth century, had triumphed over all opposition. One more dissenting voice, however, was heard at Naples in the following year, 1593. This was in the *De Refractione Optices Parte Libri Novem* of Giambattista della Porta (1543–1615), the last book of which is on the rainbow. Throughout the century knowledge of refraction had remained very imperfect, and portions of Della Porta's work show this very clearly. Some of his diagrams show light passing obliquely across the boundary between glass and air without undergoing a refraction, others grossly exaggerate the deviation; and in at least two cases his diagrams of light refracted through a prism show the rays bent in the wrong sense! [40] And yet Della Porta's *De Refractione* was significant for its attempt at a theory of lenses.

The author was one of the first to try to give a mathematical explanation of the telescope, the invention of which he claimed.[41] The next century was to discover that the telescope is less subtle and less difficult to explain than the rainbow, but, ignorance being bliss, Porta rushed into print with one of the weirdest explanations of the bow ever given. He seems to have read widely, but perhaps not too systematically or sympathetically. Certainly there was a regrettable lack of orderly thought in his theory of the rainbow. He evidently had rummaged about in ancient sources, uncovering sources not generally quoted. To Metrodorus and Parianus he ascribed the idea that when the sun shines across a cloud in forming the bow, the cloud imparts the blue color and the light the red color.

Nicolaus Peripateticus held that the diversity of color is from the diversity of the parts of the cloud; but Albumasar believed the blue was from the water, red from the sun, and green and purple from the mixture. It is to Albertus Magnus that Porta attributed the idea that the colors in the bow are produced in the four media—moist air, cloud, earthy humor, and smoky humor—through which the rays of the sun pass. Porta apparently did not know directly of Grosseteste, but through Albertus Magnus he derived some confused general ideas on the formation of the bow. Influenced apparently by Albert's careless use of the words *reflectio* and *refractio*, Porta does not always distinguish clearly between the two terms. But on the basic phenomenon of the rainbow he is quite clear and categorical. Against Aristotle it is asserted that in the generation of the bow no reflection is necessary, only refraction.[42] The justification for this statement is an interesting combination of new ideas and old preconceptions. The old meteorological sphere, with the sun A on the periphery, is retained (Fig. 31). But the cloud on the sphere is replaced by a vertical

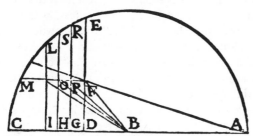

Fig. 31.—Diagram illustrating Porta's idea of the formation of the rainbow.

sheet EFD of rain; and instead of looking at the individual drops, Porta reasoned in terms of the rain-sheet as a continuum. He contended that rays from A which strike the rain at F will be reflected upward and will not reach the eye at B; and moreover, he argued, diaphanous bodies such as rain refract rather than reflect. In order that the light should reach the eye, he envisioned a bizarre sort of refraction—one which should not only incline the rays toward the perpendicular, but should bend them back almost upon themselves, introducing a deviation of not far from 180°! From his diagram (Fig. 31), one sees that the diversity of color corresponds to the varying depths to which the refracted rays penetrate the body of the raincloud FM, again thinking of this cloud as a continuum; but his ordering of the colors contradicts that observed in nature. The portion between M and O he asserted would be red, that from O to P green, and that from P to F yellow.

Porta's diagrams are completely inconsistent with the simplest observations of the phenomenon of refraction; and his explanation of the secondary bow is even more contrary to the laws of physics. After criticizing Nicolaus the Peripatetic for holding the view that the inner bow is generated from the sun, the outer from the moon; and after denying the assertion of Albertus that experience verifies the fact that the second is an image of the first, Porta proposed a confused theory in terms of two concentric spherical clouds! Rays from the sun first strike the inner more tenuous cloud (Fig. 32). Here part of the light is refracted toward the eye, while another portion continues upward toward the outer coarser cloud, from which it then is returned toward the observer. A glance at his diagram shows that this imaginative explanation accounts for the inverted order of the colors! Porta gave also a quantitative statement on the bows. He reported that in 1590 he had observed the higher bow to be 42°, the lower 38°. This would make the difference in their radii only 4°, a value quite inferior to those of Theodoric and Maurolycus.

In a day when the story of the rainbow centered about reflection, one welcomes any suggestion that refraction is important; and yet one can feel little indebtedness to Porta. His utter disregard for reflection carried him from the frying pan into the fire. One gets the impression that he was trying to capitalize on an idea which he had picked up somewhere and did not really understand. Porta ascribed the earliest refraction theory of the rainbow to a Philip who lived at the time of Plato, probably the very one whom Alexander had cited in defense of the reflection theory! Is it possible that Porta owed to Philip not only the basic idea but also some of the peculiarity of detail in the theory which he presented? Did he perhaps have access to a copy of an ancient work on the rainbow which is no longer extant? His explanation certainly shows little sound scientific knowledge; and in one respect it represents a very definite retrogression. Some of his predecessors had begun to associate the bow with the optical properties of spherical drops; but Porta had carried the theory back to the older nebulous cloud-as-a-whole stage. And when one recalls that Porta, author of the fabulously popular *Magia Naturalis*, was one of the most popular science writers of his time, one is prone to speculate on whether or not the "popular science" of today will appear as ridiculous in the year 2325 as Porta's does today.

Many a myopic scientist feels that he is above reading about the silly ideas which his predecessors entertained concerning the world about them. He

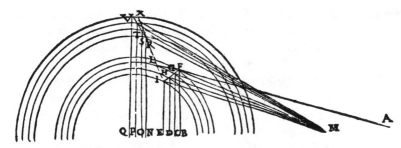

Fig. 32.—The generation of the secondary rainbow, according to Porta.

wishes to have his history of science expurgated of all but those portions which are consistent with the latest views within his field. Such a man desires a highly oversimplified and perverted account of the rise of science in which an exaggerated hero-worship is the central theme. He can dispense with the background of thought except for that of a score or so of "makers of science"; and even in the work of these paragons he overlooks all but those portions which strike him as modern or as evidence that the hero was "ahead of his time." Newton's interest in alchemy, Galileo's views on the tides, Kepler's astrology, are family skeletons to be forever closeted. But such a picture is not only a distortion of man's normal intellectual development; it is entirely inconsistent with the method and attitude of science, in which trial and error and suspension of judgment are important elements. One of the darlings of modern scientists has been William Gilbert (1544–1603), physician to Queen Elizabeth I and King James I, and father of the science of electricity. He is familiar to all as the author of (presumably) a single book, the *De Magnete* of 1600, a model of independent thought and experimentation which stands in marked contrast to most of the works of his day.

Few ancient scientists have earned so persistent and deserved a reputation as did Gilbert. As Dryden wrote:

> Gilbert shall live till loadstones cease to draw,
> Or British fleets the boundless ocean awe.

But a well-launched reputation all too easily grows beyond the limits of sober valuation. More often than not, in the case of great scientists, the truth that men write lives after them, the error oft is interred with their books. One is tempted to add, "So let it be with Gilbert," were not truth, in the history of science, a nobler virtue than adulation. To illuminate only those ideas which eventually led on to further success, passing over in silence those which posterity has judged to be of no value, can foster a totally erroneous impression of the way in which science has developed in the past—and, presumably, is likely to continue to grow in the future. This is not to say that one should admire less such a great book as the De Magnete; it is simply to call attention to the fact that Gilbert wrote also the De Mundo Nostro Sublunari, a work of a different stripe. In this book one finds an aspect of Gilbert's thought which appears to be virtually unknown today—his explanation of the rainbow.[43]

In a recent history of physics one reads that "in optics he [Gilbert] gave an explanation of the formation of the rainbow by the refraction of light in the drops of water, which recalls that of Maurolycus." [44]

There is an inadvertency here, for Maurolycus did not attribute the rainbow to refraction; but the thought occurs that in Gilbert one might find, for the theory of the bow, a worthy precursor of Descartes.[45]

The first two books of De Mundo are on "physiology," by which the author meant the constitution of the earth and the heavens; the last three are on meteorology. Gilbert's physiology and meteorology both are directed Contra Aristotelem. And is this not just what one should expect from reading about the author? His biographer writes of him, "Whilst the fantastic philosophies of the schoolmen still prevailed, he calmly worked out the inductive method of reasoning from the known to the unknown, trying all his arguments by the touchstone of experiment."

It is midway in the meteorology that Gilbert presents his views on the rainbow, beginning with Chapter X of Book IV. The illumination of the rainbow, he wrote, is generated in a vapor and is checked or brought back by a dark object, just as the image is not reflected by a mirror unless tin or a similar substance is placed behind the glass. Thus the rainbow, an image of the sun, is not transmitted to the observer, unless there is an object nearby, such as a cloud, a cliff, a mountain, or a building. The primary rainbow is formed by the vapor lying nearby and surrounding us, being reflected by a dense cloud or a high cliff. Nevertheless, the bow is not generated within the cloud, just as, in the case of glass mirrors, the light which appears to be reflected by the glass, and is in reality reflected by the tin or pitch placed behind the glass, is not in the tin or pitch. The secondary rainbow is formed by vapor near to, or within, the cloud by whose darkness it is reflected; and hence it is appropriate that here the brighter colors should arch around nearer to the primary bow, whereas in the case of the primary rainbow, which is reflected from a distant cloud, the order of colors is inverted, with the brighter outermost. Where the illumination is smaller, due to a greater refraction of the rays, there the color is more intense. Thus, there is an analogy between the generation of the luminous bow in the vapor and its colors. However, the light will not be reflected to the eye unless there is some dense object placed not too far away for it to appear, and unless the proportion is just right.

The above free summary of Chapter X contains the essentials of Gilbert's theory. His ideas—vague, speculative, and wearisome—resemble closely those of Avicenna; but this does not necessarily mean that they were derived directly from Muslim sources, or indirectly from somewhat similar views expressed by Albertus Magnus. Gilbert was predisposed to immoderate attacks upon medieval thinkers, both Latin and Arabic; and it is possible that his explanation was either independently arrived at or else was suggested by other scholars of the sixteenth century. Gilbert's account does indeed at one point include the word refraction, but he seems to be thinking solely in terms of reflection. All in all, his explanation strikes one as an artless version of medieval philoso-phizing, and this impression is not dispelled by the following chapters.

In Chapter XI Gilbert undertakes to refute an ancient error which had been maintained more or less frequently throughout the sixteenth century and which Gilbert mistakenly attributed to Aristotle—the idea that the secondary rainbow is a mirror-image of the primary. In particular, this error had been espoused not only by Cardan, but also by his older mathematical contemporary Clichtovaeus, against whom Gilbert directed his attack. Inasmuch as the position and curvature of the secondary are the same as for the primary, he pointed out, then the colors, too, ought to be similarly ordered, if the one is a mirror-image of the other. To explain the inverted order of the colors in terms of images in a mirror, it would be necessary to have the tips of the bows bending in opposite directions, something which is never seen to happen in the atmosphere. Then, too, the secondary would have to be brightest when the primary is at its brightest, whereas, Gilbert thought, the former is more likely to appear when the latter begins to disappear. Gilbert here labors to demolish a view which had been similarly refuted in ancient, medieval, and early modern times.

In Chapter XII the author considers more closely the shape and colors of the rainbow. The bow is a sort of reflection not of the form of the sun, but of a representation of the illuminated vapor carried to the eye from the air which is opposite. If the air is dense and sufficiently smooth and placed directly opposite, it will reflect the undistorted form of the sun (as happens in the case of glass mirrors). However, the rays do not cling together but are diffused in the vaporous air the illumination of which has been increased and concentrated by the force of the light from the spherical drops. Then an image, with a variety of colors produced when the luminous rays are weakened in travers-ing the moist vapor, is reflected if something dense, like a cloud or mountain is interposed, just as images are not reflected by glass unless it is backed by tin or asphaltum or some such substance. Moreover, it is not truly a repro-duction, but a figure arising from a mixture of light and vapor. It is a repro-duction of light rather than of form and color, just as in the calcination of lead by a slow fire there is at first a tincturing with a green color, then with yellow, and finally, after long and intense heating, with red, as it is converted into oxide of lead. Thus in the rainbow the stronger light is of a red color, the weaker is yellow, and the still feebler is green.

The distances between the objects and the eye, or the great spaces be-

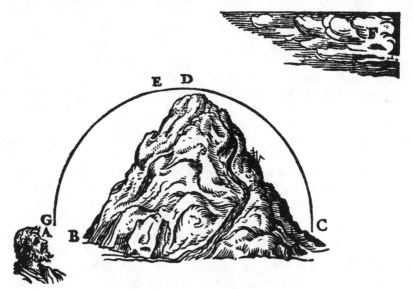

Fig. 33.—Illustration showing the vagueness of Gilbert's idea of the relationship between the dewy vapor and the dark cloud in the formation of the rainbow.

tween bodies, do not affect the proportions or shape of the bow. If the eye is at A (Fig. 33) and if the interposing mountain (against the side of which the rainbow is perceived) is at B, a distance of a stadium about 606 English feet or less, and if a cloud is at F, two thousand paces away, the rainbow GDC nevertheless will be uniform, and not broken or distorted—that is, if between D and E there is no cloud seen in the sky but only pure, thin, and empty air. Chapter XIII is devoted to another assault upon the already prostrate form of the Clichtovaean hypothesis, and the old argument is repeated in superfluous detail. When one views a reflection in water the portion which is higher above the water appears lower in the reflection, and vice versa. Consequently, if the point of a knife is directed toward the water with the handle toward the sky, in the image in the water the pointed end of the iron portion will be uppermost and the handle will be lowest. And so let it be assumed that a log of wood is divided into three parts: a, the highest, red; b, the middle, green; and

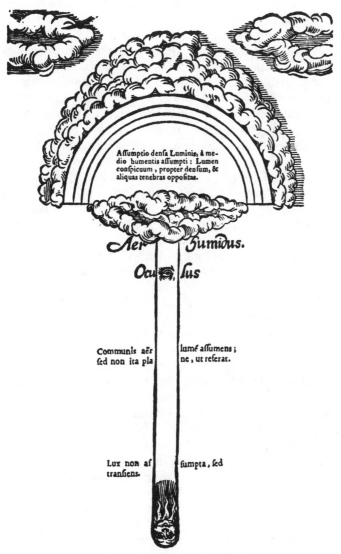

Aſſumptio denſa Luminis, à me-
dio humentis aſſumpti : Lumen
conſpicuum , propter denſum, &
aliquas tenebras oppoſitas.

Aer   ʒumidus.

Ocu   lus

Communis aër   lumé aſſumens ;
ſed non ita pla   ne , ut referat.

Lux non aſ   ſumpta , ſed
tranſiens.

FIG. 34.—Diagram illustrating the formation of the rainbow, according to Gilbert.

c, the lowest, purple; and let it be placed above clear water. Then in the image appearing in the water, the highest part, c, will be purple; the middle part, b, will be green; and the lowest part, a, will be red. This lofty demonstration, says Gilbert, is inappropriate, inept, and wholly worthy of Aristotelian subtlety. "But what do you say, Clichtovaeus. In the general ordering of things should not the entire proportion and form be inverted? See how your own demonstration refutes you." And Gilbert again points out that under this hypothesis the primary and secondary bows should be convex toward each other (not concentric), with the horns of the secondary pointing upward. Hence the secondary is not an image of the primary, either in air or in the clouds.

The infelicity of Gilbert's account of the rainbow must be admitted by his most sedulous admirers, for here he departs widely from the methodology and scientific habit of thought which are presumed to be so characteristic of his work—from the "new style of philosophizing" of which he boasts in the preface of De Magnete. The above passages, if placed beside comparable sections of Grosseteste or Witelo, will be found to be no less medieval in spirit and method; and a comparison with the fourteenth-century work of Theodoric will be positively devastating to Gilbert's vaunted reputation as "the great father of experimental philosophy." It is not without reason that the De Mundo has been described as "a non-experimental 'philosophia contra Aristotelem,' a typically Renaissance treatise which attacked Aristotelian doctrines in perceptibly peripatetic style." Were the De Mundo a juvenalium, it would not be fair to compare it with the product of the author's mature thought; but the evidence indicates that it was prepared after 1591, i.e., considerably later than De Magnete. One might argue further that the posthumous publication of De Mundo suggests that the author did not take this speculative work seriously; but here, too, evidence points to the conclusion that Gilbert clearly attached great importance to his cosmological and physical speculation. It appears, in fact, that Gilbert undertook his work on magnetism largely in order to apply it to his theories of the heavens. The fact that De Mundo appeared only in 1651, long after the author's death, suggests also the possibility that the material was not in final form; but this factor is mitigated by the circumstances that the editor (the brother of the author) collated two manuscript copies of the work. How, then, does one account for the fact that Gilbert's study of the magnet differs toto caelo from his approach to the problem of the rainbow? One is tempted to hold that the former lent itself more easily to experimental investigation, whereas the latter called for quantitative analysis, a field in which Gilbert was relatively less competent. Nevertheless, the work of the fourteenth century would seem to belie such an explanation. Just as the idea of the earth as a huge magnet led Gilbert to experiments with a "terrella," so the exactly analogous idea of a raindrop as a minute sphere had led Theodoric and Qutb al-Din to experiments with a "globosum"—a large glass globe filled with water. Was the idea of a magnified raindrop more elusive than that of a miniature earth? And Theodoric seems to have been as weak on the quantitative side as was the great "electrician"; his success in the theory of the rainbow resulted from the very same factors that served Gilbert so effectively—from the care with which he made qualitative observations. There seems to be little room for doubt that what Gilbert did for magnetism and electricity he might similarly have accomplished also for the rainbow, had he but applied to the latter phenomenon the same experimental philosophy which he had exploited so successfully in connection with the former. By a strange coincidence it appears that one of Gilbert's acquaintances, the mathematician Thomas Harriot (1560–1621), was perhaps even then experimenting with globes of water in an effort to decipher the rainbow problem. But this work is properly associated with the following century. Had Gilbert lived only a few years longer, his theory of the rainbow inevitably would have been completely different. The obvious inference is that, in some respects at least, Gilbert definitely was not "ahead of his time," as is frequently maintained. It is not given to any one man to make all the discoveries or to transcend all error. If Gilbert's dexterous experiments on electricity strike sparks of modernity, let his devious arguments on the rainbow be a reminder that his mental habits were those of an earlier day. Taken together, *De Magnete* (or *Physiologia Nova*) and *De Mundo* (or *Philosophia Nova*) show Gilbert to have been a typically Janus-faced physician of the sixteenth century, modern in his advocacy of laboratory precision, yet medieval still in his predilection for fuzzy speculation.

VII

Kepler and His Contemporaries

THE FAILURE OF THE SIXTEENTH CENTURY IN GENERAL, AND OF GILBERT IN PARticular, in the story of the rainbow illustrates the importance of ideas, as compared with instruments, in the history of science. No unusual equipment was required for an investigation of the rainbow. The one appropriate gadget was a spherical globe of water, and this had been readily available to experimenters at all times. It had, in fact, served Alhazen effectively in some of his optical studies. The one thing that was needed above all else for the geometrical solution of the rainbow problem was a rediscovery of the idea that a large sphere is not unlike a magnified raindrop. The seventeenth century often is known as the age of genius—and this for at least two reasons. The century effectively invented far more than its share of scientific instruments: the thermometer, the telescope, the microscope, the pendulum clock, are but a few of these. But, more than this, the age of genius also produced more than its just measure of ideas: among them, the circulation of the blood, the wave theory of light, and the law of gravitation. To some extent, it is true, the instruments and ideas had been adumbrated by earlier periods; but it probably is safe to say that in no century, with the possible exception of our own, was the interplay of instruments and ideas more effective than during the age of genius. And when one adds to this the penchant of the time for formulating and solving problems mathematically, it becomes obvious that the time was ripe for a fresh advance in the theory of the rainbow. And yet the advance was not achieved immediately nor without difficulty. Few aspects of science illustrate more vividly than does the rainbow the tremendous difference between hindsight and foresight. How often has a brilliant scientist, such as Alhazen, appeared to be perversely oblivious to the solution which lay before his eyes!

Johann Kepler (1571–1630) represented an unusually happy conjunction of the factors making for success in science. He had a high appreciation of the value of scientific instruments; he was confident that the laws of nature were mathematically ordained; and he was just bubbling over with ideas. Moreover, he lived during the very middle of the period of renewed interest in the theory of the rainbow, a time when whole volumes were published on the subject. Could he have ignored these works devoted to one of the most striking of optical phenomena? After all, what Gilbert was to electricity, or Galileo to dynamics, or what Boyle was to pneumatics, that was Kepler to dioptrics. And yet historical works on the rainbow make virtually no mention of his views. A search of his works, however, reveals that Kepler did indeed make a determined effort to explain the rainbow—a study almost as earnest as his efforts, far better known, to discover the law of refraction.[1] And one of the striking characteristics of his search is that the development of his own ideas as an individual follows closely the stages found in the wider history of the theory as developed by mankind at large. So close, in fact, is the resemblance that one is almost tempted to see here the operation of a general principle of mental development somewhat akin to the biogenetic law of recapitulation.

Kepler's interest in the rainbow seems to date from about the time of his *Mysterium Cosmographicum* (1596), when faith in the mathematical harmonies of the universe appeared to have been strikingly vindicated. Shortly after the publication of this work had launched Kepler upon a successful astronomical career, he sought to extend the harmonies of astronomy and music to include the phenomena of color. In marginal notes to the *Mysterium*, written apparently toward the very close of the century, the range of colors in the rainbow is compared to the infinity of tones in the musical octave. Yellow is taken as a sort of mean; and from this one passes, outward, through red into black as the solar influence diminishes and the admixture of the crass material in the cloud increases. From yellow inward one passes through green, blue, purple, and violet into black, and this transition is due to quite another cause, namely, refraction. Kepler added that often he had considered whether or not the proportion of the angle of refraction determines the limits between the colors green, blue, etc. Direct vision, in which the angle of refraction is zero, results in yellow light; and when the angle of refraction is a right angle, all light ceases, so that this corresponds to blackness. But, added Kepler, how the greatest angle of refraction is to be subdivided in terms of color, it is difficult to say. Nevertheless, he believed that if the right angle were divided into parts corresponding to the simple unit fractions 1/6, 1/5, 1/4, 1/3, and 1/2, the five colors would be yellow, green, blue, purple, and violet respectively. "And lo, is not the magnitude of the rainbow always about 45°, which is the measure of half a right angle?" Kepler added cautiously, however, "But these may be notions." [2]

This, Kepler's early explanation of the rainbow, is a strange mixture of Aristotelian color theory and Pythagorean numerology. Had he thought of Occam's razor, he would scarcely have postulated two distinct causes of the rainbow. It will be noted further that the explanation is Aristotelian in the tacit assumption that it is the cloud as a whole which causes the phenomenon. Kepler seems to have been completely unaware of earlier theories explaining the bow in terms of individual spherical drops. It is astonishing how poor scientific intercommunication of the time appears to have been. One wonders whether the radius of 45° which Kepler, throughout his life, accepted for the bow, was determined independently or was borrowed from others—perhaps Maurolycus. Kepler cited no authority for it.

The early views of Kepler on the rainbow were immature in the extreme, and yet he hastened to publish them. In an astrological treatise of 1602, *De Fundamentis Astrologiae Certioribus*, he reiterated the Aristotelian distinction between reflection from the surface of a mirror, in which form is preserved, and reflection from an uneven surface, in which the light is imbued with color. Pursuing this thesis, he held, somewhat as had Seneca, that the colors of the rainbow fall into two classes: those arising from the darkening or privation of light; and those from refraction or tincturing. Kepler here seems to use the term refraction in the Aristotelian sense of reflection. Of the former class, the first color is the light of white heat itself, which cuts the circle of the rainbow as if in two. On the one side the intensity of illumination diminishes, producing first yellow, then red, a darkish color, and finally black; on the other side it is refracted and reflected, resulting in green, blue, purple, and, finally, dark violet.[3]

Kepler's views of 1602 still were strongly tinged with Peripateticism; and two years later he published, in his classic commentary on the optics of Witelo —*Ad Vitellionem Paralipomena Quibus Astronomiae Pars Optica Traditur*— a crude qualitative modification, based upon refraction, which resembles one suggested by Grosseteste and other medieval commentators on Aristotle. Here Kepler again reiterated the idea that the colors of the rainbow result from two distinct causes—the attenuation of light and the injection of aqueous material —and he asserted categorically that the diameter of the rainbow is always 90°. Then he said that the bow is due to the refraction of the rays of light by rain or aqueous material *between the spectator and the sun*. It is therefore not true, he argued, that the rainbow is caused by the reflection or refraction of the rays of the sun or of vision in that portion of the cloud in which the bow appears—repeating a proposition which had long been argued pro and con in scholastic circles. In this same treatise Kepler naively cited the colors in the rainbow as supporting his quasi-medieval belief that comets and novae are aqueous in origin and nature. In a description by Cornelius Gemma Frisius (1535–1577) of the variations of color in the nova of 1572, Kepler had read that the new star was at first red in color, then became brilliantly yellow, then green, and finally disappeared after having taken on a violet color. But this is precisely the order of the colors in the rainbow, which results from humidity in the air; and from this Kepler felt that one may reasonably conclude that the star was engendered by humidity! [4]

Kepler apparently was soon convinced that his views of 1604 were untenable, as his correspondence in the following year indicates. David Fabricius (d. 1617) had written him asking why the radii of lunar haloes are always half the apparent radius of the rainbow, i.e., half of 45°. Kepler replied that he could not say more of the rainbow and halo than he had given in the *Paralipomena* of 1604, except that he had erred in placing the source in the sublime region far above the cloud, and that the cause lay in the drops or vapors in which the bow appears.[5] In this same year Kepler made his first crude attempt to explain the rainbow geometrically in accordance with his revised views and to answer the question raised by Fabricius. Writing to Johann Georg Brengger, professor of medicine at Kaufbeuren,[6] Kepler said one sees clearly that for the formation of the rainbow it was not so much rain which was necessary as it was the disposition of the air to collect into drops. His explanation, nevertheless, is based, not upon individual drops, but upon the spherical shape of the cloud as a whole. Let the rays of the sun, regarded as parallel, be given by the lines HDC, IE, and KFAB, and let the points B, C, D, F lie on a circle with center at A (Fig. 35). If the eye of the observer is at A, then the angle of the rainbow, angle BAC, is 45°. Kepler apparently followed Aristotle in assuming that in reflections causing color, the law of the equality of angles need not be satisfied. Therefore, Kepler argued, inasmuch

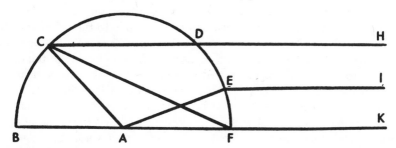

Fɪɢ. 35.—Kepler's early explanation of the production of the rainbow by reflection from a spherical cloud.

as the rainbow and the halo are both due to refraction, were the eye to be placed at C, the sun would be observed directly along the line HDC and the halo would be seen along the ray CFK. But from geometry the angle DCF is precisely half the angle BAC, so that the radius of the halo is 22½°. Years later Pierre Gassendi (1592–1655) likewise was unable to resist the temptation of seeing in the relation between central and inscribed angles the fact that the apparent size of the primary rainbow is double that of the ordinary halo.[7]

Kepler realized that his explanation gives rise to many questions. Why, for example, should one not see a rainbow of radius 45° about the sun [presumably between the lines HDC and KAC], or a halo of radius 22½° opposite the sun [allowing the ray EI to be extended to touch the circle a second time and be reflected to A]? Or does the surface of the air really have the spherical figure assumed in the explanation? The nebula in which the bow is seen is usually agitated by rapid wind and does not preserve any particular form, whereas the rainbow is always circular. Moreover, the water and drops descend swiftly in paths which are not always straight lines. Kepler was unable to explain why such irregularities do not manifest themselves in the shape of the rainbow, except to suggest that it must be the body of the aqueous air, rather than the surface, which serves as the agent. Incidentally, he would have been greatly pleased to know that there is indeed a halo about the sun, not often seen, of radius about 46°!

Kepler, in his correspondence with Brengger, made a number of other suggestions, some of which are wildly imaginative. Inasmuch as haloes are generally seen through fine mists, whereas rainbows appear during a downpour, Kepler asked whether refraction in air in which drops are just beginning to be formed may not be half as great as that in air in which rain is already falling. Perhaps, on the other hand, the double radius of the rainbow is due to the combined effect of refraction and reflection, the halo resulting from refraction alone. He seemed to be unable to decide whether it is water or a shower of raindrops or simply aqueous air which causes the rainbow; and he asked whether the bow is peculiar to each observer or is due to the coloring of the whole mass of the air in question. Of one thing he was certain, however, and that is that such phenomena are not optical illusions, but come to the eye by real rays; for he himself had admitted into his room light from parhelia and had seen the colors on the wall. Then Kepler brought his feet to earth and suggested to his correspondent that he examine the refraction of solar rays through a spherical globe of water. Is it not true that rays traversing near the center of the globe are not colored while those passing near the edge are? Color therefore seems to depend upon the magnitude of the angle of incidence. Kepler's letter closes, however, with a frank admission of failure. What can one say about the origin of haloes and the rainbow? "I don't know," is his answer.

It was almost three years before Brengger replied to the letter of 1605, and in this interval Kepler developed his ultimate explanation of the rainbow.[8] On October 2/11, 1606, Kepler wrote to Thomas Harriot (1560–1621), an Oxford mathematician and scientist, proposing problems in optics and mechanics. Harriot is one of the earliest scientists to have visited America, for in 1584 he accompanied Sir Walter Raleigh on his expedition to survey and map the Virginia territory. Upon his return to London he wrote a description of his journey in the *Briefe and True Report* of 1588, a book which became very widely known. He found a patron in Henry Percy, Earl of Northumbria, but this proved to be a doubtful blessing. He was indeed paid a handsome salary—300 pounds a year. But Henry was suspected of complicity in the gun-powder plot and in 1606 was jailed. Harriot remained with his patron, and illness and political turmoil prevented him from completing the promising projects he had undertaken. Given more favorable circumstances, he might have become known as the inventor of analytic geometry or as the one who solved the rainbow problem.

Kepler had heard of Harriot's work in chemistry and optics (as well as his criticism of astrology!) and hence he requested the latter's views on the *Paralipomena*. If Harriot would but tell him the cause of color in refraction and send him the measures of refraction in his experiments, Kepler believed that the explanation of the rainbow would be much expedited. Then followed a lengthy description of Kepler's views on the rainbow, views contrasting sharply with those of the previous year. The demonstration of the rainbow, Kepler now held, depends not on the cloud as a whole, but on the smallest elements of it—tiny drops of rain which are exactly round. Kepler finally had reached the point at which a few of his predecessors had arrived three centures before. Just how he came to this view is not made clear, but it appears likely that he had followed his own advice to Brengger to study refraction in a spherical globe of water—a step which later led Descartes to success. Inasmuch as rays near the center of the sphere are not colored, while those passing near the edge are brightly tinted, Kepler made the natural but erroneous assumption that the solar rays which produce the rainbow are those which strike the drop along a line of tangency. If, for example, S is the sun and O_2 the eye of the observer (Fig. 36), he assumed that the tangent ray SA would be refracted along the line AB, then reflected from the concave surface of the drop along BC, and finally at C leave the drop refracted along the tangent line CO_2. Now Kepler assumed that the radius of the rainbow is 45°, and hence the arc AC must be 135°. Therefore AB = 67½° and RAB = 33¾°. Here one has a beautifully clear and simple explanation of the bow; but Kepler noted at least one fly in the ointment. The angle 33¾° is too small, for tables of refraction indicated that the angle should be at least 37°. (For such a refraction the radius of the bow, under Kepler's explanation, would be about 32°.) Kepler rather half-heartedly suggested that perhaps lukewarm rainwater, being less dense than our standing waters, may cause a lesser degree of refraction. He anticipated scepticism with respect to the tangency requirement for the solar rays, but he reiterated that only in this way are colors formed.

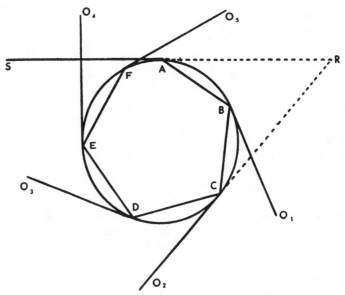

Fig. 36.—Kepler's later explanation of the production of the rainbow by refractions and reflections within the spherical droplets.

The faith Kepler placed in his plausible explanation was confirmed by the fact that, with some modification, it served equally well for haloes. Not all of the ray AB is reflected at B, for a portion of it passes through the transparent surface, undergoing refraction along the tangent line. If the eye were at a point O_1 on this line (Fig. 36), it would see a halo of radius 67½°, whereas experience shows that the actual radius is only 22½°. To account for this one could assume that the angle SAB is 168¾°, but then how much less dense would be the substance causing the halo than that which produces the rainbow! Even Kepler hesitated here; and so he devised an alternative explanation for the halo which is based upon the same refractive index as is that for the rainbow. A portion of the ray BC, which causes the bow seen from O_2, is again reflected internally along CD and leaves the drop, after undergoing another refraction, along DO_3 (Fig. 36). This would cause a halo or rainbow of radius 22½° in opposition to the sun. The fact that this is never seen Kepler ascribes, rather inadequately, to the fact that it would be visible only when the sun is near the horizon—i.e., within 22½°. Now a portion of the ray CD will undergo a third internal reflection, followed by a refraction at E along EO_4. This should cause a bow or halo of radius 90°, but this likewise is never seen. Finally, part of the ray DE is reflected a fourth time and leaves the drop at

F, undergoing a refraction along FO_5. This ray does indeed make, with the solar rays, the required angle of 22½°!

Kepler for several reasons hesitated to accept his beautiful pinwheel theory. In the first place, he thought it incredible that the colored light, following so many reflections, would reach the eye in sufficient strength to cause an impression of color. Then, too, he found it difficult to reconcile the mobility of the falling drops with the constancy of the circular arch of the bow. Perhaps, he suggested, the bow is caused principally by dew. After all, there are, besides the common bow, also many kinds of extraordinary rainbows; and, after describing one of June 10/20 seen at Mogontia, Kepler closed with the challenge: "Thou, then, oh excellent priest of the mysteries of nature, tell the causes."

Harriot replied from London on December 2/11; but the reply must have been a disappointment to Kepler. The bulk of the letter is made up of a table of refractive indices of more than a dozen substances and an attempted explanation of the simultaneous reflection and refraction of rays by transparent media. Of the rainbow Harriot said only that when he writes on this he will give the proximate and immediate causes, for these are not correctly explained by the Peripatetics. He asked Kepler to be patient, adding: "I would say this of the rainbow just now, that the cause is to be demonstrated in a droplet through reflection on a concave surface and refraction on a convex. However, I have said nothing in consideration of the mysteries which are concealed." [9]

Kepler answered from Prague on August 2/11, 1607 that he was filled with eagerness to see Harriot's works on colors and the rainbow. The two men were in apparent agreement that the arc is caused by the individual drops and that the colors are produced by reflection on the concave surface and refraction on the convex; and so Kepler hoped to receive a more adequate reply to his first letter. It was almost a year before Harriot wrote again, on July 13/22, 1608. He pleaded lack of time for writing and philosophizing. He reported on some arguments for the existence of a vacuum which had been directed by Gilbert against the peripatetics; but he said nothing further about the rainbow. Kepler in turn delayed over a year before sending a reply on September 1, 1609, in which he argued against the possibility of a void. The correspondence between them appears to have been broken off here, and we have no further information on Harriot's theory of the rainbow. Kepler meanwhile had resumed correspondence with Fabricius on November 10, 1608, giving some of his latest thoughts on the rainbow. Contrary to his previous letter of some four years before, he cited it as certain that the cause is to be found in the individual drops, and that the halo likewise is caused by the very smallest dew drops. He added that he can give a beautiful cause for the fact that only the arc, and not its interior, is colored. Here he clearly had in mind the explanation which had been sent to Harriot, for he said that colors arise only at places where the refraction is a maximum, i.e., where the angle at the drop between the incident ray and the visual ray is 135°. In a work defending astrology published in 1610, *Tertius Interveniens. Das ist, Warnung an Etliche Theologos, Medicos und Philosophos Dass sie Nicht das Kindt mit dem Badt Ausschütten*, Kepler again asserted that the colors of the rainbow are due to solar rays passing through round droplets of rain; but he did not explain his views further.[10]

Most assuredly one would have expected Kepler to go into further detail in his second classic treatise on optics, the *Dioptrice* of 1611. Nevertheless, one finds here only the familiar statement that the colors of the rainbow arise where refraction is great.[11] In the preface Kepler said that he had thought of adding a little book on the rainbow but that satisfactory causes of extraordinary rainbows were to be desired, and these at present have failed to appear. Kepler apparently never did write this projected book, even though he seems to have maintained confidence in his ultimate explanation. In correspondence with Remus in 1619 he declared that certain things concerning the rainbow and halo are clear. The radius of the primary bow is 45°, that of the secondary being 11° greater; and the radius of the halo is 22½°. The reason is to be sought in the maximum refraction in the round drops of water, for where the refraction is greatest, there do colors arise. The same explanation is given by Kepler in notes accompanying his translation from Greek into Latin of Plutarch's *De Facie in Orbe Lunae*. Here he says that some Aristotelians too readily conceded that the secondary bow is a mirror image, upon an outer cloud, of the primary bow which appears on an inner cloud; yet Kepler's failure to explain the secondary rainbow is a serious deficiency in his own work. His lack of success here, as well as in the case of the primary, may be ascribed, at least in part, to his failure to observe his own admonition to observe and measure accurately. He had the necessary mathematical imagination, but he lacked the needed experimental data.

The failure of Kepler to publish his definitive views on the rainbow was most regrettable. He was very close to the solution, and yet the important ideas found in his correspondence were not printed until they appeared in an edition of his *Epistolae* in 1718, long after they were outmoded.[12] One wonders whether he was discouraged from publishing by his acknowledged failure to discover the second of the two laws basic to the theory of the rainbow. The law of reflection had been known since Aristotle; but the law of refraction,[13] the cornerstone of optics, had eluded discovery, in spite of the best efforts of Ptolemy, Alhazen, Grosseteste, Witelo, Maurolycus, and Porta. The law was of particular concern to Kepler because of the astronomical importance of atmospheric refraction, and hence he made a concerted effort to discover it.

In this work he made an unfortunate start, for he followed his predecessors in the designation of the angles of incidence and refraction. Nowadays these are the angles between the normal to the surface and the incident and refracted rays; but Kepler took the angle of refraction to be that between the refracted ray and the rectilinear prolongation of the incident ray. That is, if r and r' are the modern and the Keplerian angles of refraction, and if i is the angle of incidence, then $r' = i - r$ or $r = i - r'$. For small angles of incidence —i.e., angles less than 30°—Kepler used the simple proportionality $r' = ki$, equivalent to $r = k'i$. He knew of Alhazen's work (but not of Ptolemys) through the Risner edition, and hence he was aware that this proportionality was an approximation only, and that it gives poor results for larger angles. He therefore tried, both in the *Paralipomena* of 1604 and in the *Dioptrice* of 1611, to find a trigonometric relationship between i and r'. How could he fail, one now asks, when he was thus on absolutely the right track? And yet he did fail completely. He tried more than half a dozen trigonometric combinations: $i - r = k \sec i$ and $i - r' = k \sin i$ and $\tan i = k \tan r$ and $\tan i = k \sin r'$ and $1 - \tan i \cot r' = k \tan i$ and $1 - \tan i \cot r' = k \sin i$, $r' = k_1 i + k_2 \sec i$, and $1 - \tan i \cot r' = k_1 + k_2 \sin i$. He devised a simple instrument for testing these equations experimentally, and he found everyone of them incorrect. Kepler was no experimental physicist and the data at his command were unbelievably slight. Is it any wonder that he became utterly discouraged and abandoned the search for the law of refraction and, perhaps consequently, the solution to the rainbow? His interest in optics seems to have waned quickly. And yet he was so very close to the truth! Had he only tried $\sin i = k \sin (i - r')$, and had he only shifted his rainbow ray slightly, he might have experienced pleasures of discovery comparable to those which he lyrically described in connection with his great astronomical discoveries. He might thus have done with Witelo's observations on refraction what he had successfully accomplished with the astronomical observations of Brahe in becoming the law-giver of the heavens.

The year 1611 was an active one from the point of view of optical publication. There was Kepler's *Dioptrice*; an edition of the *Photismi de Lumine* of Maurolycus; Ambrosius Rhodius' *Optica*; and—last but most important—the *De Radiis Visus et Lucis in Vitris Perspectivis et Iride Tractatus* by Marco Antonio de Dominis (1564-1624). What a cross-section of optical thought of the time these books furnish! The *Optica* of Rhodius is a long book of over four hundred pages, with at least eight of these devoted to the rainbow.[14] The author was a doctor of medicine and philosophy at Kembergensis and professor of mathematics at Academia Leucorea, qualifications which lead one to hope for something worth while. The result, however, is keenly disappointing. Rhodius presented a theory derived from Aristotle, Witelo, and—*mirabile dictu*—Porta. First he gave the familiar Aristotelian theory in terms of reflection, including (to his credit) the correct explanation of the circularity of the bow. Next he introduced, à la Witelo, the idea that there are infinitely many little drops in the dewy vapor and that in the rainbow there is an admirable mixture of reflection and refraction. He added the quaint belief that thus a double image of the sun is seen, one by reflection alone, the other by both reflection and refraction. But Rhodius found difficulty with the colors, and so finally he had recourse to the explanation of Della Porta. He supplied a figure virtually identical with those to be found in Porta's *De Refraction* (although he cited the *Magia Naturalis*). He felt that this explains satisfactorily why the arc is low when the sun is high, and vice versa, without apparently realizing that it is in obvious disagreement with the familiar properties of reflection and refraction.

The last of the four works of 1611 cited above shows considerably more understanding. De Dominis was an ambitious Dalmatian Jesuit who rose to become Archbishop of Spalato. (Goethe wrote of him that "he discovered the solar spectrum while saying mass.") Upon renouncing his faith, however, he became a convert to Protestantism and, through the influence of the English ambassador at Venice, was appointed Dean of Windsor by James I. Ultimately, vexed by doubts and tempted by the prospect of a cardinal's hat, he sought reconciliation with the Pope at Rome, only to die in a dungeon of the Inquisition in St. Angelo. Later the Inquisition tried him posthumously and found him guilty. His corpse was exhumed and burned, together with his writings. There has been a tendency to regard him, in view of his optics, as a martyr to science; but evidence points to religious heterodoxy, not scientific views, as the factor leading to his imprisonment.

The license for *De Radiis Visus et Lucis* is dated January 27, 1610, but the preface declares that the work is based upon notes prepared for the author's lectures at Padua and Brescia some twenty years before. Whatever truth there may be in this claim, it is quite obvious that in this case much new material must have been added just before publication, for the first part is a not unimpeachable treatise on the properties of the then recently discovered telescope. In fact, the little book of some four score pages carries in Latin a subtitle, "In which, among other things, there is shown the theory of a certain instrument which I have thought up for viewing clearly things which are very far away." By a curious coincidence Kepler and De Dominis both, within the same year, published explanations of the telescope which Galileo had so effectively publicized a year or two before.[15] Is it also merely a coincidence that their explanations of the rainbow are so nearly alike? It would appear so, for Kepler's later ideas had not yet been published; but De Dominis does not make known the path by which he hit upon his theory. The section on the rainbow includes frequent and extensive allusion to medieval and early modern precursors, constituting almost a little history of the subject; and yet his views depart radically from theirs. He described in particular detail the opinions of Albertus Magnus, Witelo, Piccolomini, and Cardan—men who were outstanding proponents of the philosophical current. There is no reference to others who represented the tendency toward geometrical optics, such as Theodoric, Fleischer, Maurolycus, Porta, or Kepler. The theory of De Dominis definitely belongs, nevertheless, in the second category. Did he somehow have access to a manuscript of Dietrich's work, or had he noticed Themo's unobtrusive reference to a reflection within the raindrops? Did he learn of the correspondence of Harriot and Kepler, or was his idea a sudden flash of insight? Was his theory merely a combination of notions taken from Maurolycus and Porta, or was it the result of painstaking observation with a globe full of water? De Dominis gives the last as the source of his work; but if this is so, he must have taken precious little pains in his experimentation, for he made several egregious errors.

The explanation of the rainbow given by De Dominis is a remarkable combination of "pearls and sour dates" (a phrase Al-Biruni had used long before to describe Hindu contributions to science). He demolished most of the antecedent theories by "proving" the impossibility of reflection as the sole cause. His "proof" probably was taken from Porta, although he doesn't mention this fact. He assumed that a ray of sunlight strikes a vertical sheet of rain, and he concluded that, according to the law of reflection, the ray should be reflected upward, rather than downward toward the eye of the observer. The bow therefore must be caused in part by refraction. But where Porta had abandoned reflection entirely and had not made use of the sphericity of the raindrops, De Dominis came almost as close to the modern theory as Theodoric had. Rays from the sun, he said, strike the outer convex surface of a spherical raindrop and are refracted into the drop. Some of the light passes unrefracted through the rear wall of the drop; but at the rear concave surface some of the light is reflected toward the eye of the observer, where the conjunction of such rays from many different drops produces the rainbow. The attentive reader will note that De Dominis failed to mention the second refraction which is bound to occur as the ray leaves the drop. He distinctly says that the bow is caused by a reflection after a refraction. Was this a casual oversight on his part? The answer would appear to be a categorical "No." The accompanying diagram (Fig. 37), which is repeated several times throughout the book, shows no sign of this second refraction.[16] Moreover, in about a dozen other figures in the book, showing light passing through lenses of various types, rays are portrayed as passing obliquely across the boundary between media without refraction. And in the explanation of the halo he followed the

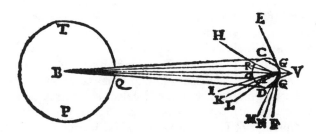

FIG. 37.—Diagram used by De Dominis to explain the formation of the rainbow.

habit of earlier writers in ascribing the phenomenon to refraction but picturing a single change of direction as in reflection.[17]

De Dominis correctly represented the primary bow as caused by rays striking the upper portion of the drop; but again he erred on rays striking the
lower portion. The latter he pictured as refracted into the drop and reflected
in part at the rear of the drop. He represented them as leaving the drop, unrefracted, in an upward direction, and hence not reaching the observer. Once
more, at the rear of the drop, the unreflected portions of the rays are pictured
as emerging unrefracted at the concave surface to converge to a focus. The
author was in possession of the key to the rainbow, but so badly did he fumble
the optical problems inherent in it that one wonders whether he truly understood his own explanation. Such doubts become more pronounced with his
"explanation" of the secondary bow. The generation of this, he held, differs
from that of the primary only in the fact that the rays causing it strike the
drop more obliquely. The diagram is even more confusing on this point, for
it implies that the secondary bow should lie within the primary. The more
sharply inclined rays in the direction F (Fig. 37) he thought formed the
primary bow, those less inclined and emerging in the direction I he believed
produced the secondary arc. (In another figure which does not show the refractions and reflections within the drops, De Dominis correctly pictured the
rays for the secondary rainbow as more steeply inclined than those for the
primary.) [18] Then, too, the figure indicates that he thought the secondary bow
also was caused by rays which strike the upper portion of the drop and are
reflected and refracted only once. (At one point, however, he does casually
mention the possibility of a refraction as the rays leave the drop.) [19] He placed
among the "insoluble questions" the problem of why the secondary arc never
appears without the primary rainbow; [20] but the weakness of the higher bow
he attributed to the greater distance from the eye and to the fact that the
reflection takes place not from the anterior portion of the vapor but from
the interior portion, as well as to the tenuity of the medium at higher altitudes.[21] On the problem of multiple rainbows his position is uncompromising.

"There are two different types of reflection [one producing the primary rainbow, the other the secondary] from drops of vapor, and no more; and hence two distinct bows, and not more, can appear." [22] In view of this assertion, he doubted the validity of Witelo's report of multiple rainbow arcs.

De Dominis could not, in view of his errors, explain adequately the order of the colors. For a prism he assumed the red ray was that which passed through less of the dense medium, the violet through more of it. (This is not true, however, of refraction through parallel plates.) For the primary bow he therefore thought the order of colors was given similarly by the magnitude of the distances traversed through the sphere of water, red taking the shorter path. On the other hand the colors for the secondary were determined instead by the degree of obliquity of the rays.

The size of the bow De Dominis accurately set at 42°, but this figure he could have borrowed from any one of many predecessors. There is, however, a statement which is not easily accounted for. No one drop, he said, sends to the eye all of the colors of the rainbow; each drop is responsible for only one color. This idea had been expressed by Theodoric of Freiberg, and possible plagiarism by De Dominis has been suggested. Serious indebtedness to Theodoric would appear to be ruled out, however, by the optical blunders De Dominis made, especially with respect to the exterior bow. Borrowing from Themo, whose *Quaestiones* once had been published at Venice, where De Dominis' book was also printed, would appear more likely. Themo had described experiments with globes of water, and he had mentioned refraction into raindrops, followed by a reflection at the rear surface, without calling attention to the second refraction. But Themo then had not known that each drop contributes to but one of the colors in the rainbow. The identification of different colors with different drops could, of course, have been suggested by Porta's theory, in which diverse colors were associated with differing portions of the watery medium. There is also a possibility that somehow De Dominis had run across Harriot's statement that the bow is caused "in a droplet through reflection on a concave surface and refraction on a convex"? The explanation of De Dominis would literally be well described in precisely such terms. Until further evidence is adduced, one can assume that the theory of De Dominis was not derived from any one source. It was rather a mosaic of notions borrowed from the philosophical and optical traditions, verified or modified perhaps by direct experimental evidence.[23] His book includes not only the explanations given above, but also many of the popular notions which have enjoyed wide currency, such as the presumed greater convexity of the rainbow when the sun is lower (and hence the bow is higher). It closes on a familiar note—the role of the rainbow in weather prognostications.

It is difficult to assess the place of De Dominis in the story of the rainbow.

His explanation, with all of its faults, is superior to any other published in the interval of three centuries from 1311 to 1611. Theodoric's work was clearly of a higher order in the precision and correctness of thought with respect to what took place within the raindrops; but at least De Dominis correctly followed Maurolycus in abandoning that old incubus, the Aristotelian meteorological sphere.

One should bear in mind, nevertheless, that in several respects De Dominis' views are quite inferior to the unpublished opinions of Kepler. In the first place, Kepler's explanation was consistent with the elementary principles of geometrical optics, for he recognized the inevitability of the second refraction. Then, too, De Dominis made no attempt to account for the size of the bow, a problem which Kepler essayed, albeit unsuccessfully. To Kepler one owes the clear recognition that "to measure is to know"; and to him physics appears to be indebted for the earliest *quantitative* theory of the rainbow based upon refraction in raindrops. Had he but measured more accurately, he might have anticipated the theory that Descartes gave half a dozen years after Kepler's death. Yet the explanation of Kepler has been universally overlooked by historians of physics whereas that given by De Dominis somehow has captured their imagination. Emphasizing its merits, they pass over in silence its deficiencies. One writer goes so far as to call De Dominis "one of the clearest heads of the century"! Newton, Leibniz, Musschenbroek, Goethe, and others virtually accused Descartes of plagiarism from De Dominis; [24] and in many an authoritative treatise on physics one can read that "the elementary theory of the rainbow was first given by De Dominis." [25] Indebtedness of Descartes to De Dominis is not an a priori impossibility.

The Dutch scientist Willebrord Snel (1581–1625), in a work on the comet of 1618, referred to haloes and rainbows as caused by reflection and refraction. He promised that on another occasion he would comment more fully on this idea; but his commentary, if written, is not extant. One wonders if it might not have served as a connecting link between De Dominis, with whose work Snel (or Snell) apparently was familiar, and Descartes, who had access to some of Snel's manuscripts. Snel wrote, in a marginal note in his copy of Risner's edition of the *Optics* of Alhazen and Witelo, that De Dominis made the greatest height of the bow 42° and explained the bow by two refractions, one on entering and one on leaving, as well as a reflection within.[26] The fact that Snel corrected De Dominis on the second refraction might indicate that he had come upon the explanation independently or through other sources. There also is a possibility that Snel operated as a link between Kepler's work on refraction and that of Descartes. Kepler had cited a suggestion of Alhazen and Witelo that light which falls obliquely upon a surface can be regarded as a composite motion, the refracted ray being bent toward the normal because the perpendicular component is strengthened.

Descartes later used precisely this type of argument in justifying the law of refraction, and it was long thought that, through Snel, he may have derived the suggestion. Snel was engaged in writing a book on refraction, and in this connection he had made many laborious measurements—an aspect in which Kepler had not excelled. Was it on the basis of Kepler's ideas or of his own measurements that Snel discovered the law? It may be that in the law of

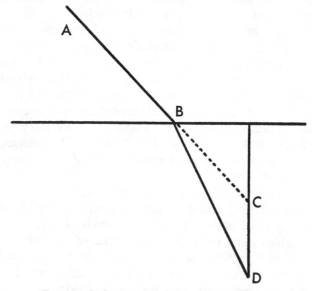

Fig. 38.—Snell's formulation of the law of refraction.

refraction one finds the first discovery of a physical law through precise measurement. It is known that by 1621 Snel knew the correct law of refraction, but just how he discovered it can not now be determined. The author died a few years later, leaving his manuscript unedited. This manuscript, which may have been available to Descartes, apparently was last seen by Christiaan Huygens (1627–1695), and it now appears to be irretrievably lost. From Huygens' report it is known that Snel expressed the law as follows: If the prolongation of the incident ray AB and the refracted ray BD (Fig. 38) meet a normal to the surface in points C and D respectively, the segments between the foot B of the two rays and that normal are in constant ratio—that is, BC:BD = k. This is, of course, equivalent to a proportionality between the cosecants of the angles of incidence and refraction. Huygens questioned whether Snel really understood the nature and importance of his discovery inasmuch as it had been assumed that a ray perpendicular to the surface undergoes a refraction. Snel appears to have been led to this error by the peculiarity of his approach. He concentrated upon the apparent raising of the level of an object seen under water, attributing this to a contraction along the vertical. Huygens correctly pointed out that the apparent lifting of the bottom of a pool is in reality due to the numerous oblique rays which meet the eye, not to a foreshortening of vertical rays.

Whether or not Snel fully appreciated his discovery,[27] the fact remains that it did not influence the story of the rainbow before the days of Descartes.

Refraction and the individual-drop theories of Maurolycus, Kepler, and De Dominis did not immediately outmode older approaches. In 1615, for example, Josephus Blancanus, or Giuseppe Biancani (1566–1624), in a book on mathematics in Aristotle, gave a clear exposition of the Aristotelian interpretation. The author referred to Maurolycus and Porta as authority for a radius of the bow of 45°, but he applied this only to reflection from a cloud on the meteorological sphere.[28] In 1621 Bartholomew Keckermann, in a section on "Scientiae opticae pars prior" of his Systema Mathematices, followed Witelo in describing the rainbow as a combination of rays reflected and refracted in a cloud.[29]

Kepler's greatest contemporary undoubtedly was the incomparable Galileo Galilei (1564–1642), a man of wide scientific accomplishments, with whom he corresponded frequently. And yet, in spite of a strong community of interests, Galileo seems not to have been attracted especially to the work of Kepler. The mathematical laws of astronomy, for example, did not rouse in him the enthusiasm they deserved. Each man worked on his own problems, with little apparent spirit of cooperation. Kepler wrote to many others about the rainbow, but not to Galileo. Through Remus, who corresponded with both, Galileo may possibly have heard of Kepler's ideas; but if so, he did nothing about them. His references to the rainbow are entirely casual. A contemporary Jesuit, Father Horatio Grassi, writing under the pseudonym Lothario Sarsi, had compared the bow with comets, whereat Galileo, in the important Il Saggiatore of 1624, pointed out that the rainbow has no physical objective reality such as that of a comet.

Galileo was misunderstood to say that the rainbow is not formed unless there is a plane surface opposite the sun, whereat Galileo expostulated: "Temerario bestiuolo! When did I say that?" It appears that Galileo had merely pointed out to Sarsi that whereas the halo appears in the direction of the sun, the rainbow appears opposite; and comets don't necessarily appear in either of these directions.[30]

It is a great pity that Galileo never really tackled the rainbow problem, for he was even better equipped for it than was Kepler. He had an intense curiosity; he was quite unafraid of new ideas; he invented and used effectively many scientific instruments; and he believed strongly in the mathematical operation of nature. His famous contemporary, Sir Francis Bacon (1561–1626) enjoyed at best only the first two qualifications. But then Bacon, who was raised to the nobility as Baron Verulam, has been heralded as "the prophet of the new science." Could he see what others had overlooked? His goal was the reorganization and advancement of all learning to the end that, by understanding nature, the lot of mankind might be improved. By pursuing a relentless program of technocracy, he believed, the time ultimately would arrive when the operations of nature no longer concealed any secrets. The story of the rainbow would be at an end. But either Bacon over-rated human ingenuity or else he grossly underestimated the subtlety of natural laws. Since that day advances in science have indeed been such as would stagger even his bold imagination. And yet this vast accumulation of knowledge has not served to confirm the optimism of Bacon; it has operated rather to restore a wanting sense of intellectual modesty. The problem of the rainbow has turned out to be far more complex than one would have dreamed. Bacon himself did not make a systematic investigation of the problem—the rainbow does not, after all, have a high degree of technological utility! But even had he applied his mind to it, he probably would have had little success, for his inductive, essentially taxonomic, method was quite inadequate. The medieval period had gone quite as far as one would expect, in studying the rainbow, on the basis of Bacon's advice. First of all we must prepare a natural and experimental history. Tables and instances of rainbows must be constructed without preconceived ideas. Then from among the collections of facts one must look for the causes or forms in order to acquire command over nature through knowledge. In this search one must ascend, through the selection and rejection of particular instances, toward ever broader generalities—that is, by Baconian induction.

It is to be regretted that Francis Bacon did not apply even the first step of his method to the rainbow. There are only a few scattered references to the bow, and these show virtually no comprehension or interest. In the *Historia Ventorum* he wrote: "The rainbow, which is the lowest of the meteors and generated nearest the earth, when it does not appear entire, but broken and only with the ends visible, is resolved into winds, as much if not more than into rain." [31] And in the *Advancement of Learning* one reads only that "The rainbow is made in the sky out of a dripping cloud; it is also made here below with a jet of water." [32] Such woefully inadequate statements are perhaps suggestive of the fact that, more than most other physical phenomena, the rainbow lends itself very poorly to treatment by the Baconian method. In this case there is no substitute for the geometrical approach.

The period from Bacon to Bacon (i.e., from Roger to Francis) again and again had underestimated the mathematical intricacies of nature. Man could have gone on for many another century to classify observations on the rainbow and to build up a treasury of information on related phenomena without ever unraveling the secrets of the rainbow. Nor was the modern confluence of the two medieval streams of craftsmanship and scholarship of much help here, for the ethereal bow was too elusive and inconsequential to be studied by those who were concerned with the conquest of nature. But, fortunately for the story of the rainbow, modern science was not entirely modelled along Baconian lines. Empiricism had to be supplemented by a Pythagorean faith in the role of mathematics. Galileo was a disciple of just such a faith, holding that the book of nature is written in mathematical language without which it is humanly impossible to comprehend a single word. Whatever he touched was illuminated with a new light; but somehow the rainbow had not caught his fancy any more than it had Bacon's.

Francis Bacon is one of the most controverted figures in the history of science, both as to character and intellectual accomplishment. He has been cuttingly described, with perhaps some exaggeration, as the greatest, wisest, meanest of mankind. In science he made no great discovery, nor was any outstanding contribution the direct result of application of the Baconian method. Moreover, he failed to appreciate the discoveries of others. The brilliant achievements of Copernicus, Gilbert, Kepler, and Galileo meant nothing to him. William Harvey (1578–1657), discoverer of the circulation of the blood, said facetiously of Bacon that he wrote science like a lord chancellor. On the other hand, it can not be denied that Bacon exercised a profound and stimulating influence on the development of seventeenth century science. If some irreverently point out the tragi-comic element in his scientific martyrdom—he died from a cold contracted while stuffing a drawn chicken with snow—others emphasize that he died on Easter Sunday, prophesying the dawn of a new era. In the story of the rainbow the prophecy was indeed soon to be fulfilled.

But dawn does not break suddenly. When in 1626 Bacon and Snel died, the challenge of the rainbow had not yet been taken up by Descartes; and two meteorological treatises of 1627 show that Aristotelian doctrines continued to prevail. The *Partitiones Meteorologicae* of Conrad Cellari (fl. 1619–1636) is a thoroughly weak treatment based on narrow Peripatetic views. Citations of sources are from the older philosophical stream only, and the explanation makes no reference to refraction or reflection in spherical raindrops. Not more than two bows are possible because the secondary bow is an already weakened image of the primary bow.[33]

A more sophisticated study of the rainbow, but one also antithetical to the optical tendencies of Maurolycus, Kepler, and De Dominis, was made by Liebert Froidment—or Fromondus—(1587–1653). In 1627 was published the first edition of Froidment's *Meteorologicorum Libri Sex*, a popular treatise of which over fifty pages are on the rainbow. The author, professor of philosophy and theology at Louvain, was a worthy successor of Thomas of Cantimpré who had taught there four centuries earlier. Had Thomas been able to return to the university in the early seventeenth century, he would have had little difficulty in making an adjustment to the instruction. Froidment's purely Scholastic lectures, excluding anything contrary to the teachings of the church and conforming mostly to the works of Aristotle, personified teaching in Belgium.[34] In science as in philosophy, Froidment was conservative. He wrote anti-Aristarchan works in astronomy, and he objected strenuously to placing comets beyond the realm of meteorology. Not being able to disregard Tycho Brahe's observations placing the comet of 1577 beyond the moon, Froidment insisted that at least some comets are sublunar. Celestial comets, he said, are perhaps exhalations from stars, just as terrestrial comets are exhalations of water and earth. It occasions no surprise, then, to find him adhering to traditional explanations of the rainbow.

In order perhaps to emphasize the difficulty of the subject, the book on *Meteoris Apparentibus* (the sixth and last book), opens with historical comments on extraordinary rainbows, with citations to Witelo, Cardan, Scaliger, and many others. There follows a lengthy chapter on whether the iris is caused by reflection or refraction of the sun's rays. In this the story of the rainbow is reviewed from Aristotle, Posidonius, and Seneca to Vimercati and Maurolycus, but nowhere in the account is Theodoric's name mentioned. Froidment seems to be swayed especially by Witelo's argument on rays concentrated and diffused by refraction through drops, and so his ultimate verdict is that the rainbow is caused by refraction into the medium. Because the angle at which the rays of the sun strike the drops will vary, so also will the angle of reflection of the rays from subsequent drops vary; and hence Froidment believed that the radius of the bow necessarily varies with the altitude of the sun. Although he admired the ingenuity of Maurolycus, he did not accept the latter's explanation. He knew of Kepler's published cloud explanation, but not of his unpublished raindrop theory. It is significant that although he cited Themo and De Dominis, he found nothing of particular value in their work. The important idea of a reflection within the drop he evidently regarded as not especially worthy of mention. One thing which particularly held Froidment's interest was the question of the altitude of the rainbow. He cited Themo's assertion that the distance from the eye to the bow is a scant half a thousand Parisian feet, and Cardan's belief that the bow can be generated 772 thousand feet from the earth. Froidment proposed a compromise figure of from 750 to 1500 feet.[35]

Unable to give answers to fundamental questions concerning the rainbow, authors were tempted sententiously to dwell on insipid and often trivial details, and Froidment was no exception. He devoted a whole article or chapter to "Various Motions of the Rainbow." [36] Here he catalogued six "motions": (1) as the sun gets higher, the rainbow sinks lower; (2) as the sun moves lower, the rainbow rises; (3) as the sun moves across the sky from east to west, the bow conversely moves from west to east; (4) as the vapor condenses or rarefies, the radius of the bow increases or diminishes, "as Vitello and Themon say"; (5) as the observer moves forward or backward, so also does the bow; (6) as the observer moves from side to side, so does the bow move likewise. One article is given over to the traditional comments on the times and places at which the bow can be seen, and another concerns the presumed effects and signification of the rainbow as recorded in Genesis and the classical writers. The lunar rainbow is the subject of another weak section in which one reads the not very significant fact that Americus Vespuccius observed a white lunar bow in the middle of the night.

Froidment, an assiduous compilator, touched upon most of the favorite questions on the rainbow, including the collinearity of the sun, eye and center of the bow (which he accepted), the reality of the colors (which, contrary to Albertus, he denied), and the number of colors (on which he held to no particular theory). But of all the old problems the one which he treated at greatest length concerned multiple rainbows. He devoted three chapters to the secondary bow and another to the possibility of additional arcs. Froidment ran the gamut of ancient and modern views. He found neither Maurolycus nor De Dominis satisfactory on the secondary rainbow, saying that "Antonius sweated to bring forth his explanation," but that it in no way explained the inverse order of the colors. Maurolycus and Vimercati he described as "clamoring" against the reflection theory of the secondary bow; but in the end Froidment himself adopted a curious casuistical compromise. Rays from the sun fall obliquely upon the drops, are refracted, and then are reflected. Of these some go to the eye to form the primary iris, others are diffused; and some of these latter rays, scattered above the primary bow, find a suitable angle for reflection to the eye as the secondary rainbow. In a sense, then, the secondary is no more a mirror reflection of the first than the first is a mirror image in the lower cloud, Froidment held; but his erroneous assumption that the drops causing the primary bow play an essential role in the formation of the secondary arc places him in a category with the Clichtoveans nevertheless.[37]

Froidment believed, after reading Witelo, Themo, and others on multiple rainbows, that four, five, or more concentric irises can appear with colors alternating in the same order, but that a secondary bow with red inner band "scarcely ever shows a third bow with a red outer band." [38]

Perhaps nowhere in the story of the rainbow was the need for a precise mathematical theory more pronounced than in connection with the question of a tertiary rainbow. A theory should not only "save the appearances"; it should serve also as a guide to further observation by predicting what more to look for. No theory yet proposed had been able to predict with any confidence where one should search for the elusive third and fourth bows. The colored bands within the primary rainbow were not generally identified as bows of higher order, but rather as somehow related to the primary bow. Hence premature extrapolation had led to the impression that if a third proper arc should appear, it would be above the second; and as Froidment inferred, the appropriate order of colors was presumed to be that found in the primary bow—that is, opposite to that in the secondary. But Froidment's *Meteorologicorum* was, in a sense, the last stand of the imprecise tradition in the story of the rainbow. The great mathematicians of antiquity had virtually boycotted the bow, but those of the seventeenth century, notably the inventors of analytic geometry and the calculus, were about to expropriate, under protest, to be sure, the philosophers' hold on the theory of the rainbow.

VIII

The Cartesian Theory and Its Reception

THE YEAR 1637 IS NOTABLE IN THE HISTORY OF THOUGHT, FOR IN THAT YEAR was published one of the most celebrated works of all times. Hundreds of thousands of different postage stamps have been issued to pay homage to political rulers, both good and bad; a few hundred have appeared with the object of honoring great intellectual figures. But it is rare indeed for a stamp to be printed for the sole purpose of commemorating the publication of a philosophical and scientific treatise. The appearance in 1937 of two French stamps celebrating the tricentennial of the *Discours de la Méthode* of René Descartes (1596–1650) is a worthy tribute to a great book. (The second of the two stamps was issued to correct an error in the first in which the title was

FIG. 39.—French stamps (Nos. 335-336) of 1937 celebrating the three hundredth anniversary of the *Discours de la méthode.*

inaccurately given as *Discours sur la Méthode.*) How many philatelists are aware, one wonders, of the fact that this stamp commemorates also the greatest single contribution to the story of the rainbow?

   In 1603 the Jesuits had obtained from Henry IV authority to set up schools in several cities, and of these the Collège Royal at Flèche achieved a reputation as one of the best in Europe. Descartes entered the school in 1604, the year it was founded, and here for eight years he studied logic, ethics, mathematics, physics, and metaphysics. In spite of the admitted soundness of this education, Descartes felt that he had derived little advantage from his studies, and he resolved to acquire first-hand knowledge from thought and experience rather than from books. He took service with Prince Maurice of Nassau, and

200

later with the Duke of Bavaria, travelling extensively throughout Europe. Wearying of these aimless expeditions, he resolved in 1628 to seek out a retreat where he could work on what he regarded as the much-to-be-desired reform of philosophy. He found the atmosphere of Holland so congenial for this that he spent a score of years there. A few years later Descartes was prepared to publish his first work—a vast treatise on physics entitled *Le Monde, Ou Traité de la Lumière*—when he learned of the condemnation of Galileo. Fearing recrimination, he withheld publication; but in 1637 he yielded to the importunities of friends to the extent of publishing, anonymously at Leyden, the celebrated *Discours de la Méthode Pour Bien Conduire Sa Raison et Chercher la Vérité dans les Sciences.*

Descartes, like Roger Bacon, took all knowledge for his province, for he was convinced that "all of the sciences are interconnected as by a chain." But whereas Aristotle had placed his faith in logic as a means of unifying knowledge, Descartes was profoundly dissatisfied with such a basis. No branch of learning satisfied Descartes' standards of certitude save mathematics alone, and he set about to determine whether the methods of geometry could not be employed with comparable success in other branches of learning. This is, essentially, the theme of the *Discours.*[1] Man is endowed in equal measure, he believed, with good sense or the capacity to reason; it is failure to employ this endowment aright which leads to error and diversity of opinion. In mathematics alone was a conflict of views almost unknown, and the reason for this seemed to lie in the fact that it was built up from a small number of basic principles through a chain of step-by-step reasoning—the deduction of one truth from another. Consequently he laid down four precepts to be followed in the acquisition of knowledge and the avoidance of error:

(1) To accept as true only such conclusions as are clearly and distinctly known to be true and to exclude all possibility of doubt.

(2) To analyze problems under consideration into as many parts or elements as possible.

(3) To reason correctly from the simpler to the more complex elements.

(4) To adopt a comprehensive view which should omit nothing essential to the problem.

Descartes rejected provisionally all opinions until he could find something which he regarded as certain, a course of methodical doubt which ended in the enunciation of the celebrated postulate: *Cogito, ergo sum.*

The *Discours* has been highly esteemed, and it represents a cornerstone in the reputation of Descartes as the father of modern philosophy; but one wonders if the author would have agreed with the traditional evaluation. It appears that the essay was written somewhat as an introduction to three treatises which Descartes had composed earlier and of which he was rightfully very proud, notwithstanding the fact that they were published as appendices to the *Discours*. This accident of publication had serious consequences for the history of science. Many casual readers are preface-skippers, and an even greater number never get to an appendix. As a result, the introductory *Discours*, the work dealing with the generalities of scientific method, stole the thunder of the appendices, the valuable and technically more difficult investigations which were to have served as specific illustrations of the method. Then, too, publication of the opening philosophical portion was cheaper than the setting of type and the illustrations required in the mathematical and scientific material in the appended parts of the book. It was not long before the *Discours* was republished without the three supplementary treatises; and, in fact, few editions since the first contain the entire work. Yet all three of the appendices have a bearing on the story of the rainbow, and one in particular is of crucial relevance.

One of the appendices, entitled *La Géométrie*,[2] is a landmark in the history of mathematics, for it is in reality the first published treatise on analytic geometry. This in a sense represented the "royal road to geometry" for which Alexander the Great is said to have yearned. Ancient classical Greek geometry had centered about the synthetic study of the relationships between the only curves then known—about half a dozen in all, including the straight line, the circle, the conic sections, the spiral of Archimedes, the quadratrix, the cissoid, and the conchoid. The algebra of the ancient Babylonians, the medieval Arabs, and the early modern Europeans had been concerned primarily with the solution of *determinate* equations (that is, equations in a single unknown). One morning Descartes suddenly had a brilliant thought. This idea—which came independently to Pierre de Fermat (c. 1608–1665) a few years later—was that *indeterminate* equations (i.e., equations in two or more unknowns) were closely related to geometry. That is, he discovered that each equation in two unknown quantities (now called variables) is a curve—the locus of points whose coordinates (with respect to a given coordinate system) satisfy the given equation. Conversely, geometry became related to algebra in a new way; to a given curve, there corresponds (with respect to a given coordinate system) an equation in two unknown quantities. The discovery of analytic geometry appeared at the time to be quite unrelated to the rainbow problem; but such an impression was corrected almost precisely two centuries later.

The second appendix, *La Dioptrique*,[3] had more immediate relevance for the rainbow. Here Descartes first published the optical principle for which physicists had sought in vain from Ptolemy to Kepler. Snel had discovered the law of refraction some sixteen years earlier, but he died before he was able to prepare his work for publication. Snel's uncompleted manuscript still was in existence during the years when Descartes lived in Holland. Had Descartes seen the manuscript and was he guilty of plagiarism? The question frequently has been argued, with some bitterness, since that time. Huygens believed that Descartes had seen Snel's manuscripts on refraction, concluding that it was "perhaps from this work that he has deduced that ratio expressed in sines which he has employed so ably in explaining the rainbow." Recent studies indicate that Harriot, too, knew of the law,[4] and this suggests still another possible source of inspiration, especially in view of the fact that Harriot was interested in the problem of the rainbow and that Descartes once was believed to have appropriated some of his algebra. Nevertheless, the weight of the evidence exonerates Descartes. He would appear to have known of the law a couple of years (or more) before his trip to Holland in 1629, although one does not know how he was led to it. It was partly Descartes' lack of frankness with respect to the time and manner of his discovery that had heightened suspicion he had borrowed it. He expressed the law in a geometrical form equivalent to $\sin i = m \sin r$—from which it sometimes is known as the "law of sines" (not to be confused with the law of sines in the solution of triangles). This is slightly different from the form Snel had used, but suspicious critics suggested that Descartes had modified Snel's form in order to cover up his indebtedness. Descartes explained that refraction must be measured by lines (i.e., by trigonometry) and not by angles—and especially not by the angle which Maurolycus and Kepler had called the angle of refraction. When one recalls that trigonometric functions, before the eighteenth century, were regarded as lines, one sees in this statement the assertion that the law involves trigonometry. Now every student of elementary mechanics is familiar with the fact that the vertical and horizontal components of a force are found by multiplying the force by the sine and the cosine respectively of the angle of inclination of the force. Hence the key to Descartes' explanation is found in the idea that light is a force the vertical and horizontal components of which are independent of each other. The reflection of light from a mirror is like the rebound of a rubber ball from a wall—as Aristotle had suggested two thousand years before—and Descartes explained the equality of the angles of incidence and reflection by postulating that the velocity component horizontal to the surface remained unchanged, while the vertical component was reversed in direction. In the case of refraction he made a somewhat similar assumption—that when a ray strikes the surface of the denser medium, the vertical component of velocity is increased (i.e., multiplied by a constant factor k greater than unity), the horizontal component remaining the same. Thus, if V_i and H_i are the horizontal and vertical components of the incident ray (Fig. 40) and V_r and H_r the corresponding components of the refracted ray, then $H_r = H_i$ and $V_r = kV_i$. From these equalities Descartes deduced the law $\sin i = k \sin r$.

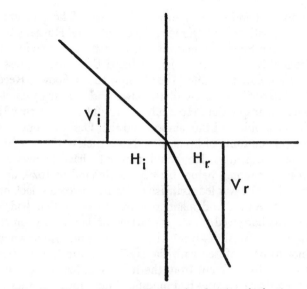

Fig. 40.—Diagram illustrating Descartes' explanation of refraction.

The law of refraction, one of the cornerstones of optics, had long been sought for; yet when the discovery was published by Descartes,[5] it seems not to have been widely appreciated. In many cases it was overlooked completely; in others it provided material for attacks on Cartesian thought. There seemed to be an inconsistency in the explanation of the "law of sines," for on the one hand Descartes made use of the idea of velocity components, yet on the other he insisted on the instantaneity of the transmission of light. Whether or not light was propagated in time had been argued from the days of the Pythagoreans, with opinion somewhat evenly divided. Among the ancient Greeks there was a general belief, supported by Aristotle, that the speed of light was infinite, although Empedocles had taken exception to the opinion. During the Middle Ages the theory of finite velocity was supported by Avicenna, Alhazen, and Roger Bacon; but Kepler, in the early seventeenth century, reverted to the doctrine of infinite speed. In 1620 Francis Bacon was tempted to think, in the *Novum Organum*, that light might be transmitted in a perceptible time interval; but in the posthumous *Sylva Sylvarum* he categorically maintained the instantaneity of propagation. Galileo described an experiment which he interpreted as deciding the issue in favor of an infinite speed, for no time interval was observable in the transmission of visual signals over distances of two or three miles.[6] In reality, however, the experiment merely established a lower limit for the velocity of some sixty miles per second; but Descartes maintained his confidence in the instantaneity of light. He compared the transmission of light with the use by a blind man of a cane by means of which the impact of his stick upon a stone is felt immediately. Yet in his derivation of the law of refraction, Descartes reasoned that light travelled faster in a dense medium than in one less dense. He seems to have had no qualms about comparing infinite magnitudes! In 1676, however, the Danish astronomer Ole Roemer (1644–1710) was to announce to the Paris Académie des Sciences that the speed of light, while of very great magnitude, was nevertheless finite. From observations of the eclipses of the satellites of Jupiter he had calculated that it takes light about twenty-two minutes to traverse a distance equal to the diameter of the earth's orbit about the sun— known at the time to be about 180,000,000 miles. Later work confirmed Roemer's conjecture and placed the speed of light at thirty billion centimeters per second! Descartes was wrong in insisting that light travels instantaneously; yet so stupendous is the velocity that the discovery of Roemer did not vitiate the Cartesian experiments on the rainbow. Were the speed of light not so great, the rainbow could not have been studied by Theodoric's method, in which a stationary globe of water was regarded as analogous to a falling raindrop.

Cartesian science seemed calculated to raise controversies, and one of Descartes' severest critics was Fermat, the coinventor of analytic geometry. He agreed with the statement of the law of refraction (and incidentally did not suggest plagiarism), but he felt that the Cartesian explanation was far-fetched. Why should the component perpendicular to the surface be increased when light enters the denser medium? Should one not rather assume that the greater density would offer greater resistance? Fermat instead had recourse to the principle of the economy of nature. Heron and Olympiodorus had pointed out in antiquity that, in reflection, light followed the shortest possible path, thus accounting for the equality of the angles. During the medieval period Alhazen and Grosseteste had suggested that in refraction some such principle was also operating, but they could not discover the law. Fermat, however, not only knew (through Descartes) the law of refraction, but he also had invented a procedure—equivalent to the differential calculus—for maximizing or minimizing a function of a single variable. That is, given a function $f(a)$, he solved the equation $f(a + e) = f(a)$ after he had simplified the equation, divided through by the common factor e, and finally had set $e = 0$ in the resulting equation. This procedure is tantamount to setting the derivative of $f(a)$ equal to zero. (The notation, except for the letters a and e, is an anachronism, but the thought of Fermat is not thereby misinterpreted.) Fermat then set himself the following problem. Let C be an object in water and A the eye of the observer in air (Fig. 41).[7] Assuming (1) that

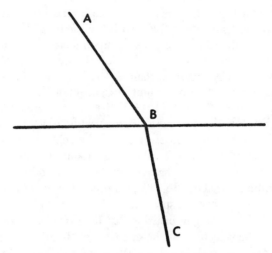

Fɪɢ. 41.—Fermat assumed that in refraction light travelled from C to A in the shortest possible time.

light travels more rapidly in air than in water, and (2) that light travels from C to A in the shortest possible time, what will be the path ABC of the refracted ray? Fermat applied his method of maxima and minima in this case and discovered, to his delight, that the result led to precisely the law which Descartes had enunciated. But although the law is the same, it will be noted that the hypothesis contradicts that of Descartes. Fermat assumed the speed of light in water to be greater than that in air; Descartes' explanation implied the opposite. Not until two centuries later was a crucial experiment devised to determine whether Descartes or Fermat was right—whether the light producing the rainbow speeds faster through the raindrops or in the surrounding atmosphere. Fortunately, however, a definitive decision on this question was not needed for the next steps in the story of the bow.

Descartes was fully aware of the importance of the law of refraction in connection with the grinding of lenses and in the construction of telescopes, a study which led him to the well-known curves called the "ovals of Descartes." But he realized that the discovery of the exact law of refraction was significant also in connection with the rainbow problem. Huygens, in fact, wrote that Porta, Maurolycus, and Kepler tried to find the law of refraction "seeing that it was necessary to know this in order to explain the refractions of the atmosphere, of the rainbow, and other meteors, as well as to explain the effect of convex and concave lenses." The attention of Descartes had been drawn to the rainbow very shortly after he had become familiar with the law of refraction, a topic which had held his thoughts for about a dozen years. On

March 20, 1629 there had been an unusually striking parhelic display at Rome, and Descartes had been asked by that universal scientific correspondent, Marin Mersenne (1588–1648), for his opinion on this. Descartes eagerly seized upon this opportunity to prove to a potentially doubting world the efficacy of the method he was to announce.

The rainbow is such a remarkable natural wonder and its cause has been so zealously sought by able men and is so little understood, that I thought that there was nothing I could choose which is better suited to show how, by the method which I employ, we can arrive at knowledge which those whose writings we possess have not had. [8]

For a while, therefore, Descartes broke off his metaphysical meditations to study meteorological phenomena, and the result was a third appendix to the *Discours de la Méthode*, one to which he gave the title *Les Météores*.

Descartes habitually undervalued the work of his predecessors; and, with respect to the rainbow, it is likely that he was encouraged in his depreciation by a popular work published in 1624 and 1628—the *Récréations Mathématiques* of the Jesuit teacher, Jean Leurechon (ca. 1591–1670).[9] It is known that this charming book had been read by Descartes with considerable care; and his emphasis on the wonder and lack of understanding of the rainbow, as well as his experimental approach to its study, may well have stemmed from the comments of Leurechon. Problem 46 of the *Recreations* explained "how to represent down here different irises and rainbow figures." The author eulogized the rainbow as "the richest piece in the treasure of nature" and cited a passage in *Ecclesiastes* referring to the bow as "the chief work of God." [10] But then he pointed out that mathematicians have brought the bow down to earth to produce the same mixture of colors and the same ingredients as on high. It is first suggested that the reader, standing with his back to the sun, produce irises by spraying droplets of water into the rays of sunlight. The only trouble, Leurechon pointed out, is that they don't last, any more than does the rainbow. For a more stable rainbow he advised exposing to the sun's rays a prism or crystal or glass of water or bottles of soap suds. Instead of proceeding to make an experimental or mathematical study of the rainbow, the author closed with an abject apology.

I fear that you will ask me further about the production, disposition, and figure of the colors. I shall reply that they come by reflection and refraction of light, and that is all. Plato has very well said that the iris is a sign of admiration, not of explanation. And he hit the nail on the head who said that it is the mirror in which human nature has had a full view of its ignorance; for all the philosophers and mathematicians who for so many years have been engaged in discovering and explaining its causes, have learned nothing except that they know nothing but a semblance of truth.[11]

Here was a challenge which Descartes could not resist.

The three appendices to the *Discours de la Méthode* centered about three problems which Descartes believed his contemporaries incapable of solving: the problem of Pappus, the law of refraction, and the cause of the rainbow. The first two had indeed been solved (the first probably by Apollonius, the second by Snel), and the third was not so far from solution as Descartes had supposed. It is natural to think of his work on the rainbow as a consequence of the views of Kepler, Harriot, De Dominis, or Snel; but Descartes rarely admitted indebtedness to others. Newton and Leibniz virtually accused Descartes of plagiarism from De Dominis; and Huygens wrote that the Cartesian theory of the bow "has been taken from a place of the incomparable Kepler"; yet there is no adequate evidence of any such filiation of ideas. So far as is known, his work was a fresh attack upon the problem. And where the ingenuous Kepler had a habit of recording for posterity even his fumbling attempts, the sophisticated Descartes published only the evidence of his ultimate triumph. He wrote in French rather than Latin, so that those who use their natural reason, rather than depend on ancient books, should judge his work.

The *Discours* and the three appendices are not intended to teach the author's method, but simply to tell enough about it so that its efficacy might be judged—"to show those who examine them whether my method is perhaps not worse than the method which is commonly used." [12] Meteorology formed a traditional part of scholastic philosophy, and hence Descartes was especially eager here to show the superiority of his approach.[13] Only in discussing the rainbow, he wrote a year later, had he let slip some sample of his method.[14] And yet his investigation of the rainbow violated some of the basic canons of his methodology. Descartes, like Galileo, preferred ordinarily to use deduction from indubitable first principles, rather than to amass and correlate, in Baconian fashion, a large number of specific instances. But it was out of the question, in that day and age, to have any clear and distinct ideas on the rainbow; and hence Descartes perforce began as would any other scientist, with patient inductive observation.[15] But if patience alone were sufficient, Kepler would indeed have been the legislator of the rainbow, as well as the law-giver of astronomy. Ingenuity and freedom from earlier preconceptions were also needed, and here Descartes excelled.

Les Météores opens with the observation: "We naturally admire more the things which are above us than those which are on the same level or below us." But Descartes preferred understanding to admiration, and so he proposed, in Book VIII of *Les Météores*,[16] to explain the causes of the rainbow so that there no longer will be occasion to wonder at this most admirable display. So thorough is his avoidance of the supernatural that he used *arc-en-ciel* instead of the name of the goddess Iris which had been taken over into the romance languages. He brought the rainbow down to earth, in fact as well as in name, by studying the passage of light rays through a magnified raindrop.

Considering that this bow appears not only in the sky, but also in the air near us, wherever there are drops of water illuminated by the sun, as we can see in certain fountains, I readily decided that it arose only from the way in which the rays of light act on these drops and pass from them to our eyes. Further, knowing that the drops are round, as has been formerly proved, and seeing that whether they are larger or smaller, the appearance of the bow is not changed in any way, I had the idea of making a very large one, so that I could examine it better.[17]

Descartes believed, apparently, that he was the first one thus to study the bow experimentally; but we now realize how mistaken he was in this. He was but repeating the observations which Theodoric and Kamal al-Din had made long before, and he drew the conclusions his predecessors had reached.

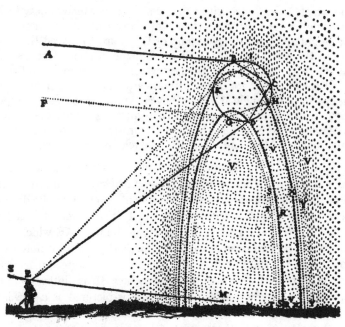

FIG. 42.—Descartes' diagram to illustrate the formation of the rainbow arcs.

I found that if the sunlight came, for example, from the part of the sky which is marked AFZ (Fig. 42), and my eye was at the point E, when I put the globe in the position BCD, its part D appeared all red, and much more brilliant than the rest of it; and that whether I approached it or receded from it, or put it on my right or my left, or even turned it round about my head, provided that the line DE always made an angle of about forty-two degrees with the line EM, which we are to think of as drawn from the center of the sun to the eye, the part D appeared always similarly red; but that as soon as I made this angle DEM even a little larger, the red color disappeared; and if I made the angle a little smaller, the color did not disappear all at once, but divided itself first as if into two parts, less brilliant, and in which I could see yellow, blue, and other colors. . . . When I examined more particularly, in the globe BCD, what it was which made the part D appear red, I found that it was the rays of the sun which, coming from A to B, bend on entering the water at the point B, and pass to C, where they are reflected to D, and bending there again as they pass out of the water, proceed to the point E.[18]

Descartes found that he could cover all of the globe except the points B and D (Fig. 42) without in any way interfering with the red ray. In other words, it was this ray alone, of all those striking the sphere, which produced the red in the rainbow.

From which I clearly perceived that if all the air which is near M is filled with such globes, or instead of them with drops of water, there ought to appear a bright red point in every one of the drops so placed that the lines drawn from them to the eye at E make an angle of about forty-two degrees with the line EM, as I suppose those do which are marked R; and that if we look at all these points together, without any consideration of their exact position except of the angle at which they are viewed, they should appear as a continuous circle of a red color; and that something similar ought to appear at the points marked S and T, the lines drawn from which to E make more acute angles with EM, where there will be circles of less brilliant colors.[19]

The formation of the secondary rainbow Descartes explained with equal clarity. He found that there is a uniquely defined ray which strikes near the bottom of the drop at a specific point G, is bent on entering the drop to proceed along the path GH, is reflected at H to I, is again reflected at I toward K, and at K is refracted to reach the eye along the path KE which makes an angle of about fifty-two degrees with the line EM. It will be noted that the deviation within the drop of the ray which produces the secondary bow is in the sense opposite to that causing the primary bow. If, as illustrated in Descartes' diagram, the deviation in the former case is pictured as clockwise, then the deviation in the latter case will be counterclockwise.

Descartes often was neither so original nor so definitive as he presumed.

For example he claimed to have been aware of only one earlier estimate of the size of the bow, a value of 45° given "by common opinion" and which he ascribed to Maurolycus; and he unjustly concluded, "This shows what little confidence one can put in observations which are not accompanied by correct reasoning." Yet the figure of 42° had appeared in a dozen manuscripts and printed works from 1269 to 1611. If Descartes was unaware of any of these anticipations (which appeared in some of the most popular books of the time) one can only conclude that he had a remarkable facility for overlooking in the works of his predecessors anything which might be of value in connection with his own discoveries. Is it any wonder that few scientists have been more frequently subjected to charges of plagiarism? In fairness to Descartes, however, it must be pointed out that he was the first one to improve on the traditional value. The greatest radius of the bow he placed at 41° 47′, the smallest at about 40°. The limits for the secondary bow he gave as 51° 37′ and 54°, definitely more accurate than the 56¼° of Maurolycus.

Up to this point *Les Météores* has afforded a rediscovery of earlier ideas, with some improvement in accuracy of measurement. Many an historian has credited Descartes with this, and nothing more, on the rainbow.

But Descartes' explanation did not end with the account given above. As he himself expressed it,

> The principal difficulty still remained, which was to determine why, since there are many other rays which can reach the eye after two refractions and one or two reflections when the globe is in some other position, it is only those of which I have spoken which exhibit the colors.[20]

Remembering the refraction of light in a prism, he realized that it was not the curvature of the drop nor the size of the angle of incidence nor the reflection nor the plurality of refractions. He decided that for the production of color at least one refraction is necessary, and, in the case of two refractions, the effects should not be contrary, so as to destroy each other, as in the case of light passing obliquely through a window. But even after carefully comparing the rainbow with the spectrum from a crystal prism, Descartes still was in the dark about the reason for the critical rainbow angle of about 42°. Accurate observation was not sufficient. Then the mathematician in him came to the rescue.

> I took my pen and made an accurate calculation of the paths of the rays which fall on the different points of a globe of water to determine at what angles, after two refractions and one or two reflections they will come to the eye, and then I found that after one reflection and two refractions there are *many more rays which can be seen at an angle of from forty-one to forty-two degrees than at any smaller angle; and that there are none which can be seen*

at a larger angle. I found also that, after two reflections and two refractions, there are many more rays which come to the eye at an angle of from fifty-one to fifty-two degrees than at any larger angle, and none which come at a smaller angle.[21]

Through these simple but profound statements the fundamental problem of the size of the rainbow was solved for the first time. Without a knowledge of the law of refraction, this calculation would have been much more difficult, although not necessarily impossible; and *Les Météores* consequently is, in a sense, a corollary of *La Dioptrique*. In connection with the computation one sees a facet of Descartes' character which all too frequently is overlooked. Far too often science is pictured as developing from the facile inspirations of the great men of genius—a gross and dangerous oversimplification.

If one accepts the traditional view of Descartes as a man of brilliant ideas but with little persistence in experimentation and calculation, then the work on the rainbow must be taken as an exception. Patience and persistence are the essence of scientific success. When Newton was asked how he discovered the law of gravitation, he is said to have replied, "By thinking about it incessantly"; and much the same can be said for Descartes and the rainbow. Too often one thinks of the Descartes who slept until ten in the morning—who invented analytic geometry while keeping warm in bed during a cold Bavarian winter. But of one thing we can be sure; and that is that Descartes did what every mathematician has done and will continue to do—keep a large wastebasket and keep it busy. And in science also the results of Descartes were not sudden inspirations; he did what every good scientist must do—patiently carry out experiments and calculations virtually without end. Descartes often concealed this prosaic side of his life, but in *Les Météores* he candidly reported the detailed tables of calculations he had made on the passage of light through his globe of water. One historical account mistakenly avers that Descartes calculated the paths of ten thousand rays! [22] This is an exaggeration; but it was the very accumulation of his multitude of calculations which led him to his important discovery—the discovery for which Theodoric had looked in vain. Nature, in setting the angle of the rainbow at about 42°, had not acted arbitrarily or capriciously. Descartes did not simply rediscover the fourteenth century theory, as many histories hold; he gave it true scientific status by showing the quantitative agreement of theoretical calculations with the results of observation.

In tracing the paths of rays which traverse the drop, Descartes made the critical observation that the figure of about 42° was the maximum value of the (acute) angle θ (angle DEM in Descartes' diagram—Fig. 42) between the rays of the sun and the rays which emerge from the drops after two refractions and one reflection. Kepler had believed that the angle θ increased monotonically from a value of 0° for a ray through the center of the drop to a maximum value for his tangential ray. Descartes found instead that θ at first increases from 0° to about 41° 30' and then decreases rapidly to 13° 40' for the tangential ray. As is characteristic of maxima and minima, there is at the greatest angle a clustering of rays which emerge nearly parallel to each other; and it is this accumulation of rays which makes the beam of refracted and reflected light especially effective for an angle of about 41° 30'. Descartes expressed this angle as 41° 47', allowing 17' for the apparent radius of the sun.

The calculations of Descartes were essentially trigonometric, but they were expressed in terms of geometric lines, as was customary in his day; and, in order to avoid fractions, the radius of the drop is taken as 10,000 units. The angle of incidence of the ray EF (Fig. 43) is not given in degrees; it is determined by the magnitude of the distance FH between this ray and a

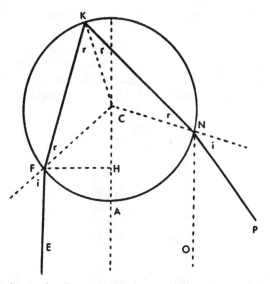

Fig. 43.—This diagram, based upon p. 263 of Les météores, retains the lettering used by Descartes, but has been somewhat modified for purposes of exposition in terms of modern trigonometry.

parallel ray passing through the center of the drop. The ratio FH/FC is, therefore, the sine of the angle of incidence i of the ray. For FH = 0, the ray EF coincides with the central ray AH, and the angle of incidence is zero; for FH = 10,000, the ray EF is just tangent to the drop, and the angle of incidence is 90°. Descartes now calculated, for values of FH from 1,000 to 10,000 (at intervals of 1,000), the corresponding angle ONP—that is, the angle between the emergent ray and the direct rays from the sun. This computation

is easy if one notes that at F the deviation is $i - r$ (where r is the angle of refraction), at K there is a further deviation (in the same sense) of $180° - 2r$,
and at N there is a final deviation (still in the same sense) of $i - r$. The total
deviation therefore is $D = 180° + 2i - 4i$; or, since $\angle ONP = 180° - D$, the
angle ONP is $4r - 2i$. For $FH = 1{,}000$, for example, i is about $5° 44'$. The angle
r is found from the law of refraction, $\sin i / \sin r = m$ (where Descartes took m
as $4/3$, or, more accurately, 250 to 187) to be about $4° 17'$. In this case the
angle ONP therefore is about $5° 40'$. In the same way one computes the angle
ONP for other values of FH to obtain the following entries from Descartes'
table [23] (Table I below).

TABLE I

HF	1000	2000	3000	4000	5000	6000	7000	8000	9000	10000
ONP	5° 40′	11° 19′	17° 56′	22° 30′	27° 52′	32° 56′	37° 26′	40° 44′	40° 57′	13° 40′

From this chart of values one notes that the maximum value of the angle is
close to 41°, but the clustering of rays is not obvious; and so Descartes calculated the paths of other rays in the vicinity of the critical ray, his values being
those in the accompanying Table II: [24] Here the conjunction of many rays in

TABLE II

FH	ONP
8000	40° 44′
8100	40° 58′
8200	41° 10′
8300	41° 20′
8400	41° 26′
8500	41° 30′
8600	41° 30′
8700	41° 28′
8800	41° 22′
8900	41° 12′
9000	40° 57′
9100	40° 36′
9200	40° 4′
9300	39° 26′
9400	38° 38′
9500	37° 32′
9600	36° 6′
9700	34° 12′
9800	31° 31′

the neighborhood of 41° 30′ is quite apparent. The values in Descartes' charts explain clearly not only why the bow appears at an angle of about 41° 30′; they account easily, and at one stroke, for another characteristic of the rainbow which Theodoric had noticed but could not justify. The outer boundary of the primary rainbow is more sharply defined than is the inner edge; and the space above the rainbow, Alexander's dark band, is palpably darker than the space below the bow. But this is precisely what is to be expected from Descartes' calculations. No light whatever is returned, after the indicated reflection and refractions, at an angle greater than about 41° 30′, but there are an appreciable number of rays at angles slightly less than this angle. An exactly analogous calculation for rays producing the secondary rainbow showed that the *minimum* angle between the rays of the sun and those which emerge from drops after two reflections and two refractions is about 51° 54′ (corresponding to a maximum deviation of the rays of 231° 54′). Again allowing 17′ for the apparent radius of the sun, Descartes placed the angle at 51° 37′. About this angle there is a clustering of rays at slightly larger angles, but there are no rays at all returned to the eye at smaller angles. Hence, the inner boundary of the outer rainbow is the more sharply defined; and the space between the two rainbows is darker than the portion of the sky just above the outer bow. The story of the rainbow finally had reached full-fledged scientific status when Descartes, with justifiable pride, showed that his theoretical calculations of the rainbow radii were in remarkably close agreement with observations of the bow as seen in the heavens.

It is generally held that Descartes' discovery of the key to the rainbow problem—the reason for the clustering of rays about the angle of 42°—was made possible by the prior discovery of the law of refraction. There can be no doubt that knowledge of Snel's law facilitated the resolution of the rainbow problem, but it is not strictly correct to say that this could not have been achieved beforehand. In fact, Descartes' discovery concerning the primary rainbow could have been made by Theodoric of Freiberg in terms of knowledge of refraction current in his day. Ptolemy had bequeathed the following table of relationships between the angles of incidence and refraction for air to water:

TABLE III

i	10°	20°	30°	40°	50°	60°	70°	80°
r	8°	15½°	22½°	29°	35°	40½°	45½°	50°

(Noting that the angles of incidence increase in simple arithmetic progression, and that the angles of refraction increase in an arithmetic progression of second order, the first differences being in simple decreasing arithmetic progression [the first difference being 7½ and the common second difference being ½], it would be easy to extrapolate to find that an angle of refraction of 54° would be implied to correspond to an angle of incidence of 90°.) This table had been perpetuated, with the slight change of 8° to 7° 45′ for the first entry, by Witelo, with whose work Theodoric undoubtedly was familiar. Had Theodoric carried through the reasoning of Descartes, using the Ptolemy-Witelo refraction table instead of Snel's law, he would have been led to the values in Table A below, where i is the angle of incidence and θ is the rainbow angle (angle ONP in Figure 43). Descartes' table, slightly modified, has been set beside this as Table B below for purposes of comparison.

Table A			Table B	
i	θ		sin i	θ
10°	12°		.1	5° 40′
20°	22°		.2	11° 19′
30°	30°		.3	17° 56′
40°	36°		.4	22° 30′
50°	40°		.5	27° 52′
60°	42°		.6	32° 56′
70°	42°		.7	37° 26′
80°			.8	40° 44′
90°	40°		.9	40° 57′
	36°		1.0	13° 40′

A glance at these tables shows that the conclusion to which Descartes was led by Table B stands out no less sharply in Table A. That is, in Table A, as the angle i increases, angle θ at first increases, it reaches a maximum value of about 42°, and then it decreases; and there is an obvious clustering of emergent rays at an angle of 42°, the very measure which Bacon and Witelo had given for the radius of the rainbow! (A similar comparison of Theodorican and Cartesian values for the secondary rainbow, however, would disclose that the Ptolemy-Witelo refraction table would not have been sufficiently precise to show the clustering of rays which Descartes, using Snel's law, had noted.)

The diagrams and tables of Descartes make it possible here to justify the statement made above that the theory of the Theodorican arc (the great-circle arc on the raindrop between the points of incidence and emergence of the rays) is false. That this arc (the arc FAN in Figure 43) does not play the decisive role becomes clear from the observation that, inasmuch as it measures twice the inscribed angle FKN, it is equal to four times the angle of re-fraction. It therefore is given by the equation FAN = 4 arcsin (¾ sin *i*). Theodoric's arc FAN does not take on a maximum value within the range of the tables above, but it increases monotonically from almost 30° for *i* = 10° to more than 194° for *i* = 90°. At the incidence *i* = 65°, close to the critical value for the rainbow, it is somewhat over 171°. Rays returned to the eye consequently do not correspond to any particular value of the arc of Theodoric. Moreover, this arc is not the determiner of color, as Theodoric mistakenly thought. From Table B above, for example, one sees that light which reaches the eye at an angle of elevation of say 40° to produce the blue of the primary rainbow can correspond to an angle of incidence of either 50° or 80°, with Theodorican arcs respectively 140° and 190°. Consequently, the direct correspondence which Theodoric thought he saw between his arc and the colors of the rainbow turns out to be a chimera.

Descartes had discovered his effective ray through patient observations and laborious calculations applied to large numbers of individual rays; but at that very time geometers were developing a new and powerful tool, the differential calculus, ideally suited to handle problems concerning maxima and minima. Descartes was aware of these, but after 1637 he seems to have taken little further interest either in mathematics or in the rainbow. It therefore remained for the age of Newton to simplify the tedious computations of Descartes.

The Cartesian geometrical solution of the rainbow problem was indeed a triumph of scientific method, harmonizing as it did mathematical deductions from optical theory with precise observation in experimental verification. With respect to the physical explanation of the rainbow colors, however, Descartes was singularly unsuccessful. In spite of his numerous observations, he accepted the traditional view that light suffered a qualitative change in passing through a raindrop. It will be recalled that Descartes thought of light as a movement or action of the subtle matter which occupied space, and this view led him to explain color formation through variations in rotary motions of the spherical particles responsible for the transmission of rays. When the particles of the subtle matter tend to turn more vigorously than to travel in a straight line, the result is red light; where the rotary motion is less pronounced, one has yellow; and when rotation is relatively slight in comparison with the speed of direct transmission, green or blue is produced. The inversion of the colors in the secondary bow, he said, was easy. In a thoroughly ad hoc manner he attributed the reversal to the fact that in this case the little prisms (of which the drop can be thought of as composed) are arranged in an inverse order.[25] So well did he feel his arguments on color accorded with experience that he regarded it as impossible to entertain any other supposition. Is it not true that motion is of two kinds only, rotary and translatory? The latter accounts for the angle of refraction, and hence it must be the former which explains the sensation of color. Because there can be no diversity in movements other than those mentioned, and inasmuch as there are no sensations connected with the rainbow except those of color, Descartes concluded that his explanation must be the true one. One of his weaknesses was the wholesale postulation of microcosmic particles with complicated qualitative properties modeled on the observed properties of the macrocosm. But how false his emphasis on matter and motion here played him! There was a characteristic of the rainbow which Descartes largely disregarded—its width —and it was in this narrow band, rather than in rotary motion, that the secret of color lay all unnoticed. Had he measured what was measurable, instead of speculating on a motion which could be neither seen nor measured, he might have discovered the secret.[26]

Descartes was all too prone to assume that his work constituted the last word on a subject, and his arbitrary impatience frequently kept him from getting the most out of his brilliant ideas. On the age-old problem of multiple rainbows Descartes might have found a beautiful confirmation of his geometrical explanation of the bow. One of the acid tests of an effective scientific theory is its fruitfulness in predicting phenomena which subsequently are verified through experience; and the Cartesian theory could have done what no earlier hypothesis could possibly have done—predict where one should expect to find the tertiary rainbow, if it exists. But Descartes missed this marvelous opportunity; he seems not to have seriously entertained the possibility of a proper rainbow of third order. It had been reported to him that a third rainbow had been seen above the two ordinary bows, much paler and about as far from the second as the second is from the first. Descartes held this to be impossible unless there were mixed in with the raindrops particles of hail, very round and transparent, and having a notably greater index of refraction than that of water—a remark which reminds one strongly of Themo's views on multiple bows back in the fourteenth century. In this case, Descartes pointed out, there should have also been a fourth bow within the primary, although he admitted that this might have escaped attention because of the brightness of the adjacent arc, or because it might have been mistaken for a part of the primary bow.[27] To provide a display of natural magic, Descartes suggested mixing (with the water in fountain sprays) liquids of varying refractive power, thus multiplying the number of rainbows visible to an observer. He also called attention to the formation of rainbows by reflection in a body of water; but of the possibility of a tertiary bow from ordinary raindrops, there is no hint.

Les Météores closes with the confident expectation that "those who have understood all which has been said in the treatise will no longer see anything in the clouds in the future of which they will not easily understand the cause, or which will evoke their admiration." [28] Descartes should have known better! In the first place, he had not really answered all the problems connected with the rainbow, as future generations were to find out. Then, too, he should have realized that his work was not exempt from the rule that new ideas do not meet with immediate acceptance. He had of course sent a copy of the book to Froidment, whose *Meteorologia* had appeared only a decade earlier. The two works naturally had much in common. It must be realized that the explanation of the rainbow constitutes only a small portion of *Les Météores*. Descartes presumably had been introduced to meterology through the Conimbricenses *Commentarii* on Aristotle's *Meteorologica*, which he had used as a textbook at La Flèche. His own treatise covers much of the same ground in much the same order. It deals with such diverse topics as winds, clouds, thunder, salts, vapors, and exhalations. *Les Météores* would be far from notable were it not for the theory of the rainbow,[29] and yet the section on the rainbow is not typical of the treatise as a whole. Of the three appendices to the *Discours*, the author described one, *La Géométrie*, as pure mathematics, another, *La Dioptrique*, as mixed mathematics and philosophy, and the third, *Les Météores*, as pure philosophy. He seems to have counted on his meteorology to introduce his philosophy into Jesuit schools; [30] and he therefore must have been eager to learn of the opinion of others on his work. The result must have been a cutting disappointment. Contemporaries judged *La Dioptrique* and *La Géométrie* highly; but *Les Météores* they regarded as arbitrary and problematical. Froidment's reaction was especially negative. He never wrote Descartes about his opinions and never acknowledged receipt of the book. After 1637 one could not well teach Aristotelian meteorology in colleges without refuting Descartes—unless one pretended to ignore Cartesianism, which is just what Froidment did! He found the philosophy of Descartes too mechanical and materialistic, and set about to counter it by a campaign of silence. He republished his own *Meteorologia* in 1639 (it reappeared again in 1656 and 1670) without so much as mentioning *Les Météores* or its author; and he made no appreciable change in the views he expressed. The work of Descartes had not suddenly overthrown the Aristotelian theory of the rainbow.

Philosophical disagreement was not the only impediment to the spread of the Cartesian explanation of the rainbow. In 1637 there were no scientific periodicals to disseminate discoveries, and news travelled slowly. In Bohemia there was a contemporary scientist by the name of Marcus Marci (1595–1667) whose interests were much like those of Descartes, yet apparently he never became familiar with the *Discourse on Method* or its three appendices. Marci (or Marek) of Kronland, a Jesuit, took his medical degree at Prague in 1625 and practiced there for some time, becoming ultimately physician to Ferdinand III and Leopold I. For years he was a member of the medical faculty at the University of Prague and served in many executive capacities, including that of Dean.[31] In spite of these activities, he wrote works on medicine, philosophy, geometry, astronomy, physics, meteorology, and anatomy, earning himself the title of "Bohemian Galileo." His best known work is the De Proportione Motus of 1639, a classic of mechanics dealing with the laws of motion and of impact. Far less known, but nevertheless of considerable significance for the attempt to explain color quantitatively, is the Thaumantias, Liber de Arcu Coelesti, published at Prague in 1648. This tediously long treatise of 268 pages shows that the theory of Descartes, eleven years after its publication, had not yet become generally known in parts of central Europe. Marci was unaware of either the law of refraction or the Cartesian geometry of the rainbow.

"Indeed," wrote Marci, "no one whom I know has either observed or demonstrated the diameter [of the bow], and yet this is of the greatest importance, for from it a great part of the demonstration could have been derived." [32] That is, he, too, was searching for a quantitative explanation. But first he outlined his lengthy project. First he will give a history of earlier explanations, next he will lay the foundations, then he will explain the essential causes, and finally he will refute errors. There is no need to follow Marci through all of the preliminaries. He cited an abundance of authors—ancient, medieval, and modern—with an indication of some of their views. These form a cross-section of the story of the rainbow up to about 1600. Marci refuted, for example, the Conimbricenses definition of the iris as a multicolored arc formed on a moist, opaque, and concave cloud by reflection of the sun's rays to the eyes of the spectator.[33]

The next section, on foundations, includes a quasi-medieval discussion on the intention and remission of qualities, the nature of fire and gunpowder, the possibility of a vacuum (in which he refers to Pascal and Roberval), and the properties of light. By modern standards much of this now is irrelevant to the rainbow. More significant is that which follows—the formation of color in refraction through a triangular prism (trigonia). Rays from the sun differ by half a degree (the apparent diameter of the sun) in their inclination on the side of the prism, and in this half-degree difference, Marci thought, is contained all the variety of color which arises in refraction.[34] That is, in refraction it is different angles, not different intensities or the amount of material, which determine the various colors. "Different kinds of colors are due to different parts of refraction." [35] There is in Marci's views a resemblance to those later expressed by Newton; and yet the similarity is essentially superficial. Where Marci reiterated that differences in color were caused by minute differences in the angle of *incidence*, Newton later demonstrated instead that color is characterized by small differences in the angle of *refraction* (Fig. 44). Marci said that the purple-producing incident ray is more nearly perpendicular to the surface of the prism than is the red; Newton showed that the purple refracted ray is less nearly perpendicular.[36] But Marci did very definitely anticipate another of Newton's discoveries. He noted that each color, after refraction, was, in a sense, disentangled from the others. If, for example, the

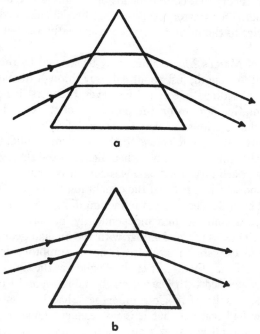

FIG. 44.—The idea of Marci (a) and Newton (b) on the formation of spectral colors by a prism.

red portion in the spectrum is refracted a second time, no further change in color occurs, although Marci believed the light to be further condensed. Each of his four colors (red, green, blue, and purple) was, so to speak, elementary or homogeneous.

The ideas of Marci are an intriguing combination of penetrating insight and confused speculation. Noting that the colors of transmitted and reflected light for thin gold sheets are different,[37] he jumped to the odd conclusion that light is made up of two types of rays, one of which (through refraction) produces colors, the other (in reflection), images. Color somehow is due to a condensation of light, but in what respects degree of condensation differs from degree of intensity is no clearer than is the difference between coloro-genic and photogenic rays. Had he recalled Occam's razor, and avoided such ad hoc explanations, he might well have anticipated Newton on the composi-tion of white light. But one must bear in mind that Newton lived a genera-tion later and, moreover, had the advantage of knowing the law of refraction. Marci depended in his work on the centuries-old tables of refraction of Witelo, believing that "the magnitude of refraction is not yet determined by any cer-tain law." [38] His theory that color depends on minute differences in the angle of inclination called for greater precision than his data afforded, and his com-putation of angles to the nearest second consequently was but specious accu-racy.

The reader of Marci's *Thaumantias* may be wearied by the tedious intro-ductory material. At length, just about half way through the book, one comes to the promising heading, "De arcu." However, the worst is not yet over, for the author promptly digresses on the properties of the eye, the operation of binary vision, the relations between light and fire, and the formation of images by prisms. Finally one comes to the problem of finding the angle of inclination of the colorogenic rays when light passes through a triangular prism or a circular cylinder—for Marci was convinced that the problems were identical. Somehow he had reached the conclusion that the colors of the rain-bow are caused by the reflection and refraction of light in the round drops of rain. Did this idea come to him independently, or was it derived through Maurolycus or Kepler? He was familiar with the Sicilian geometer's explana-tion, which he takes pains to refute. It is not too improbable that he knew also of the tangential-ray theory. After all, Prague was the very city from which Kepler had corresponded with Harriot on the rainbow; and Kepler's relations with the Jesuits, in spite of religious differences, had been most cordial. But Marci somehow had realized that it is not Kepler's tangent ray which pro-duces the rainbow, and he asked again the old question of Theodoric—what lies behind the fact that only rays at a certain angle are effective in forming the bow? His answer was rather ingenious and typical of a geometer. (He had been a student of the mathematician Gregory of St. Vincent.) Having made a careful study of prismatic refraction, he had recognized that there, too, there was a characteristic effective ray. He believed that the rays which pro-duce the prismatic spectrum are inclined at an angle of between 41° 30′ and 42° to the side of the equilateral prism, pass through the prism in a direction parallel to another side of the prism, and emerge from the third side at an angle of refraction of from 41° 30′ to 42° (Fig. 45a). If, now, one thinks of inscribing a sphere within the prism and of causing the same incident ray to strike the sphere and prism at a point of tangency (Fig. 45b), the ray will follow the same path as before, emerging at another point of tangency. Hence

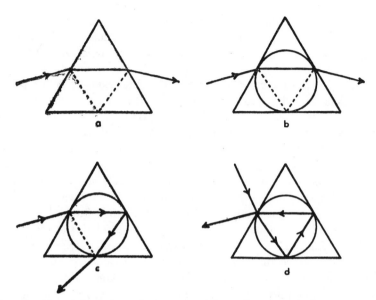

Fig. 45.—The relation between the formation of the rainbow arcs and the refraction of light through a prism, according to Marci.

one can dispense entirely with the prism, and think only in terms of the sphere. Here, believed Marci, was the key to the rainbow. He thought of each great circle of the drop as inscribed within an equilateral triangle such that the solar ray strikes the triangle both at the point of tangency and at the effective colorogenic angle for the prism, of which the triangle is a cross-section. Was it coincidence only that for his effective ray he used the familiar 42° angle? Perhaps, for this was not Marci's estimate of the radius of the bow.[39] Rays at other angles, he said, would not leave the prism at the same angle at which they had entered. If one removes the triangle, leaving only the circle, the color-producing angle will remain the same. The incident and refracted rays will cut off from the circle an arc of 120°. In the formation of the rainbow, however, the rays of the sun are reflected within the drop. In this case they follow a path (within the great-circle cross-section) given by two sides of the inscribed equilateral triangle, emerging at another vertex (Fig. 45c). The emergent ray likewise makes an angle of about 42° with the tangent to the circle; and hence, the angle between the incident and emergent rays, that is, the apparent radius of the rainbow, is about 36° (Marci gave, as a more precise value, the angle 36° 52′). Apparently Marci realized that his calculated radius was not in close agreement with observation, and so he took refuge in the conjecture of Maurolycus. Because the falling raindrops are changed to ellipsoidal form, the inclination in reality is somewhat greater than that which his theory predicts.[40] For the halo or corona Marci calculated, on the basis of his inscribed-circle theory, with two refractions but no reflections (Fig. 45b), a diameter of 38° 45′ 12″. The discrepancy between this and the observed value of 45° he again ascribed to the fact that the diameter of the rainbow and the corona are subject to variation. Some such escape became even more necessary for Marci's secondary bow inasmuch as he calculated that the emergent ray, after two internal reflections (Fig. 45d), would make with the incident ray an angle of 83° 8′, a value far in excess of the radius of the bow given by experience.

The width of the rainbow arc (which he set at 50′) and the formation of the colors Marci attributed to the nonparallelism of the incident rays. The apparent size of the sun is about 30′, and in this small interval he believed the secret of color lay. Convinced of the correctness of this assumption, he went into tedious and intricate calculations of the refractions of rays from opposite limbs of the sun.[41] There is little to be gained in following him through the labyrinth of calculations, for without a knowledge of the law of refraction, one could not hope to determine angles to the requisite degree of accuracy. Marci also paid considerable attention to the calculation of the relative distances of the rainbow arcs, but here his theoretical preconceptions, with resulting huge errors in the radii of the bows, vitiated his work. From the radii of 36° 52′ and 83° 8′ for the primary and secondary arcs respectively, he computed (using the law of sines) that the exterior rainbow should be eight times as far from the eye of the observer as the interior, assuming that the drops producing the two bows lie in the same vertical plane.[42] This is greater than the actual ratio by some six hundred per cent!

Marci's theory should have been just as suggestive of a tertiary rainbow as any which had yet been proposed. The extension to cover one more internal reflections should have been quite natural, and the result would not have been less in agreement with later experience than were those he obtained for the first two bows. But Marci, like his worthy predecessors, did not take such a step. He was inclined to agree with Aristotle that not more than two bows are possible—unless one counts the red and purple arcs as separate rainbows. Witelo's multiple bows he thought must have been just part of the main bow; but possibly in subpolar regions there can be others.[43]

Marci's *Thaumantias* closes, after a long section on the opinions of others, with a strange theological conjecture. The rainbow observed in the drops of rain does not seem to be the same as that seen by Noah. The former is unique for each observer—there are, in other words, as many bows as there are eyes, and if there were no eyes, then there would be no bow. But this could not be true of the bow of God! The ordinary bow apparently was too intangible to serve as a sign of divine covenant. And so the author spends a score of pages on fruitless speculation concerning atmospheric conditions which would make possible the formation of a *Great or Universal Rainbow*, that is, one which should have more than subjective existence.[44]

The explanations of Marcus Marci are bound to strike the modern reader, familiar with the Cartesian theory, as tedious and misdirected. Tedious they undoubtedly are, but not misdirected; for his object was the same as that of Descartes—to give a quantitative theory which should be consistent with experience. His ingenious theory came closer to the truth than any with the exception of Descartes'; but it did not accord well with experience and could not hope to compete with the Cartesian explanation. It has been held that Marci "knew the correct explanation of the rainbow";[45] but such a statement becomes true only when one discards all of the quantitative element, to which Marci gave so much attention, and retains only the imprecise qualitative skeleton. On the physical side, however, Marci made at least a beginning where Descartes was weakest—in a quantitative study of colors. What was most needed at the time was a combination of geometrical study of the rainbow with a quantitative physical analysis of color. But central Europe remained for some time unfamiliar with the works of Descartes, as is apparent from correspondence between Huygens in Holland and G. A. Kinner von Löwenthurn, tutor in the household of the emperor Leopold I.

In 1653, Kinner wrote from Prague that *La Dioptrique* was scarcely heard of there, adding that what Descartes had written on the subject of color failed to satisfy him. Kinner asked Huygens also to settle a controversy which had arisen many years before in connection with the theory of the rainbow. In 1634 Balthasar Conrad (1599–1660), a Jesuit professor of mathematics at Olmütz, Prague, and Gratz, had written a treatise entitled *Propositiones Mathematicae de Natura Iridis*; and in this he had raised the question as to whether or not rays which strike the spherical drops tangentially are refracted in the same manner as those which travel along secant lines. It would appear from this query that the theory of Kepler had lived on in spirit, if not in fact, for half a century. Huygens merely referred Kinner to Descartes on this matter, asking him to have Marcus Marci send observations on parhelia and halos, for he is going to try to show the true causes of these.[46]

Bohemia is far from France; but even at home Descartes was a prophet without honor in certain quarters. In 1650, the very year Descartes died, there appeared at Paris a book on the rainbow rivaling that of Marci in length and prolixity. This work, a volume of well over three hundred pages entitled *Nouvelles Observations et Coniectures Sur l'Iris*, was written by Marin Cureau de La Chambre (1594–1669), physician to Louis XIII and Louis XIV and member of the Académie des Sciences from the year of its foundation in 1666. La Chambre, an associate of Cardinals Richlieu and Mazarin and author of other works on physics and medicine, wrote as though Descartes had never lived. As a matter of fact, his book reads almost as though it had been written some four hundred years earlier. The rainbow, one reads, can not be due to reflection alone because strong reflection is possible only from opaque, hard, and united surfaces. Nor is the bow formed in the drops of rain, for these are like atoms in the cloud—too small to allow a sensible angle to be formed at the eye. Most philosophers ascribe colors to refraction, the color being more or less clear according as the rays penetrate parts which are more or less opaque; but La Chambre protests. For the formation of prismatic colors three things are needed—sunlight, a glass prism, and a wall—and so he postulated three corresponding agents for the rainbow: sunlight, a cloud, and a medium between the sun and the cloud. The intervening medium, where the marvelous change takes place, is more dense than a vapor and more continuous than the cloud of raindrops. The rays of the sun are refracted in this medium and then are reflected to the eye by the rain cloud. But there are two types of reflection—one on bodies with smooth polished surfaces (in which case the angles are equal), the other from a rough uneven surface. The rainbow can result from each of these; the cloud is like a wall, rough and uneven; but the surfaces of the drops are smooth, and hence one sees the bow only along certain lines which are the result of both reflection and refraction, for neither reflection nor refraction alone can produce color.[47] This curious opinion would appear to be in conflict with observations on the spectrum produced by refraction through a prism; but La Chambre ingeniously interpreted prismatic colors as a combination of refracted rays with other rays which had undergone

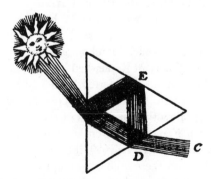

Fɪɢ. 46.—La Chambre's idea of the way in which prismatic colors are produced by a combination of rays which are both refracted and internally reflected.

iterated internal reflections at the surfaces of the prism (Fig. 46). He recognized seven pure colors—black and white, red and blue, yellow, green, and purple; and he believed that the colors of the rainbow are models for all colors.[48]

The medieval character of his views notwithstanding, La Chambre had not rushed into print without reading the newer works on optics and the rainbow. He cited Maurolycus, Porta, Kepler, Froidment, and others. There is even, in his work, some indication of independent observation. Recalling that "all ancient philosophers" had set the maximum altitude of the bow at 42°, that Maurolycus and Kepler had accepted instead a value of 45°, and that Porta had found a radius of only 38°, La Chambre reported that he himself had found one in 1625 which was about 60°. Admitting that he had not measured it carefully, he nevertheless asserted that at least it exceeded considerably the greatest measure which had been given. In view of this he was confident that not only the radius of the primary bow, but also the interval between the two rainbows, is subject to fluctuation.[49] When the scientific observation of a charter member of the Académie des Sciences could be so slovenly, one can only wonder at the rapidity with which seventeenth-century science progressed!

One would be inclined to dismiss with an impatient shrug the hopelessly antiquated notions of La Chambre were it not for the authority of his position in the world of science. In retrospect it is so easy to praise the Cartesian solution and to damn the vagaries of unappreciative contemporaries. How could any but the most obtuse fail to see the superiority of Descartes' work, one asks? And yet the *Nouvelles Observations* was sufficiently successful to warrant a new edition a dozen years later. The author at the same time (in 1662) published a related book (of over 400 pages) with the title *La Lumière*.

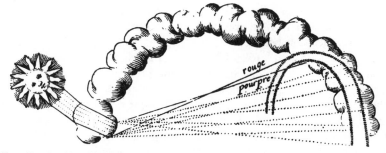

Fig. 47.—La Chambre's idea of the formation of the rainbow. Rays of the sun are refracted by a subtle material between the sun and the observer and are then reflected by a cloud mass opposite the sun.

This work illustrates well the fact that throughout the century there were two types of treatises on the subject—those in the geometrical style of Descartes and others along the philosophical tradition of the scholastics. La Chambre's treatise abounds with metaphysical subtleties, frivolous distinctions, and stultifying notions. He imagined four kinds of light in nature, two sorts of transparency, and two kinds of opacity! [50] When light changes to color, there is no change in its nature or essence. Light has more essence than all sensible things, and it is more independent of its subject than any other accident. "Finally, howsoever one wishes to describe the production of light, it is enough to know that it comes directly from God." [51] In the midst of such metaphysics there are occasional passages indicating that he had heard something of geometrical optics, but very little. He knew of the equality of angles and minimum distance in reflection, holding that there is nothing analogous in refraction. (Apparently he rejected Fermat's idea.) He was aware that refraction is greater for media of greater density and for rays of greater obliquity, and that it is not proportional to these. (Either he was unaware of the correct law or else he was oblivious of its significance.) The Cartesian explanations of reflection and refraction in terms of components he opposed on the grounds that light in indivisible.[52] Refraction he explained as due to "a natural antipathy for matter." There is one point on which De la Chambre sounded a quasi-modern note, for he believed that in the world of color there is a counterpart to the harmonies of sound.[53] But without any apparent knowledge of the law of refraction, his project for a quantification of color, however well taken, could scarcely lead to a satisfactory conclusion.

One who is not familiar with the historical development of science may be prone to assume that the work of La Chambre was but an instance of the survival of isolated ignorance and reaction. After all, an intellectual laggard can be found in every age—and even within the inner circle of scientists. But the appearance of a second edition of the Nouvelles Observations et Coniectures indicates that La Chambre's explanation of the rainbow was finding probably more readers than was that of Descartes. Even among top-level scientists the decision between the two explanations was far from clear-cut. Jean Baptiste Duhamel (1624–1706), the French physicist and perpetual secretary of the Académie des Sciences, found little to admire in the work of Descartes, while he had high praise for La Chambre. His De Meteoris et Fossilibus Libri Duo of 1660 (reprinted in 1681 in his Operum Philosophicum) included an extensive and far from impartial account of the rainbow. This is in the Galilean form of a dialogue in which the views of La Chambre are put in the mouth of Simplicius, those of Descartes are presented without conviction by Menander, and Theophilus represents an inquiring mind. But whereas Galileo's Simplicius was made to appear ridiculous, the Simplicius of Duhamel is expected to be admired. At considerable length Simplicius recounts the favorite medieval problems which had been argued ad nauseam—Whether the bow is formed as an image on a concave cloud, buffeted as it is by winds and constantly changing in form; whether it is formed by reflection or refraction, and whether in a continuous medium or in discrete drops; whether or not the sun, the eye, and the center of the bow are collinear. It would lead to ennui to review here the details of the answers which Duhamel, through Simplicius, gave. In general he agreed with La Chambre that the colors of the rainbow, produced by refraction in a continuous medium between the observer and the sun, are projected upon the dark continuous cloud where the bow is seen. Still following La Chambre, he held that the diversity in colors is in conformity with the various degrees by which the light has been weakened; if the strength of white light is taken as 24, then that of yellow is 18, red is 16, green is 12, and purple is 8—"and this is almost the same as that in sounds." [54] After Theophilus has presented, in none too favorable a light, the Cartesian doctrine of rotating particles as the key to color, Menander protests that they have not been entirely fair to Descartes, the one to whom they are indebted for the modern spirit in philosophy. He therefore suggests, in a single sentence, that the colors of the bow arise from solar light reflected and refracted in globes of water, as Descartes had skilfully shown. He says that he will dispense with the arguments prompted by the secrets of mathesis and will cite in support only the observation that one can get a beautiful rainbow by spraying water by mouth into the light of the sun.[55]

Not all of the treatises of the mid-seventeenth century were as unfavorably disposed toward Cartesianism as was that of Duhamel. The exponent of atomism, Pierre Gassendi (1592–1655) represented in a way a transition from pre-Cartesian to post-Cartesian views. He evidently had made a thorough study of ancient opinions, for the chapter in his *Opera* which treats of the rainbow, the halo, and parhelia cites classical views from Anaximenes to Olympiodorus. Strangely, little is said on medieval or early modern ideas; but Gassendi accepted the Cartesian theory as definitive. He reported that Descartes had established by experiments with a glass globe of water, that the reflection producing the bow is in the spherical drops; it is not from the anterior surface, but from the posterior surface, with two refractions, one on entering and the other on leaving the drop.[56]

Gassendi was also more accurate than La Chambre in his observation of the rainbow. Accepting the usual 42° measure of the primary bow, he estimated that the amount by which the radius of the secondary arc exceeds 45° is double the amount by which the radius of the primary arc falls short of 45°. He estimated the width of the rainbow as about 3°, being double that of the halo. Gassendi seems to have been unduly impressed by the fact that the two-to-one proportion enters also in the ratio between the radius of the primary rainbow and that of the halo. He tried, like Kepler, to see some relationship analogous to that between the measures of central and inscribed angles in a circle, suggesting that for the halo the eye is on the circumference of a circle with the sun at the center, while for the rainbow the eye is at the center with the sun on the periphery. Gassendi was on more solid ground in holding, like Descartes, that the halo is caused by a double refraction of solar rays through particles between the earth and the sun, even though he was wrong in assuming these particles to be spherical.

The astronomer J. B. Riccioli (1598–1671) gave some independent rainbow observations in his celebrated Almagestum Novum of 1651. The rainbow, strictly speaking, is a part of meteorology rather than astronomy; but Riccioli became interested in the altitude of clouds and, hence, in the height of the rainbow.[57] If one will note carefully the place where the bow rests, he said, he will find that the distance of the bow from the eye usually does not exceed a thousand paces. But the apparent altitude of the bow is not more than 45°, from which it follows that the height of the bow (and hence of the cloud) is not greater than a thousand paces. Riccioli was aware of the estimates of the radii of the primary and secondary bows given by Maurolycus and Descartes, values which he checked against experience. In a table of observations made in the years 1643, 1644, and 1647 he cited radii of half a dozen primary rainbows (varying from 45° 50′ to 42°), and of one secondary bow (50° 20′), his values agreeing quite well with those calculated by Descartes. These observations supplied whatever confirmation of theory by experience sceptics may have believed to be needed.

Riccioli mentioned several seventeenth-century authors on the rainbow, although he gave no theory. This is understandable in a work on astronomy; but one is struck by the utter inadequacy of treatments given in contemporary treatises on optics. Nicolaus Zucchi of Parma (1586–1670) in 1652–1656 published a popular two-volume work entitled Optica Philosophia Experimentis et Ratione a Fundamentis Constituta, yet he gave no evidence of valuing the achievements of Descartes. He knew that there is "some sort of proportion" connecting angles of incidence and refraction, but he added that this law commonly is omitted! He realized that the colors of the rainbow are produced in the same manner as those formed by a prism; but he gave no quantitative explanation, being satisfied to report that the bow is due to moisture dispersed in the form of drops.[58] A decade later J. C. Kohlhans (1604–1677) rector at Coburg, published a Tractatus Opticus (1663) which leaned heavily upon the Optica of Zucchi and added only the banal comments that the angle of refraction is not proprotional to the angle of incidence, and that the rainbow is a mixture of reflection and refraction.[59] Even the noted Scotch mathematician, James Gregory (1638–1675), seems to have been unfamiliar with the *Dioptrique* of Descartes when, also in 1663, he published his *Optica Promota*, for he presented the law of refraction as an independent discovery. Unfortunately, he did not include a treatment of the rainbow, perhaps still feeling as the ancients did that it belonged to meteorology more than to optics.

The triumph of Cartesianism, in science as in philosophy, was far from precipitate. In some quarters the work of Descartes remained unappreciated until well into the second half of the century, in others it was actively opposed. Symptomatic of this was the publication in 1669 of yet another edition, intended for students, of the *Margarita Philosophica* of Reisch, and the reissue in 1670 of Froidment's *Meteorologicorum*. More striking still is the appearance in 1678 of a new volume on the rainbow in which the author, apparently in full awareness of the Cartesian theory, nevertheless preferred a thoroughly quantified Aristotelian explanation to that of Descartes. This book, *L'iride: Opera Fisicomatematica*, by Giuseppe Antonio Barbari, is made up of two parts. The first section, "Discourse on the iris," is a sort of overgrown preface of thirty-one pages in which some of the problems concerning "the most beautiful question of the nature of the rainbow" are enumerated, together with comments on the proper way of philosophizing about the subject. The half-dozen basic questions remind one strongly of those posed during the medieval period; but Barbari's general attitude appears to be favorably modern. He objected to the blind credulity of leading Peripatetics who fail to question the dicta of the Philosopher, and he praised the experimental philosophy of Galileo. Most encouraging are several statements: (1) that Aristotle is wrong in assuming that the rainbow is caused by an ordinary reflection; (2) that nothing is needed for the bow except rays of the sun (or moon) and vapor in the form of drops, and (3) that the colors of the bow are produced by refraction as in crystal globes full of water.[60] This section closes, however, with the disconcerting remark that those who are not satisfied with the type of explanation to be given can find an opposing view in Descartes, Gassendi, and Grimaldi.

The second section, the body of the book, is entitled "Commentary on the Text of Aristotle," and it consists of over a hundred pages of elaboration on a few lines from the ancient *Meteorologica*. Whereas earlier Aristotelian commentators had been attracted by the non-mathematical portions of the master's explanation, Barbari focussed attention almost entirely on the quantitative aspect. By means of an exposition based upon more than a score of geometrical diagrams, he tried to make the Aristotelian explanation quantitatively acceptable. Just as the ancient astronomers' circles had in later hands become crystal spheres, so the Aristotelian meteorological hemisphere (Fig.

5) had become, among some commentators, a concave cloud. Barbari accepted implicitly the idea that the Aristotelian hemisphere is a cloud, from the inner surface of which rays of the sun are reflected to the eye of the observer. Recognizing, however, the incongruity of placing the sun and the cloud at equal distances, he placed the source of the rays beyond the periphery of the hemisphere. Moreover, in order to reconcile the Aristotelian explanation with the equality of angles called for in the law of reflection, Barbari followed a suggestion which had been made before and displaced the eye of the observer from the center of the hemisphere.[61] In most respects the tedious commentary of *L'Iride* is not unlike those written by Alexander, Olympiodorus, Vimercati, and many others whom Barbari cited. Having made the necessary adjustments, the author appears to be more, rather than less, confident in the basic soundness of Aristotle's theory of the rainbow than were his pre-Cartesian predecessors!

The story of the rainbow shows clearly that, for the development of science, more is needed than a few great discoveries. Descartes, in his appendices of 1637, had hit upon three capital advances; and yet not one of the three was integrated into scientific thought for several decades. The fault lay in part with Descartes. In the case of each of the appendices—*La Dioptrique* and *La Géometrie*, as well as *Les Météores*—the author was primarily boasting of the efficacy of his methodology. He was not explaining, with the meticulous care required in new situations, three great contributions to science. New scientific ideas, before they become common property, must be explained over and over again; they should be surrounded with an aura of proselytizing enthusiasm. They must be explored further; their implications should be determined, their relationships to other phenomena investigated. Descartes himself, however, promptly lost interest in analytic geometry, the law of refraction, and the rainbow; nor had he groomed a successor. Is it any wonder that his work, even though published again in 1656 in Latin, the universal medium of communication, was slow to achieve the recognition it deserved?

IX

The Age of Newton

ON CHRISTMAS DAY OF 1642 WAS BORN A PREMATURE CHILD WHO LATER BE-
came the greatest scientist of his day—perhaps of all time—the one who made
the greatest post-Cartesian discovery in connection with the rainbow. Sir Isaac
Newton (1642–1727) in childhood gave no particular sign of great promise,
either intellectually or socially; but shortly after he had graduated from Cambridge University, his mathematical genius became clearly apparent. This was
particularly evidenced in the invention (in 1665–1666) of the method of
fluxions—i.e., the calculus, invented independently by Leibniz during the
following decade. This great discovery, later to become so important in the
story of the rainbow, was but the beginning in the long chain of Newton's
contributions to mathematics and science, the best known being the verification of the law of gravitation. But whereas Descartes had been inclined to
overrate his own originality, claiming that he seldom read the works of others,
Newton always maintained a high degree of modesty. If he had seen further
than others, he said, it was because he stood on the shoulders of giants. And
among these giants he had singled out for particular recognition Christiaan
Huygens (1629–1695), the brilliant Dutch scientist. In the story of the rainbow, as in other respects, in science and mathematics, Huygens played a role
as a transition figure between the age of Descartes and that of Newton. He
was one of the earlier scientists to appreciate Cartesian science, although he
was sharply critical of certain aspects of Descartes' physics; and he had particularly high regard for the explanation of the rainbow. Of it he wrote: "The
prettiest thing which he [Descartes] found in physical matters, and in which
alone perhaps his view was well taken, is the cause of the double arc of the
rainbow, i.e., the determination of their angles or apparent diameters." With
respect to the physical explanation of the colors, on the other hand, Huygens
opined that there was "nothing less probable" than the Cartesian doctrine.[1]

The chief contribution of Huygens to the theory of the rainbow was indirect, and its influence was not felt until well over a century later. In the
Cartesian geometrical theory it matters little what light is, or how it is transmitted, so long as propagation is rectilinear and the laws of reflection and
refraction are satisfied. Even the old visual-ray theory would have been com-

patible with the Cartesian explanation. But rainbow developments of the nineteenth century were to hinge closely on the nature of light, and here Huygens introduced a major change. Before the middle of the seventeenth century it was generally believed that light was a steady emanation, either material or immaterial, corpuscular or fluid, from luminous objects. In 1665, however, Robert Hooke (1635–1703) reverted to a quasi-Aristotelian suggestion that light is an activity conducted in a medium. There had been assertions in medieval times, and more recently by Francis Bacon, that heat was related to motion, and Hooke held that light was an exceedingly quick motion, somewhat similar to the more "robust" motion producing heat. The motion he assumed to be vibrative, for he could not see how a circular motion (such as that which Descartes had proposed to explain color) could be impressed upon the pellucid medium. Finally, he held this vibratory motion to be very short, for the transmission of light through diamond, the hardest body then known, seemed to imply this.[2]

Hooke's *Micrographia* is especially important in biology for the rich microscopic observations it contains; but it contains also much of interest to the student of physical optics, especially with respect to the study of color. Attention is called to the colors which appear in thin sections of mica; and Hooke showed that similar color fringes and rings can be produced by pressing together plates of glass. He inferred correctly that the colors somehow were produced by the interaction of light which was reflected from the front and back surfaces of the plates; and although in his day the principle of interference was unknown, he, nevertheless, had recourse to a somewhat similar principle of superposition of effects. To explain the colors produced in refraction Hooke had used an ingenious geometrical argument. Let $A_1 B_1$, $A_2 B_2$, $A_3 B_3$, . . . , $A_n B_n$ be the wave-fronts of the light in the less dense medium (Fig. 48), and $A'_1 B'_1$, $A'_2 B'_2$, $A'_3 B'_3$, . . . , $A'_m B'_m$ the wave-fronts in the more dense medium. Hooke assumed that for white light the wave-front is perpendicular to the direction of propagation. But when light strikes the surface and enters the denser medium, Hooke said, its velocity increases (an assumption agreeing with that of Descartes but diametrically opposed to that of Fermat). Because the points of the wave-front do not experience the increased speed simultaneously (when the wave-front is oblique to the surface separating the media), the resulting wave-front in the denser medium takes on the form of a straight line oblique to the direction of propagation, as illustrated in the accompanying diagram (Fig. 48). Hooke assumed that such a wave-front corresponded to a weakening of the light, resulting in its coloration. When the weakening affected the more advanced portion of the oblique front, he believed the light was blue, when the weakening affected the rearmost portion, the confused and weakened pulse was red. Other hues were

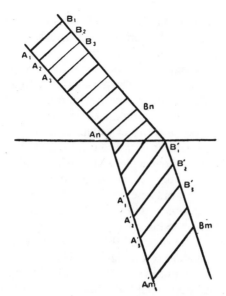

FIG. 48.—Hooke's idea of the relation of color to the orientation of the wave-front of light.

regarded as combinations of these two basic colors. To explain the colors produced in thin plates Hooke assumed that the combination of the light reflected from the front surface with that reflected from the rear was a super-position of such relatively weak and strong components. The colors of the rainbow presumably were to be explained in a similar fashion; but Hooke, no mathematician, was unable to give his generally cogent views a quantitative form.

Huygens had read without conviction the comparison of color and sound published by De la Chambre, but Hooke's theory, appearing in the *Micrographia*, spurred him to search for a mathematical theory of light. The result was the first systematically developed wave-theory of light with a convincing geometric background. Huygens began with the essentially Cartesian idea of an all-pervasive medium. Each particle of this luminiferous ether serves as a source of pulses or waves which are communicated to neighboring particles in much the same way as one billiard ball strikes another. Light thus travels outward from a source in somewhat the same manner as a disturbance is transmitted across the surface of a smooth pond when a stone is dropped into the water. Each point of the wave becomes in turn the source of a new disturbance of secondary waves; but the effective wave-front is the envelope of these secondary waves, and what one sees on the surface of the water is this advancing wave-front.[3] From this idea, known as "Huygens' Principle," the author derived with impeccable geometry the basic laws of optics upon which the theory of the rainbow rests. The derivation of the law of refraction led to the conclusion that light travels more readily in air than in water, contrary to the conclusion of Descartes and Hooke, but Huygens was unable to verify this inference. Nor was the author of the wave theory able to make use of his principle to explain the colors of the rainbow. The reason for this lay in the fact that he disregarded the most important characteristics of waves—their periodicity, and the possibility of waves of varying length and frequency. In view of such deficiencies, it might be better to call his hypothesis a "pulse theory" rather than an undulatory theory; and with this in mind it is easier to understand why it met with so little approval at the time it was proposed. Huygens presented his ideas to the Académie des Sciences in 1679 and published them in 1690 in Traité de la Lumière; but it was to be more than a century before the wave theory of light became a factor in the story of the rainbow.

Huygens probably never dreamed that his theory of light some day would revolutionize the explanation of the rainbow. He seems to have felt that, except for the question of color-formation, the work of Descartes was definitive; and Huygens never really accepted the challenge which the problem of colors presented. Huygens was satisfied to correspond with others on the rainbow, to amplify the Cartesian explanation, and to occupy himself with certain peripheral aspects. This is, however, just what keeps scientific problems alive, pending more definitive advances.

In 1652, while he was working on a treatise on refraction and the telescope, Huygens (then only 23 years old) considered the following problem: For a given index of refraction, to find the angle of the rainbow, and conversely. Where Descartes had found that for spheres of water (with an index of refraction of 250:187) the angle is 41° 30', Huygens computed that for spheres of glass (with refractive index about 20:13) the corresponding angle is about 21° 52'. That is, in a downpour of little spherical drops of glass the radius of the bow would be little more than half that of the arcus pluvialis. The radius of the secondary arc, for spheres of glass, Huygens calculated to be 89°, as compared with that of 51° 54' for water.[4] As Descartes had recognized earlier, the larger the index of refraction, the smaller is the radius of the primary bow. If, for example, turpentine is sprayed into the atmosphere, the primary bow will be found to lie within the water bow because of the greater refractive power. For the secondary arc the situation is reversed, and the turpentine bow will be seen above the water bow. Huygens' computations were carried out by means of Fermat's method of maxima and minima, a procedure equivalent to the use of the calculus. Huygens has been pictured as a mathematician who was too old to appreciate and to use the new tool invented by Newton and Leibniz; but this characterization can not be carried too far, as the rainbow calculations show. Nevertheless, the notations of Huygens are so far from those now in use that the details of his computation will not be given here, in view of the fact that work of a similar type was carried out later by others who published before Huygens did. Huygens had been urged by the mathematician Frans van Schooten (1615–1660) to publish his ideas on refraction, the telescope, and the rainbow in a Latin edition of *Discours de la Méthode* which Schooten was preparing; and Henry Oldenburg (1615–1677), secretary of the Royal Society, in 1669 pressed for publication. But Huygens was not ready, and this work appeared only after his death, in 1703.

The calculations of Huygens simplified considerably the work of Descartes on the primary and secondary rainbows. Whereas Descartes laboriously had calculated the paths of innumerable rays, one by one, Huygens expressed the deviation of the emergent ray as a function of the angle of incidence and then calculated, by the method of Fermat, the values for which this deviation is a maximum or a minimum. It would have been a relatively simple problem for him to have answered the old problem of the tertiary rainbow, but he missed the opportunity as thoroughly as had Descartes. He, too, had received reports that a third bow had been seen, one from his brother Lodewijk. He was interested, and he wrote asking his brother to send him further information; but his reaction was similar to that of Descartes. Recalling his calculations on the relation between index of refraction and the radii of the two rainbow arcs, Huygens tentatively attributed the third arc to drops which fell in the lower part of the valley. Such drops, he reasoned, were warmer than those at a greater elevation.[5] He wrote his brother that he would have to make some experiments on refraction in warm and cold water; but there is no record that he ever carried these out, and he died in 1695, a few years before the riddle of the tertiary rainbow was resolved.

Huygens corresponded freely with the scientists of his day, and so in 1657 one finds him writing to the mathematician René François de Sluse (1622–1685) on his rainbow calculations. Ten years later Sluse wrote to Oldenburg that he had, a good many years before, shown how to find the radius of the rainbow for a given fluid. Had he shown this, one wonders, prior to the correspondence with Huygens? Sluse was *au courant* with the newly developed methods leading to the calculus, and so one can assume that he too had used the usual rules for maxima and minima. Oldenburg sent on to Robert Boyle (1627–1691) a report of the letter of Sluse, with a request that the contents be communicated to the mathematician John Wallis (1616–1703).[6] There is no evidence that these British scientists pursued the problem further, in spite of the fact that Boyle was much interested in the problem of color and described a number of experiments with a "prismatical iris."[7] It may also be that Huygens was the original inspiration for a little-known treatise on the rainbow, *Stelkonstige Reeckening Van Den Reegenboog*, composed by Benedict Spinoza (1632–1677) and published posthumously in 1687, the year of Newton's *Principia*. To judge from the "Introduction" (which, however, was not written by Spinoza), the author evidently felt that there was no simple "algebraic computation of the rainbow" available. Hence he wrote this little work as a good example of Hudde's rule of maxima and minima (a modification of the method of Fermat), as well as of the works of Huygens and Sluse. Using Descartes' index of refraction, the calculations lead to radii of 40° 57' and 54° 25' for the two bows.[8] This little tract of a score of pages probably exerted little if any influence on later developments, but it indicates that, while contemporaries and successors of Huygens were becoming more keenly aware of the Cartesian rainbow theory, there still was need for an elementary treatment suitable for textbooks.

The lack of appropriate textbook material notwithstanding, the Cartesian explanation of the rainbow was becoming pretty generally known in the top scientific circles of the Newtonian Age. The well-known treatise of Francesco Grimaldi (1618–1663), *Physico-Mathesis de Lumine, Coloribus, et Iride*, which appeared posthumously in 1665, the year of Hooke's *Micrographia*, contained an unusually thorough-going treatment of the rainbow along Cartesian lines. Because the author does not name Descartes here, some have assumed [9] that Grimaldi, a Jesuit scholar, discovered the explanation independently of Descartes, who had been educated in Jesuit schools; but it seems to be more probable that hints of the geometrical rainbow theory had reached him somehow. After all, his book was published more than a quarter of a century after the *Discours de la Méthode*. The *Physico-Mathesis* has more than a dozen propositions on the primary and secondary rainbows. The author explains that he has experimented with a spherical phial of water; and his explanation is accompanied by a table, similar to that of Descartes, in which angles of incidence and emergence are calculated, using the sine law of refraction. His calculations are less satisfactory than those of Descartes, for they led him to a maximum angle (the angle for which the emergent rays are nearly parallel) of close to 46°; but Grimaldi says that he had frequently observed in experiments that the angle is about 41° or 42°. He evidently was a better experimenter than mathematician. In the case of the secondary bow he calculated from his table of values an angular radius of "more than 45°." He excuses the discrepancy between his result and experience (with the globe he found. an angle of about 50°) on the ground that there are other factors involved, such as the loss in vigor of the light after the double reflection and refraction, and the "fluidity of colored light." That not everyone had yet accepted the Cartesian theory is evidenced by his remark that some object to this explanation on the ground that the raindrops are too small to distinguish between the upper and lower hemisphere (used, respectively, for the primary and the secondary bows); but Grimaldi asserts that there is no basis for such an objection.[10] Grimaldi's treatment of the secondary rainbow is prefaced by the remark that if the bow is formed in spherical drops of water by means of suitable refractions and reflections, "it can appear double, or even triple, etc." This is one of the earliest suggestions of the possibility of a bow produced by a triple internal reflection; but the author did not carry the idea further, and it remained unsupported either by observation or by calculation.

The work of Grimaldi, like that of Huygens, was destined to affect the story of the rainbow more by indirection than by immediate contributions, for he was the effective discoverer of diffraction. Ever since antiquity it had been assumed that light could be transmitted in three and only three ways; but Grimaldi's *Physico-Mathesis* opens with "Proposition I. Light is propagated or diffused not only directly, by refraction, or by reflection, but also in a fourth way, by diffraction." To justify this assertion he described experiments, reproduced in modern textbooks, to show that light rays are deviated from the straight-line path on passing through a small opening or near a sharp edge. Shadows in these cases, he found, are not clear-cut; they are ill-defined, and edged with color.[11] The shadow cast, for example, by a small disc placed in a cone of sunlight from a small aperture was slightly broader than it should have been if the propagation of light were truly rectilinear; and light was similarly deflected on passing through a narrow slit. Moreover, he observed colored fringes bordering the shadow, arranged in a sort of periodic distribution. Perhaps the fringe patterns reminded Grimaldi of ripples in water, for although he died before the works of Hooke and Huygens were published, he asserted that the phenomenon of diffraction "does not favor atomism." To him light seemed to be a very rapid undulating fluid diffused throughout diaphanous bodies. Had he pursued the idea of periodicity which his results suggested, he might have solved the knotty problem of colors, but he spent his efforts instead on the age-old questions of whether light is substantial or an accident, and whether color is inherent in light or is a property of the medium through which it passes. His work on color was of no permanent value, for he accepted a sort of Aristotelian version in which red rays are dense and violet rays sparse. His discovery of diffraction, on the other hand, was later to be significant in the theory of the rainbow by removing one of the chief objections to the undulatory hypothesis of light. It was to be argued that light, if it is a vibration, should bend around an obstruction in the same way that sound travels around a corner. The strict rectilinearity of the propagation of light seemed to be an insuperable argument against a wave theory. Had the Newtonian Age paid more attention to the experiments of Grimaldi, they would have noticed that light does indeed, albeit almost imperceptibly, bend around corners.

At the very time that the books of Hooke and Grimaldi were being published, young Newton in England was performing some simple optical experiments. As he put it, "I procured me a triangular glass prism to try therewith the celebrated phenomena of colours." [12] But then many a scientist, from Seneca to Hooke, had tried his hand at this celebrated riddle, and without success. Boyle had compiled a history of all the theories of color, but he found none of these acceptable. Men praised Descartes' geometry of the rainbow, but they ridiculed his color theory. Would the callow young Cambridge graduate fare better than the greatest philosopher of the century?

The key to the mystery of color now appears to be so simple and obvious that it is almost incredible that it should have remained hidden so long. For thousands of years men had looked at colored spectra produced by light passing through spheres and prisms of water and glass; but Newton looked at the spectrum more carefully than had any one of his predecessors. One of Newton's first observations was that when a beam of solar light from a small circular opening passes through a triangular prism, the spectrum produced is not circular but very distinctly elongated, with straight sides and semicircular ends. This in itself was not a new discovery. Diagrams accompanying Theodoric's theory of the rainbow show a slight divergence after being refracted; and Marci had pointedly called attention to the divergence of the rays which had passed through a sphere or prism. Diagrams in Descartes' *Dioptrique* show a very distinct divergence of rays emerging from a prism; and Grimaldi too had noted the spreading out of the spectrum. But Newton's predecessors (and also, at first, he himself) attributed the divergence to differences in the angles of inclination of light from the opposite limbs of the sun. That is, they in essence attributed the width of the rainbow band to the fact that the sun is not a point source of light. The angles of incidence of rays from opposite sides of the sun differ by as much as half of a degree. But Newton noticed that the divergence of the spectral colors from an equilateral triangular prism was far too great to be accounted for in this manner, for the length of the oval spectrum (which was about five times its width) subtended at the prism an angle of about 3°. There seemed to be only one explanation, and that was that some of the light rays were refracted more than others. To confirm this suspicion, Newton passed through a second prism light from each of the colors in the spectrum. He noted, as Marci had before him, that each color remained the same under the passage through the second prism; but he noticed also something much more significant. He saw that rays of differing color, for like angles of incidence, were refracted by differing amounts.[13] Ever since antiquity it had been realized that the amount by which light was refracted depended on the angle of incidence, as well as upon the media in question; but Newton first showed that it depends also on the color of the light involved. Descartes, when he calculated the paths of the light through the raindrop, had used but a single index of refraction, 250:187. Newton's discovery showed that each color has its own characteristic index of refraction or, as Newton called it, degree of refrangibility, from 108:81 for red rays to 109:81 for violet. Each color corresponds to a definite degree of refrangibility, and conversely; and the color and index of refraction are not changed by further refractions or reflections, or even on transmission through plates of colored glass. Thus for the first time was color reduced to an orderly quantitative basis, and, also for the first time, an adequate explanation was possible for the width of the rainbow. As Newton phrased it, "The Science of Colours becomes a speculation as truly mathematical as any other part of *Opticks*." Computing the radii of the primary and secondary rainbows from the index for red light, Newton found values of 42° 2' and 50° 57' respectively; for violet rays the corresponding radii were found to be 40° 17' and 54° 7'. The difference between these radii, 1° 45' for the primary and 3° 10' for the secondary, would be the widths of the two rainbow bands if the sun were a point. Allowing half a degree for the apparent diameter of the sun, Newton gave the actual widths of the bows as 2° 15' and 3° 40'. "And such are the Dimensions of the Bows in the Heavens found to be very nearly, when their Colours appear strong and perfect." [14] Newton's calculations gave a value of 8° 25' for the interval between the two bows, this figure being somewhat smaller than the estimates made from Theodoric to Descartes.

Newton's 1666 experiments with the prism showed that light is constituted of dissimilar components which can be separated out by refraction; only colored light is pure and homogeneous. This is contrary to belief which had been almost universally held since the time of Aristotle. Traditionally it had been understood that white light was pure and homogeneous, and that color, such as that in the rainbow, was the result of a loss in strength or purity. (An exception must, of course, be noted in the case of Marcus Marci, who had postulated the composite nature of white light—a combination of photogenic and colorogenic rays.) Newton's experiments indicated, however, that when sunlight passes through a raindrop, color is not a result of weakening; it is simply a winnowing out, by the drop through dispersion, of the components of the light, sending each color along a slightly different path. This explanation vindicated an assumption which Theodoric had made in the fourteenth century: for a given observer the drops which furnish the red of the rainbow are different from those which supply the blue or the violet. Each color is produced by a set of drops at a position appropriate to the formation of that color by virtue of the unique correspondence between color and index of refraction. Unlike most of his predecessors, but in agreement with Ptolemy, Newton believed that there are seven distinct colors in the rainbow—red, orange, yellow, green, blue, indigo, and violet—each having its own distinct degree of refrangibility.

Newton realized that such a drastic departure from previously accepted views would not be accepted by his contemporaries without strong supporting evidence. If, as he claimed, light is a congeries of components of varying refrangibility, it should be possible, by combining the colors in the proper proportions, to reconstitute the original white compound. Newton had analyzed white light; could he complete this by a synthesis? Newton showed that he could; and in so doing, he proved that he could dispense entirely with the superfluous photogenic component which Marci had postulated. Having produced, by means of a convex lens, a beam of sunlight in which the rays are parallel, Newton decomposed the beam into its colors by passing it through a triangular prism; then passing the colors of this spectrum through a similar prism the axis of which had been rotated through an angle of 180° (See Fig. 49), and interposing a second convex lens, Newton found that the colored

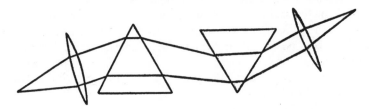

FIG. 49.—Newton's apparatus for analyzing and synthesizing light.

bands did indeed merge again into white light. It was not often that Newton permitted himself the luxury of enthusiasm about his accomplishments; but the success of this *Experimentum Crucis* was one such occasion. He wrote in 1672 to Henry Oldenburg:

I desire that in your next letter you would inform me for what time the Society continue their weekly meetings; because, if they continue them for any time, I am purposing them to be considered of and examined an account of philosophical discovery, which induced me to the making of the said reflecting telescope, and which I doubt not but will prove much more grateful than the communication of the instrument, being in my judgment the *oddest if not the most considerable detection* which hath hitherto been made in the operations of nature.[15]

Such language, coming from the one who had then but recently discovered what usually is regarded as the greatest of all laws of nature (the law of gravitation) and invented the most powerful of mathematical methods (the calculus), shows the value Newton placed on his optical experiments. This was one of the few instances in which Newton showed an eagerness to get an audience for his ideas. In 1672 he presented to the Royal Society a paper describing in broad outline his important discovery of the composite nature of white light. He did not include computations on the rainbow, even though he had presented these to students at Cambridge University in his optical lectures of 1669–1671. Newton did, nevertheless, show in a general way how his discovery solved the age-old question of the formation of the colors in the rainbow:

Why the colours of the rainbow appear in falling drops of rain, is also from hence evident. For, those drops which refract the rays disposed to appear purple, in greatest quantity to the spectator's eye, refract the rays of other sorts so much less, as to make them pass beside it; and such are the drops on the inside of the primary bow, and on the outside of the secondary or exterior one. So those drops, which refract in the greatest plenty the rays apt to appear red, towards the spectator's eye, refract those of other sorts so much more, as to make them pass beside it; and such are the drops on the exterior part of the primary, and interior part of the secondary bow.[16]

The reception accorded Newton's great discovery was to the young author a great disillusionment. Half a dozen scientists, including Hooke and Huygens, criticized his work, with the result that the over-sensitive Newton in 1675 wrote to Leibniz, "I was so persecuted with discussions arising from the publication of my theory of light, that I blamed my own imprudence for parting with so substantial a blessing as my quiet, to run after a shadow." [17] From that time on Newton was most prudent indeed. He withheld from publication anything further on optics until 1704, the year after Hooke, his sharpest critic, had died. Meanwhile his ideas went pretty much unnoticed, with credit for similar work sometimes ascribed to others. The Mémoires of the Académie des Sciences for 1679 announced that Edme Mariotte (1620–1684) had discovered the true cause of the colors of the rainbow, and his Essay de la Nature des Couleurs was published two years later. He, too, had found out that red and yellow rays have a smaller index of refraction than blue and violet. Hence the breadth of the spectrum is explained and the width of the rainbow band is accounted for. The author's diagrams show that he, like Newton, correctly interpreted the important observation that for a circular beam of light the spectrum is oval. Mariotte's explanation is given in an article, "L'Arc-en-Ciel," published in his Oeuvres.[18] Following reference to the ideas of Fleischer and De Dominis, the author wrote that "There are few scientists who are not satisfied by the explanation of Descartes." The Cartesian geometrical theory seems finally to have been generally accepted, but scepticism on Descartes' theory of colors continued. Mariotte himself [as also his contemporary Franciscus Linus (1595–1675)] raised a number of objections. One was that Descartes should not have recognized green as a principal color inasmuch as it is a mixture of blue and yellow. (Here Mariotte and Linus confused the mixture of pigments with the composition of spectral colors, for they suggested a study of color mixture by cutting up colored ribbons.) A second point is much better taken, for Mariotte points out that Descartes had not noticed that red rays undergo a refraction which is different from that for violet.

Mariotte reports in complete detail the experiments on the rainbow which he and Philippe de La hire (1640–1718) carried out. They placed on top of a stick a perfectly round thin glass globe full of water, and through it they passed a beam of sunlight. One of them then stood two or three feet away and measured the angular radius of the primary rainbow. At first the radius was 42° 40′; an hour later it was more than 43°, and still another three-quarters of an hour later the angle was 43° 30′. In nature, Mariotte said, the radius is 41° 14′. The discrepancy he ascribed to differences in density of the water and air in the laboratory as compared with the elements out of doors. In a second experiment with the phial, the angle for red rays at first was about 41° 20′; but after heating the water to almost the boiling point, the angle was 44° 44′. By extrapolation for lower temperatures he concluded that when the rainbow has a radius of only 41°, the refractive index is less than 4 to 3 by a sixtieth; and this can happen for very cold drops of rain and for greater rarefaction of air at high altitudes. Hence the angle of the bow can change; and the rainbow need not be a perfect circle because of the differing elevations of the drops. Mariotte calculated also the radius of the bow for substances other than water. For glass beads, with a refractive index of 3/2, he found a radius of 22° 48′ for the inner bow and more than 80° for the radius of the exterior bow. It is possible that this work was suggested by Huygens who, while he had not published on the subject, nevertheless had spent many years at Paris as a member of the Académie.

One wonders how much of Mariotte's work was discovered independently. His "Essay de la nature des coleurs" was read to the Académie some half dozen years after Newton had presented his discovery to the Royal Society; and yet Mariotte cited Newton's theory of white light only to criticize it. He reverted instead to a theory resembling somewhat that of Marci. White light is made up of rays which follow the law of refraction; the rays of colored light diverge either more or less than the law would indicate. In the brightest portion of the rainbow—that toward the middle of the band—the rays of differing divergencies destroy each other and are also concealed by the strength of the white rays. (Mariotte might have been well advised to recall that entia *non sunt multiplicanda praeter necessitatem.*) And, too, he was not convinced that the spectral colors were homogeneous. Contrary to Marci and Newton, he believed that, for a ray undergoing a second refraction (in the same sense), the color depends upon whether the angle of incidence is greater or less than the angle had been in the case of the first refraction.

Mariotte's work on the rainbow is of significance as illustrating the continuing interest, experimental as well as theoretical, in this and other related problems. Haloes are not rainbows, for they are not caused, as the ancient and medieval scholars had thought, by mist or raindrops; and yet the two phenomena are closely related. Kepler had at first tried to explain both phenomena in the same terms, although later he held to the opinion that the halo was caused by flat ice crystals. Descartes and Huygens, too, had attributed the halo to refractions in ice crystals, but they were unable to give the correct quantitative explanation. Here Mariotte earned credit as the one to suggest that the halo of radius 22° is caused by the refraction of light through the faces of prisms which make angles of 60° with each other. He explained the halo in terms of equilateral triangular prisms, whereas in reality the ice crystals are hexagonal prisms; but inasmuch as the refraction takes place through two non-adjacent non-parallel faces which are inclined to each other at an angle of 60°, the path of the ray is essentially that which Mariotte postulated (Fig. 50). In either case the minimum deviation is the same, being about 22°.

Following the cue furnished by Mariotte, in the next century the eccentric Henry Cavendish (1731–1810) explained the larger halo, with a radius of about 46°. This son of a lord disdained both his titles and the plaudits of his

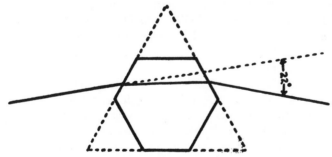

Fig. 50.—Mariotte's explanation of the halo of 22°.

fellow scholars, finding satisfaction only in his laboratory. He found that the larger halo was caused by the refraction in right prismatic ice crystals of the rays which traverse, with minimum deviation, a lateral face and a base. The parhelic circle, on the other hand, is due not to refraction, but to reflection from the vertical faces of the needle-like prisms and the vertical bases of flat crystals. The points of intersection of the smaller halo and the parhelic circle (mock suns) are bright because of the conjunction of refracted and reflected rays.

Mariotte's work on the rainbow was less significant than that which Newton had taught to his students at Cambridge; but Newton had published only the brief note of 1672, and hence Mariotte's account of colors and the rainbow was better known during the last part of the seventeenth century. There were other books of this period which contained references to the theory of the rainbow, but none of these made an appreciable contribution to the problem. Treatises on optics paid surprisingly little attention to the rainbow. William Molyneux's Dioptrica Nova of 1692, for example, has nothing on the subject. This is quite a change from the endless stream of rainbow publications which had come from the presses during the sixteenth century and the earlier part of the seventeenth. With the exception of Mariotte's work, and perhaps also that of Spinoza, rainbow references from 1672 to 1699 are quite undistinguished. There are, for example, passages on the bow in a large folio volume of 1675 by Athanasius Kircher (1602–1680) on Noah's arc. This example of seventeenth-century science fiction, by a conservative Jesuit, discussed in detail, together with floor plans, the dimensions and architecture of the arc and the stabling of the animals. Among many other curious questions, such as whether books on the sciences (especially those by Hermes Trismegistus) existed before the flood, Kircher raised the ubiquitous query about the pre-diluvian existence of rainbows. After explaining that the bow is caused by the reflection to the eye of rays from the sun which fall on a moist cloud, he concludes that rainbows had indeed existed earlier. Only after the flood did the rainbow become a sign of God's covenant.[19] Kircher's atavistic explanation of the bow seems to represent a retrogression of two thousand years; but the author refers the reader to his Ars Magna Lucis et Umbrae of 1671 in which an exposition along Cartesian lines, but without quantitative statements is included.[20] Evidently Kircher felt that the geometrical theory did not at the time appeal to the man-on-the-street, to whom his Arca noë was addressed. However, not all discussions of pre-Noachean rainbows were on this low level. In 1696 Christian Seyfried published a thesis entitled Iris Diluvii, in which an affirmative answer on this moot point was coupled with a thoroughly respectable Cartesian treatment of the rainbow.[21]

A thoroughly representative illustration of the seventeenth-century transition from earlier qualitative explanations to the modern mathematical treatment is found in an elaborate work of 1699 entitled Thaumantiados, Sive Iridis Admiranda. This Nuremberg thesis by an otherwise obscure figure, Christoph Gottlieb Volkamer,[22] is a mosaic of borrowing from remote antiquity to the author's own day. The compilation of material is fascinating but uncritical, with the outmoded meteorology of Froidement receiving little less admiration than the Cartesian ideas which overthrew it. The opinions of Scaliger and Vimercati that the bow is an image of the sun from a hollow cloud are recorded as sententiously as are the computations of Descartes and Mariotte. Volkamer is among the few rainbow writers who were more concerned about the formation and magnitude of the bow than with the enigma of color. He wrote that while Descartes' explanation of the reflections and refractions by which the colors emerge was ingenious, the Cartesian hypothesis on the nature of color scarcely meets with approval. The views of Mariotte are preferred for their simplicity. Volkamer refers in part of a sentence to the results of Newton,[23] but he was either unfamiliar or uninterested in these. He was more intrigued by the old speculative question of multiple rainbows. He cited the familiar ancient, medieval, and modern opinions and reports, including those of Aristotle, Witelo, Cardan, Snel, and Descartes. He himself had never seen three bows at a time, nor had Honoré Fabri (1606–1688), a writer on the rainbow to whom he refers with respect, and whose calculations, along Cartesian lines, he presents. Fabri seems to have doubted the possibility of a third bow; and the existence of a fourth or fifth arc he said was still more questionable. If the third and fourth bows were to appear, it was conjectured that, with respect to their radii and in the ordering of the colors, the first should correspond to the third and the second to the fourth. It is not clear just what is meant by a correspondence in the radii; but presumably this referred to some symmetry in their positions. After citing the reference in Descartes to a bow reported to have been seen as far above the second as the second is above the first, Volkamer confidently predicted [24] that the elusive tertiary rainbow should be found, if at all, at a maximum altitude of 62°. From this rash prediction it is quite apparent that neither Fabri nor Volkamer (nor, for that matter, any other of the numerous authors cited here) had carried out the necessary calculations on the third rainbow arc. It was just a year later that Halley, in the *Philosophical Transactions* for 1700, finally published the solution to the riddle of the tertiary bow.

It is well known that Newton, through the discovery of dispersion, gave the first reasonably adequate explanation of the color formation in the rainbow and of the widths of the two rainbow arcs. It seems not to be generally known that he was perhaps the first to give calculations concerning rainbows of order more than two. This work had been included in his Cambridge lectures of 1669–1671; but so great was Newton's reluctance to publish that it was not until 1704 that some of this material appeared in the *Opticks*. Meanwhile similar and more extensive calculations by his friend, Edmund Halley (1656–1742), appeared in the *Philosophical Transactions* for 1700; and analogous results were given also by Jean Bernoulli (1667–1748), probably shortly afterward and presumably independently. Still a fourth mathematician, Jacob Hermann (1678–1733), a student at Basle under Jean Bernoulli's brother Jacques, had worked out a similar formula independently before he had read a summary of Halley's article which had appeared in the *Acta Eruditorum* at Leipzig. Hermann published his short derivation of the general formula in the *Nouvelles de la République des Lettres* (in the same year that Newton's *Opticks* appeared [25]), explaining that he did so in order to point out to those who were criticizing the newly invented calculus that the infinitesimal methods were useful and valid. Halley and Hermann both expressed justified surprise that Descartes had not made use of similar short-cuts which were known in his day. Newton, Bernoulli, Halley, and Hermann all applied the methods of the calculus to the Cartesian theory in order to derive a single formula from which can be deduced the radius of the rainbow corresponding to any number of reflections within the raindrops. In this way they cut through the tedious computations Descartes and his earlier successors had carried out, one by one, for the paths of the individual rays, affording a striking example of the tremendous power afforded by the new mathematical analysis. The calculations of these four men will not here be treated separately, for they are very much alike; a single composite, in slightly more modern notation, will suffice.[26] Upon entering the spherical drop, the ray of light from the sun is deviated, according to the law of refraction, by an angle of magnitude $i - r$, where i and r are the angles of incidence and refraction, respectively (Fig. 51). A further deviation of like amount, and in the same sense, takes place as the ray leaves the drop, no matter how many reflections the ray may have undergone within the drop.

Each reflection within the drop results in a deviation, again in the same sense, of $\pi - 2r$ (where π radians represent 180°). The total deviation D in the case of a ray which undergoes two refractions and N reflections therefore is $D = 2(i - r) + N(\pi - 2r)$. For rainbows of odd order the deviation of the Cartesian effective ray is a maximum, for those of even order it is a minimum. In either case the maximum or minimum deviation is found by the calculus on equating to zero the differential (or derivative or fluxion) of D. This leads to the equation $0 = 2di - (2 + 2N)dr$ or $di = (N + 1)dr$, recalling that N and π are constants. Differentiating also the equation $\sin i = n \sin r$, where n is constant (the index of refraction), results in $n \cos r \, dr = \cos i \, di$ or, on dividing the two differential equations, in $\tan i = (N + 1) \tan r$ (recalling that the tangent is the quotient of the sine and the cosine). On eliminating the angle r from the last equation above, using the law of refraction, one easily finds that the extreme deviation takes place for those rays for

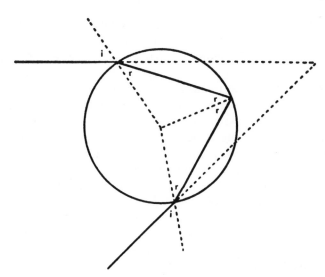

Fig. 51.—The deviations, due to refraction and reflection, of the light which produces the rainbow.

which $\sin i = \sqrt{\dfrac{(N+1)^2 - n^2}{(N+1)^2 - 1}}$ or $\cos i = \sqrt{\dfrac{n^2 - 1}{(N+1)^2 - 1}}$. On making the substitution $m = N + 1$ these take on the simpler forms

$\sin i = \sqrt{\dfrac{m^2 - n^2}{m^2 - 1}}$ and $\cos i = \sqrt{\dfrac{n^2 - 1}{m^2 - 1}}$. From either of these two equations, knowing the value of n and for a given integer $m = N + 1$, one finds the angle of incidence of the effective ray; and from this, by the law of refraction, the angle of refraction for this ray is found. Then the deviation D is calculated, and the difference between this and the nearest positive integral multiple of π is the apparent angular radius of the rainbow of order corresponding to that value of N. It is curious that Newton did not complete his calculations for values of N greater than 2. In the Opticks, he casually remarked that "The Light which passes through a drop of Rain after two Refractions, and three or more Reflexions, is scarcely strong enough to cause a sensible bow." [27] Bernoulli and Hermann likewise failed to publish the radii of rainbows of higher orders, although their formulas made such calculations a simple matter. The third bow might be visible to eagles or lynxes, Bernoulli thought, but not to human eyes. Halley seems to have been the earliest to carry through to the end the calculations on the tertiary rainbow; and the result must have been a surprise. He found that the third rainbow arc has an angular radius of 40° 20′, and that it should appear not in the part of the heavens opposite to the sun, but as a circle around the sun itself (Fig. 52). For at least two thousand years men had been looking for this arc in the wrong part of the sky! And furthermore, the reasons they had given for its failure to appear were not entirely correct. It is not so much the weakening of the light after a triple reflection that makes the bow of third order virtually imperceptible; and were it to be found where it had been expected, the light of the bow would be sufficiently strong for many human eyes. However, the light of the tertiary bow must compete against the brightness of the background in the sky in the general direction of the sun, and consequently the lack of a dark contrasting background accounts for the fact that this bow had never, so far as Halley knew, been identified. Halley found that the quaternary bow, formed after four reflections within the drops, also is to be sought on the side of the observer toward the sun. In this case the ray has undergone a deviation of 405° 33′, so that the fourth bow is a circle of radius 45° 33′ about the sun.

Halley and Bernoulli paid much attention to the converse of the old Huygenian problem: for a given radius of the primary or secondary rainbow, to find the index of refraction of the medium producing the bow. This required the solution of equations of third and fourth degree, a favorite algebraic topic of the day. Of more general interest were the definitive answers Halley gave to some of the age-old speculations concerning multiple bows. As is the case in the first two rainbows, so also for the third and fourth bows the order of colors and the sense of the deviation of the rays within the drops are opposite.

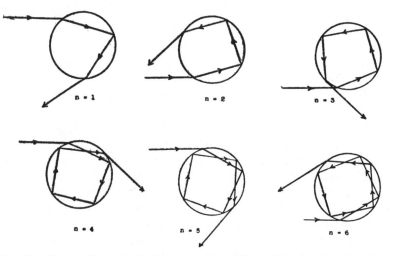

Fig. 52.—Diagrams illustrating the formation of the rainbows of the first six orders, where n, the number of internal reflections, indicates the order of the bow.

For the tertiary bow the rays enter the lower portion of the drop and are deviated in the same sense as for the secondary bow. Hence the violet band is nearest the sun, but in this case it is on the inside of the rainbow band. Rays forming the quaternary bow strike the upper portion of the drops, are deviated in the same sense as in the primary arc, and form a circle with the red rays inside and nearest to the sun. Maurolycus had pointed out that, with the sun at certain elevations, the secondary bow might appear without the primary. Halley's calculation showed that, when the altitude of the sun is more than 52°, it is theoretically possible for the tertiary rainbow to appear without the bows of first and second orders. Moreover, he and Bernoulli pointed out that for sufficiently large index of refraction the bows of lower orders may be missing for still another reason. The radicand in the expressions above for angle i may become negative, and hence the angles of these bows may turn out to be imaginary! In a downpour of genuine spherical diamonds, for which $n = 1.7$.) If the index of refraction of the drops is between three and four, be possible. (The "rainbow test" would betray synthetic diamonds, for which $n = 1.7$.) If the index of refraction of the drops is between three and four, the first two bows disappear; if it is between four and five, bows of the first three orders become impossible; and so on. Halley noted that as the index of refraction gradually increases, the radii of the third and fourth bows decrease until in turn they vanish in the sun itself. Newton and Bernoulli noted also another interesting fact. As n increases, $\cos i$ approaches zero and $\sin i$ approaches unity. This means that for rainbows of higher and higher order the incident Cartesian effective ray tends more and more toward the position for which it would just graze the side of the drop. As Bernoulli expressed it, the incident ray for the rainbow of infinite order is tangent to the drop! How pleased would Kepler have been to have lived to see that his tangential ray did indeed play a role—a quite unanticipated role—in the theory of the rainbow.

The first great landmark in the story of the rainbow should have been Theodoric's *De Iride*; but the influence of this seems to have been quite evanescent. Both before and after this volume there were dozens of books, medieval and modern, devoted entirely to the rainbow; and yet not one of these could claim to have made a significant contribution to the subject. The first truly effective milestone appeared in 1637 as one chapter of an appendix to Descartes' philosophical treatise, *Discours de la Méthode*. A second comparable contribution was published in 1704 in a more relevant context, as a section in the classical treatise on *Opticks* of Newton. The kernel idea of this, the composite nature of white light, had appeared in 1672; and through Newton's lectures (unpublished at the time), the application of this to the rainbow may have become known to those few Cambridge students who attended his lectures. Nevertheless, it was primarily through the numerous editions of the *Opticks* that the Newtonian theory became generally known throughout Europe and America. Proposition IX of the second part of Book I consists of ten pages on "Prob. IV. By the discovered Properties of Light to explain the Colours of the Rainbow." Newton began his explanation of the rainbow with the elementary statement that it is the sun shining on drops of rain which "certainly causes the Bow to appear to a Spectator standing in a due Position to the Rain and Sun. And hence it is now agreed upon, that this Bow is made by Refraction of the Sun's Light in drops of falling Rain." Newton at this point interjected historical remarks which were neither accurate nor fair. He wrote that "This was understood by some of the Antients, and of late more fully discover'd and explain'd by the famous *Antonius de Dominis*. . . . The same Explication *Des-Cartes* hath pursued in his *Meteors*, and mended that of the exterior bow." Newton presumably was here misled by the use, in Latin translations of Aristotle, of the word refraction for what in modern times was called reflection. Then, too, Newton erroneously implied that Descartes added nothing really new to De Dominis' explanation of the primary bow. This gross misrepresentation (which, incidentally, did not appear in the first edition, being added later, apparently under Newton's initiative or with his consent) could scarcely have been possible if Newton had really examined *De radiis visus et lucis*. Credit for the quantitative reasoning by which was established the correspondence between the radius of the bow and the parallelism of the effective rays belonged to Descartes and to him alone. Newton should have been aware that De Dominis had made no attempt to account for the size of the rainbow; and De Dominis would have been completely unable to do so, for he had overlooked the second required refraction. The historical lapses of Newton in this connection are a reminder that history requires every bit as much attention to detail as does science—and the history of science perhaps twice as much.

Newton then published for the first time a formula he had used in his lectures more than thirty years before—a formula, equivalent to that of Halley and Bernoulli, from which the radii of bows of all orders, and for any index of refraction, can be deduced. Details of the derivation are omitted by the author, and the reader is referred, in later editions of the *Opticks*, to the *Lectiones opticae*. It is interesting to note that Samuel Clarke, who in 1706 translated the *Opticks* into Latin, later conversely translated the *Lectiones Opticae* into English. (It was the Latin editions which did most to spread Newtonian influences abroad.) By substituting in the formula the index of refraction from air to water, first for red light and then for violet, Newton derived the radii, for the extreme colors, in the primary and secondary rainbows. The results had been confirmed, he tells us, from experience in nature and in the laboratory.

For once, by such means as I then had, I measured the greatest Semi-diameter of the interior Iris about 42 Degrees, and the breadth of the red, yellow and green in that Iris 63 or 64 Minutes, besides the outmost faint red obscured by the brightness of the Clouds, for which we may allow 3 or 4 Minutes more. The breadth of the blue was about 40 Minutes more besides the violet, which was so much obscured by the brightness of the Clouds, that I could not measure its breadth. But supposing the breadth of the blue and violet together to equal that of the red, yellow and green together, the whole breadth of this Iris will be about 2¼ Degrees, as above. The least distance between this Iris and the exterior Iris was about 8 Degrees and 30 minutes. The exterior Iris was broader than the interior, but so faint, especially on the blue side, that I could not measure its breadth distinctly. At another time when both Bows appeared more distinct, I measured the breadth of the interior Iris 2 Gr. 10′, and the breadth of the red, yellow and green in the exterior Iris, was to the breadth of the same Colours in the interior as 3 to 2.[28]

Newton closed his brief section on the rainbow by observing that light which has been thrice reflected within the raindrops is scarcely strong enough to cause a bow. So far as we know, he did not even calculate its radius, although he must have known about the then relatively recent article by Halley in the *Philosophical Transactions*. He did, however, calculate the radius of what may be called the rainbow of order zero. The formulae of Halley and Bernoulli break down for $n = 0$; but Newton showed that for rays which pass through the raindrops without being reflected (in a manner somewhat analogous to the passage of light through ice crystals to form the halo), there should be a circular band of illumination concentric with the sun and having a radius of about 26° at its brightest portion, with the intensity gradually falling off on both sides of this circle. This is never seen because the refracting spheres and the sun are in too nearly the same direction.[29] Newton incorrectly believed that the halo of somewhat less than 26° in radius was formed by rays of the sun or moon which are transmitted through slightly flattened spherical hailstones; and he thought that the halo is more strongly colored if the hailstones have opaque globules of snow in their centers.[30]

The fourth edition of the *Opticks* contained over four hundred pages, of which only one ten-page section is specifically on the rainbow; but there are other sections which, a hundred years later, were to lead in the end to a new theory of the rainbow. Among the most pertinent of these was the one with which the second book of the *Opticks* is almost entirely concerned: "Observations concerning the Reflexions, Refractions, and Colours of thin transparent Bodies." Here the discussion turns to the phenomenon known as

"Newton's rings," although such colored bands had been observed earlier by Hooke, to whose *Micrographia* Newton referred. Newton explained that he had not taken up this topic before because it is more difficult to explain these colors than those of the rainbow, and, as he thought, the explanation of the rings is not needed to account for the rainbow bands. But Newton described the colors of rings in the expectation, later amply justified, that a study of these would lead to further "Discoveries for compleating the Theory of Light."

Newton in 1675 had presented a report on his experiments on thin transparent laminae before the Royal Society, but in the *Opticks* this account was greatly expanded. In one of the experiments Newton pressed a biconvex lens against the plane surface of a plano-convex lens so that the thickness of the air layer between the surfaces (which here took the place of the space between the surfaces of a lamina) increased from the center outward. He noted a system of colored rings concentric about the point of contact; as the pressure on the lens was increased, thus reducing the distance between the surfaces, the rings moved inward, vanishing in the dark spot in the center. If the lenses were illuminated with monochromatic light, there were more rings clearly defined; and as different colors were used in succession, the positions of the rings shifted. To Newton "it was very pleasant to see them gradually swell or contract according as the Colour of the light was changed." The appearance of the rings for white (or polychromatic) light therefore was due to the superposition of innumerable monochromatic rings of varying radii. Newton measured the radii of the rings with compasses, and from the results he computed that the distances between the plates of glass corresponding to the dark rings (Fig. 53) were in the proportions 0, 2, 4, . . . , the distances for

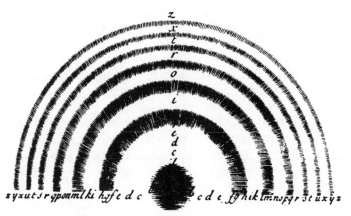

FIG. 53.—Newton's rings, from his *Opticks*.

the bright rings being in the proportion 1, 3, 5, . . . He found that similar rules applied to the rings produced by the varying thicknesses of soap bubbles, except that in this case the distance was greatest in the middle of the bottom, where the film was thickest, and decreased as the distances from this point increased. This was a second great triumph in the quantitative study of color; but although Newton sometimes referred to his rings as "Rain-bows" (or to one of them as an "Iris"), he saw no connection between these and the arcus pluvialis. A century later, though, a study of the rings was to lead to a new rainbow theory.

Just as Descartes' explanation of the size of the rainbow was independent of any particular hypothesis as to the nature of light, so also was the Newtonian explanation of colors adaptable to either one of the two chief contemporary theories of light. Whether red and violet light are streams of corpuscles or rapid vibrations in a medium was immaterial, so long as the basic fact of their differing refrangibilities was assumed. Newton at first tried to avoid committing himself to either of these theories, clinging to the celebrated phrase, "Hypotheses non fingo." But the puzzle of the rings tempted him into a more and more speculative mood. Why was it that "Rays of the same Colour were by turns transmitted at one thickness [of the space between two pieces of glass], and reflected at another thickness, for an indeterminate number of Successions?" There seemed to be some periodic relationship between the small space intervals separating the glass plates and the production of color. Had the wave theory of Huygens developed the concept of periodicity, Newton likely would have seen a connection between waves and the rings of color. Instead, he adopted a plausible but highly speculative hypothesis:

Every Ray of Light in its passage through any refracting Surface is put into a certain transient Constitution or State, which in the progress of the Ray returns at equal Intervals, and disposes the Ray at every return to be easily transmitted through the next refracting Surface, and between the returns to be easily reflected by it.[31]

That is, light near the surface of a medium has a certain periodic property which alternately predisposes it to be either easily reflected or easily refracted. Newton called these periods of easy reflection followed by easy refraction "fits."

The reason why the Surfaces of all thick transparent Bodies reflect part of the Light incident on them, and refract the rest, is, that some Rays at their Incidence are in Fits of easy Reflexion, and others in Fits of easy Transmission.[32]

Newton had opened the Opticks with the assertion, "My Design in this Book is not to explain the Properties of Light by Hypotheses, but to propose and prove them by Reason and Experiments." But the book closes with highly speculative conjectures in the form of queries, some of which appeared only in later editions.[33] Among the queries is one which must have been suggested by the experiments with colors in thin plates, or "Rain-bows."

Qu. 13. Do not several sorts of Rays make Vibrations of several bignesses, which according to their bignesses excite Sensations of several Colours, much after the manner that the Vibrations of the Air, according to their several bignesses excite Sensations of several Sounds? And particularly do not the most refrangible Rays excite the shortest Vibrations for making a Sensation of deep violet, the least refrangible the largest for making a Sensation of deep red, and the several intermediate sorts of Rays, Vibrations of several intermediate bignesses to make Sensations of the several intermediate Colours?

Such quotations make Newton appear as a supporter of the undulatory theory of light; and some of his work is tantamount to measuring the wave-lengths of various colored lights. Apparently at one point he did espouse a wave theory, as is revealed from a paper intended for the Royal Society but which was withheld from publication. In this Newton wrote:

First, it is to be assumed that there is an ethereal medium, much of the same constitution with air but far rarer; subtiler and more strongly elastic. In the second place, it is to be supposed that the ether is a vibrating medium like air; only the vibrations, far more swift and minute; those of air made by a man's ordinary voice, succeeding at more than half a foot or a foot distance, but those of ether at less distance than the hundred thousandth part of an inch. And as in air, the vibrations are some larger than others but yet all equally swift, so I suppose the ethereal vibrations differ in bigness but not in swiftness.

Had Newton persisted in such views, the story of the rainbow might have been advanced by almost a century. But Newton was obsessed by the rectilinearity of the propagation of light, and he felt that this was in irreconcilable conflict with any thorough-going undulatory hypothesis. Portions of a wave theory he did accept, for in connection with his theory of fits he assumed vibrations in an ether surrounding transparent media. He knew of Grimaldi's experiments in diffraction, and he interpreted these as indicating that light bent backward and forward with an eel-like motion in passing through narrow openings and near the edges of media. This explanation, and his belief that the polarization of light was to be explained by assuming that the rays of light had to have "sides," are suggestive of transverse waves; but Newton reached an opposite conclusion. In another of the queries he pointedly asks:

Are not all Hypotheses erroneous, in which Light is supposed to consist in Pression or Motion, propagated through a fluid Medium?

and he cites as supporting evidence a familiar comparison of light and sound.

Sounds are propagated as readily through crooked Pipes as streight ones. But Light is never known to follow crooked Passages nor to bend into the Shadow.[34]

And the following query continues in a similar vein:

Are not the Rays of Light very small Bodies emitted from shining substances? For such Bodies will pass through uniform Mediums in right Lines without bending into the Shadow, which is the Nature of the Rays of Light.[35]

These latter passages from the works of Newton underwent an apotheosis similar to Aristotle's theory of the rainbow; they became *ex cathedra* pronouncements by the High Priest of Science.

Newton's chromatic discoveries, like his law of gravitation, represented a "New Philosophy," and as such it had to make its way against established orthodoxy. Just as Aristotelianism had given ground but slowly to Cartesianism, so now the latter was displaced only gradually by Newtonianism. The physics of Descartes had become a part of the university tradition, in England as well as France, in large part through the popular textbook, *Traité de Physique*, of Jacques Rohault (1620–1675). It is of interest to note that this thoroughly Cartesian book opens with a just rehabilitation of the scientific reputation of Aristotle, calling attention to the latter's all-too-often-overlooked combination of reason and experimentation, including even the use of mathematical methods. "You will not find a great many Things in this whole Treatise contrary to Aristotle," he wrote; "but you will find more than I could wish that are contrary to most of the Commenators upon him." Rohault's *Traité* first appeared in 1671, just a year before Newton had presented his paper on the composite nature of light, and it went through a dozen editions within a generation. Before Newton's two great classics appeared (the *Principia* in 1687 and the *Opticks* in 1704), the pre-Newtonian *Traité* had established itself in British schools, largely through the Latin edition at London in 1682. Even at Cambridge, where Newton himself had lectured on the new ideas on the colors of the rainbow, Rohault's interpretation of Descartes long remained popular. Latin editions were supplemented by three English versions between 1723 and 1735.

Latin and English translations came from the pens of Samuel and John Clarke, the former of whom had been responsible also for translations of the *Opticks*. In an attempt to keep Rohault's classic up-to-date, John Clarke added extensive notes, and these were as often as not summaries of Newtonian ideas which contradicted the text itself. On the width of the rainbow, for example, the notes point out that the Rohault-Cartesian views are "a very great mistake," for the phenomenon of differential refraction was not known. (Newton published his discovery the year after the *Traité de Physique* first appeared.) Then, too, one finds among the notes almost the entire paper of Halley on multiple rainbows, as well as references to Newtonian gravitation. Changes in cosmology were more pronounced than in optics. When Voltaire in 1726 visited England, he wrote: "A Frenchman who arrives in London finds a great alteration in philosophy, as in other things. He left the world full, he finds it empty." [36] Thus, as Playfair later wrote, concerning Clarke's *Rohault*, "the Newtonian philosophy first entered the University of Cambridge under the protection of the Cartesians." At American schools much the same could be said,[37] for Clarke's versions were used as texts at Yale College as late as 1743.

The early story of the rainbow in America indicates that there was at first a considerable lag behind developments in Europe. There was, of course, no continuity between the visit of Harriot in 1585 and scientific work of a hundred years later. No traces remained of Raleigh's lost colony, in which Harriot had had a hand. The recorded history of the rainbow in America seems to begin during the period when Newtonianism was becoming a force in Europe. In 1685 Increase Mather, then acting president of Harvard College, invited Charles Morton (1627–1698), liberal founder of a "dissenting academy" in England, to come to the New World, ostensibly with the prospect of the latter's stepping into the presidency of Harvard. Morton had been educated at Cambridge and Oxford; and at the latter school he undoubtedly had come in contact, during the middle years of the century, with members of the "Invisible College," a group of scientists from which later sprang the Royal Society. In 1686 Morton settled in the Boston community, becoming pastor of John Harvard's church and teaching science at Harvard College. He had brought from England a manuscript copy of a work called *Compendium Physicae*; and during the time when Newton was publishing the *Principia*, students at Harvard were copying and recopying the notes of Morton. The *Compendium Physicae* became the adopted science textbook at Harvard, and it remained in the program of studies for some forty years. This work, which gave Harvard students their first glimpse of the "New Science," was not published at the time; but thirteen surviving manuscript copies bear witness to the important role it played in the curriculum.[38]

Morton shared the general Puritan belief that the study of natural philosophy, by enhancing admiration for God's handiwork, would stimulate piety; and hence science and theology were in a sense partners in a common enterprise. The *Compendium Physicae*, a collection of principles taken from Aristotelian, Baconian, Cartesian, and Newtonian science, consequently is spiced with moral overtones expressed in doggerel rhymes:

> Rainbow which does no native beauty want,
> Is more illustrious by Gods covenant.[39]

Morton's ideas on the rainbow are contained in a chapter, "Of Appearing Meteors," devoted to certain phenomena (including also halos, parhelia) which were categorized as meteors—incorrectly, he held, inasmuch as they are not "bodyes at all, but certain shews which bodyes by the help of light do Exhibit." The primitivity of his explanation of the bow contrasts sharply with other sections of the *Compendium*, for it shows no appreciation whatsoever of seventeenth-century contributions to the subject. It harks back instead to the conjecture of Anaxagoras.

> The rainbow therefore is nothing but a multiplyed reflection of the Sun from a dewy Cloud, or gently falling rain the Eye being placed in a direct line between the Suns body and the Center of the circle whereof the rainbow is a Segment . . . for the suns lustre is reflected to the Eye only from those points of the Cloud where the angles of incidence and reflection are Equall.[40]

The diagram accompanying this passage shows rays of the sun reflected at the concave surface of a circular cloud band to the eye of the observer. That the explanation is unintelligible seems not to be the fault of the student from whose notes this explanation is taken. Throughout the chapter one cannot avoid the impression that at least one Harvard professor of the time did not understand his own explanation. But this did not prevent him from deducing further consequences. To explain why one does not see in the rainbow many distinct Suns, he resorted to a form of Aristotelianism:

> Whereas indeed every little bubble or drop reflects the whole body of the Sun in little; the Multitude of which being Close set together seam to be but on[e] entire reflection.

Nor was Charles Morton much better at scientific observation than in theoretical explanation. He correctly noted that the rainbow could be a complete circle when he was on a hill which overlooked a long deep valley; but he followed this observation by the incorrect remark that when the cloud producing the bow is nearer to the eye, the bow is a segment of a larger circle than when the cloud is farther away, sententiously adding, "Because it makes an obtuser angle in the Eye (According to the rules of Opticks)." He was familiar with the ancient observation that the higher the sun, the lower is the arc of the rainbow; and for this reason the bow seldom is seen around noontime, except in winter. He added that, in winter, when the sun is low, then indeed a little bit of the rainbow may be seen in the north "Commonly Call'd the Dogs Eye in the N.) A Prognostick of cold stormy weather." Nor does Morton omit the more familiar applications of the rainbow to weather forecasting:

The Prognostication of Rainbow is rain if it appears in the Evening: because the nights cold will farther condense the Clouds; but in the morning it foreshews a fair day because the dayes heat will rarifye the Cloud according to the old Saying.

> " 'A Rainbow in the Eve; put thine head in the Sheave,
> A rainbow in the morrow; go take thy bow, and arrow.' "

A theologian, Morton could not well avoid the question of rainbows before the flood; and here the liberal viewpoint is espoused. "The Answer is Affirmative for the Natural cause of it was from the beginning." The Biblical passage therefore is interpreted to mean that from the time of Noah the bow also was to be a sign of God's covenant.

Morton was in England when in 1672 Newton's announcement was made of the composite nature of white light; but the significance of the paper for the theory of the rainbow seems to have made no impression on him. He taught his Harvard students a theory of color similar to that of Aristotle (except that yellow replaces green as a fundamental hue), citing Boyle as his authority.

Now that Colour of the Rainbow which is nearest the Centre and partakes more of light, or reflects it more fully to our Eye Gives the Suns Immage more red (purple or rutilus) that which is next to it may be conceived to be reflections from drops a little farther off; and Yellow is more languid than red; the utmost is blew which requires but a little light; and between the blew and Yellow is Green; which is a mixture of the blew and yellow, as we commonly see in a painters mixing of Colours.

In a later lecture "of Seeing," Morton goes further into the nature of color. The influence of Aristotle is strong here also, even though Morton includes references to whirling globules which must have come from Descartes, and to experiments with glass prisms which he ascribes to Boyle. There is even a hint of the Newtonian theory when Morton held that it is "past all doubt that Colour is made by regular reflections [of light] yet the distinct rule of reflection, and refraction proper to Each Colour is not yet determined." [41]

Morton's blundering account of the rainbow ventures to include also lunar and multiple bows. "The Lunar Rainbow is of the same nature and from like causes," he held; but he then went on to confuse this with the moon's halo. On multiple solar bows he accepted uncritically the old Clichtovean hypothesis:

If 2, or more of these bows appear the 2d is but the reflection of the first (as in the parrallius [parhelion]) and the 3d (more Languid) is the reflextion of the 2d etc: This appears by the Inversion of the Collours (according to the laws of Opticks) the Quite contrary way from that from which it was reflected.

Had he attended Newton's optical lectures more than a dozen years before, or had he read carefully the *Philosophical Transactions* of 1672, he would have found a very distinct rule for the refraction proper to each color; and he would have learned better than to teach such nonsense on the rainbow. And yet Morton was one of the best educated men of the time in the American colonies. He did not attain to the presidency of Harvard, but he was named vice-president in 1697; and he and the Mathers were members of a group which constituted a sort of Cambridge Scientific Club.[42] Cotton Mather (1663–1728) in a sense was the successor of Morton in the colonial story of the rainbow; and his advanced views on this topic are consonant with the fact that he was the first American to be elected (in 1713) to the Royal Society.[43] Cotton Mather had been educated at Harvard at a time when virtually no science was taught, but he corresponded with scientists in all parts of the world. Consequently he was far more *au courant* with recent science than Morton had been; and he made use of scientific topics in his sermons, many of which he published. (He was a prolific writer, the author of at least 444 works from 1682 to 1728.) In an anonymous polemic essay of 1711, *The Right Way to Shake Off a Viper*, he used the Newtonian theory of gravitation to symbolize the fact that "the further you fly towards *Heaven*, the more you must *Lessen*." A year later Mather published two essays, one of which he had delivered as a sermon, in which he chose the Newtonian explanation of the rainbow as his theme. These essays, *Thoughts for the Day of Rain*,[44] had been suggested by a friend; and Mather undertook them "from a perfect Concurrence I had with him, and many other Good men, in the Opinion, That among the *Engines of Piety*, wherewith our Good God has accomodated us, the *Rainbow* is one too much neglected with the Professors of our Holy Religion." In a preface to the essays Mather shows thorough familiarity with Cartesian and Newtonian work on the rainbow. He understood clearly the significance of the parallelism of rays in Descartes' explanation, both for the primary and the secondary arcs; but he nevertheless repeated Newton's depreciation of Descartes: "*Descartes*, (who don't use to betray his Tutors), took the Hints from *Antonius de Dominis*, and went on *Mathematically*, and with much Demonstration, to give us a Theory of the *Iris*." Evidently a regular reader of the *Philosophical Transactions*, Mather appreciated the way in which Halley had gone on "to cultivate the Subject with the Ingenuity proper to so accomplished a Gentleman." But his most extravagant admiration is reserved for Newton, the "Perpetual Dictator of the learned World . . . than whom, there has not yet shone among Mankind a more sagacious Reasoner upon the Laws of Nature." Mather in a brief paragraph correctly summarized Newton's account of the rainbow colors:

This rare Person, in his incomparable Treatise of Opticks, has yet further explained the Phenomena of the Rainbow; and has not only shown how the Bow is made, but how the Colours (whereof Antiquity made but Three) are formed; how the Rays do strike our Sense with the Colours, in the Order which is required by their Degrees of Refrangibility, in the Progress from the Inside of the Bow to the Outside: the Violet, the Indigo, the Blue, the Green, the Yellow, the Orange, and the Red.

In Mather's preface there is even a reference to Volkamer's Thaumantiadis, "which has not yet reached America," in which "the skilful Author lays together whatever is to be found upon this Argument, among the modern, as well as the antient Writers."

Mather's account of the theory of the rainbow is strikingly superior to that taught by Morton, not only in what is said, but also in appreciation of what remained unsaid; for the author closes his preface by pointing out that for a complete description of the phenomena it is necessary to understand algebra (which he refers to as "Oughtred's Characters"), trigonometry, and the doctrine of fluxions. Feeling that his printer "could not easily accomodate us, with the Schemes that would be needful for Dilucidation," and that very few "would be gratified with such things," he proceeded "unto more Theological and Agreeable Contemplations." Among the more agreeable contemplations with which the first essay, The Gospel of the Rainbow, begins, is an account of the names by which the bow is known in other languages. Then the author gives "the common definition" of it: celestial arc which is made by light from the sun striking upon a cloud, variously composed and tempered, in time of rain. To those educated by Morton, this must have had a more familiar ring than the sophisticated notions included in the preface. But Mather feared that even this bit of science might leave his hearers "in the clouds"; and so he hurried on to the account of the bow in Genesis and to pious meditations. He would "make some Arrows for the Bow of God, and make ready a Quiver of Good Thoughts, Which are to be shot from that Bow into the Minds of the Faithful, as often as it appears unto us." His first meditation is on the circumstances, both moral and physical, which led to the flood in which he estimated that, in terms of antediluvian longevity, 1,347,000,000 people must have perished. The second meditation concerned the graciousness of God as shown in the Covenant; and here he raised "that noted Question" of whether there were rainbows before the flood. Unlike Morton, he argued that there is "nothing in Scripture, no, nor in Nature neither," to assure us that there was a pre-Noachean bow. He felt that the catastrophic changes at the time, both in the abbreviation of human life and in atmospheric conditions, may confirm the fact that the bow made its first appearance after the flood. He admitted, however, that others, including Calvin himself, held that the rainbow was not a new thing, but that it simply was put by God to a new use, much as the prior existence of bread and wine does not prejudice their sacramental importance. Mather even goes so far as to cite others (including Aquinas, Cajetan, and Cardan) who believed that the Covenant itself has a natural basis: the appearance of a rainbow indicates that the clouds are too thin to cause a heavy downpour, and hence this is a token that the rains will not cause a flood.[45] The third meditation suggests that, just as the higher the sun the lower is the rainbow, so also the higher Christ is in us the lower will we appear to ourselves. The fourth meditation concerns the colors of the bow, and likens the blue with the waters of the flood, the green with God's patience to bring forth fruit, and the red with that fire which is to consume the world. The fifth and last meditation points out that the rainbow is not a perfect circle, a sign of the imperfection attending all sublunary things; but at the time of the resurrection the Saviour will communicate a glory beyond that of the rainbow.

The second of the two essays in The Gospel of the Rainbow, entitled, The Saviour with His Rainbow, contains only pious thoughts without anything of scientific interest. That Mather thought highly of his thoughts concerning the science of the bow is apparent from the fact that the preface to these essays of 1712 was taken over bodily as part of Essay XIII, "Of the Rainbow," in The Christian Philosopher: A Collection of the Best Discoveries in Nature, with Religious Improvements, which appeared in 1721. This is supplemented by notes on the halo which "is of so near kindred unto the Rainbow, that it claims a mention with it." He reports that Newton considered it to be caused by rays from the sun or moon which shine through a thin cloud consisting of globules of hail or water, all of the same size; and he gives also the theory of Huygens that it is formed by small round grains of hail with opaque centers, as well as the latter's modification of this theory for parhelia. The "Puritan Priest" closes the essay with a warning in characteristically vigorous language:

Though a watery Flood, which may drown the World, is no more to be feared; yet there is a fiery Flood, for the Depredations whereof, a miserable World is growing horribly Combustible.[46]

Mather's good understanding of the causes of the rainbow was but one indication of his familiarity with the scientific currents of his day. He took an active role in 1721 in advancing the practice of small-pox innoculation; and although he had not earned a medical degree, he has been called "the first significant figure in American medicine."[47] In botany he made important contributions, being acknowledged as the first to observe artificial hybridization in plants. Generally in advance of his age on scientific matters, Mather nevertheless was actively associated with the prosecution of witches.[48] When the delusion had subsided, his *Wonders of the Invisible World* (1693), with its firm belief in diabolical possession, made him vulnerable to attacks from many quarters. Failing in his desire to become president of Harvard, his influence turned toward Yale where, by a curious coincidence, there was, during these years, a young teacher and theologian, Jonathan Edwards (1703–1758), whose early scientific interests resembled those of Mather.

Students at Yale College had been warned in 1714 that the "New Philosophy," represented by the works of Descartes, Boyle, Locke, and Newton, would "corrupt the pure Religion of the Country." [49] In spite of this advice, the modern world view somehow seeped down to Jonathan from the time he entered as a boy of twelve until his graduation in 1720.[50] During these years the youth who later became known as the "Puritan Sage" was writing down notes on such themes as the ways of insects, the shape of the universe, the rumbling of thunder, and the causes of the rainbow. These *juvenilia* are not of lasting value, but they are fascinating for the bold originality of thought they represent. Young Edwards seems somewhere to have acquired an imperfect knowledge of Newton's discovery of the differing refrangibilities of vari-colored lights; but his explanation of the rainbow bears little resemblance to the Cartesian-Newtonian theory. He began with the cogent observation that the rainbow can not be a reflection of the sun's light from a cloud, as "it once was thought," for the bow can be seen among the trees close to the ground where there is no cloud, but only rain. He concluded that the rainbow is caused by "shining full on the drops of rain." This is why it is said that a rainbow in the east is a sign of fair weather, for then it must be clear in the west. And if one takes a large globular glass bottle of water, one can see that the reflection is from the concave and not the convex surface.[51] But there remains the more difficult question of why the reflection is circular. To answer this Edwards invented his own peculiar law of reflection:

If the reflecting body be perfectly reflexive, the angle of reflection will be the same as the angle of incidence; but if the body be not perfectly so, the angle will be less than the angle of incidence. By a body perfectly reflexive, I mean one that is so solid as perfectly to resist the stroke of the incident body and not to give way to it at all; and by an imperfectly reflexive, a body that gives way and does not obstinately resist the stroke of the incident body.

By "angle of inclination" Edwards meant the angle of inclination of the ray with respect to the reflecting surface, not the angle between the ray and the normal. His "law," assuming that the deviation from the rectilinear path is in proportion to the resisting power of the body, reminds one somewhat of Descartes' explanation of refraction; but Edwards gave no consistent quantitative treatment in his adaptation of this to reflection. His explanation of the rainbow consequently is ingenious, but it lacks precision, even though it is accompanied by a geometrical diagram. Drops of water being imperfectly reflexive bodies, rays are not uniformly reflected by its surface, some rays, in fact, not being reflected at all. Let the lines *ci*, *ab*, and *he* (Fig. 54) represent

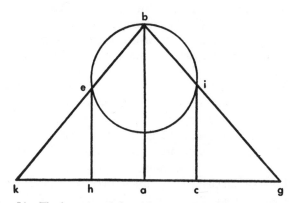

Fig. 54.—The formation of the rainbow according to Jonathan Edwards.

rays from the sun falling upon the spherical drop *ebi*. Then rays like *ab*, which fall directly on the concave rear surface, strike with such force that the surface is unable to reflect them. Others, such as *he*, after being refracted along the path *eb*, fall "with a far lighter stroke" and hence are reflected along *bg* to reach the eye of an observer. Only those rays whose degree of obliquity is adjusted to the refractive power are reflected by the rear surface and are reflected out at the sides of the drop. Hence those parts of the sky just opposite the sun (i.e., in the direction of the line *ab*) are dark, and those that are just the right amount to one side are bright. The drops must be just the right distance all around from the opposite point, and hence the bow is a portion of a circle.

Edwards closes his account with a cryptic remark:

The next grand question is what is it causes the colours of the rainbow and this question indeed is almost answered already for it is very evident.

Presumably this refers back to a statement with which the explanation opens in which the author asserted that his account "will be satisfactory to any body if they are fully satisfied of Sir Isaac Newton's different reflexibility and refrangibility of the rays of light."

It is impossible to determine to what extent the scientific writings of Jonathan Edwards were original. His explanation of the rainbow certainly is not a reproduction of any earlier extant theory; but it does bear some resemblance to that of De Dominis. In particular, his curious failure to mention the second refraction on the emergence of the ray from the drop makes one wonder if a copy of De Radiis Visus et Lucis somehow had found its way to America. Of any influence that Edwards himself might have had in science, little need be said. He has been called an "American Pascal," whose scientific ability "was nipped in the bud by a religion as lugubrious as that of Pascal." [52] But even had he not deserted science, it is not likely that he would have gone much further with work on the rainbow. Pascal was a mathematician; but when Edwards, long afterward, wrote to the trustees of the College of New Jersey (later Princeton University) apropos of his nomination for the presidency (which he accepted and filled for about a month before he died in 1758), he candidly warned them that he was "deficient in some parts of learning, particularly in algebra and the higher parts of mathematics."

The theory of the rainbow during the Newtonian Age had reached the point where no one untrained in advanced mathematics could hope to follow. In the field of poetry the change in attitude toward the rainbow was variously received. In England, some time later, on December 28, 1817 Benjamin Haydon described in his diary a dinner gathering at his home of famous poets of the day. Lamb and Keats agreed that Newton had destroyed all the poetry of the rainbow by reducing it to its prismatic colors, and the guests all drank a toast: "Newton's health, and confusion to mathematics." Not long afterward Keats composed the familiar lines in Lamia:

> Do not all charms fly
> At the mere touch of cold philosophy?
> There was an awful rainbow once in heaven:
> We know her woof, her texture; she is given
> In the dull catalogue of common things.
> Philosophy will clip an Angel's wings,
> Conquer all mysteries by rule and line,
> Empty the haunted air, and gnomed mine—
> Unweave a rainbow.

Wordsworth, too, had drunk the toast to the confusion of mathematics; but he may well have done so with mental reservations. In a description of the view of Cambridge from his window, given in Book III of *The Prelude*, he is not unsympathetic to science:

> And from my pillow, looking forth by light
> Of moon or favouring stars, I could behold
> The antechapel where the statue stood
> Of Newton with his prism and silent face,
> The marble index of a mind for ever
> Voyaging through strange seas of Thought, alone.

Shelley was not at the Haydon dinner; but had he been, he might have expressed a minority opinion on the beauty of science. Nor was he alone in adulation of his country's greatest scientist, for Pope composed the oft-quoted lines:

> Nature and Nature's laws lay hid in night;
> God said, "Let Newton be!" and all was light.

Newton himself had little interest in, or feeling for, aesthetics; and his prismatic discoveries met at first with little response in the field of poesy. At the time of Newton's death, however, interest of poets in the *Opticks* was pronounced. Eulogies and elogies of 1727 almost without exception refer to "Newton's rainbow"—in spite of Voltaire's expressed belief that Newtonian philosophy, because of its geometry, was no fit subject for versification. The austerity of the *Principia* was indeed forbidding to the layman; but it has been held that the *Opticks* gave back to poetry the beauty of light which Cartesianism had once taken away.

> Descartes thus, great Nature's wandering guide,
> Fallacious led philosophy aside,
> 'Till Newton rose, in orient beauty bright,
> He rose, and brought the world's dark laws to light.[53]

The eighteenth century became obsessed with light in all its aspects. Of it Moses Browne wrote:

> Thy Colours paint the Sky's ethereal Blue,
> And stain the lifted Rainbow's varying Hue,
> Diversify the Clouds with Tinctures gay,
> And thro' thy Depths is shed the golden Day.

Scientists to this day think of Descartes as the first one to explain the rainbow; but to the poets of the eighteenth century it was Newton and Newton alone who had unraveled its mysteries. And if some lamented that science stripped nature of her secrets, others rejoiced in the contrast between the swain who "runs to catch the falling glory" of the rainbow and the "enlightened few whose godlike minds philosophy exalts." The latter attitude was clear in James Thomson's *Spring*:

> Meantime, refracted from yon eastern cloud,
> Bestriding earth, the grand ethereal bow
> Shoots up immense; and every hue unfolds,
> In fair proportion running from the red
> To where the violet fades into the sky.
> Here, awful Newton, the dissolving clouds
> Form, fronting on the sun, thy showery prism;
> And to the sage-instructed eye unfold
> The various twine of light, by thee disclosed
> From the white mingling blaze.

And for Mark Akenside the rainbow took on greater beauty through the work of scientists:

> Nor even yet
> The melting rainbow's vernal-tinctur'd hues
> To me have shone so pleasing, as when first
> The hand of science pointed out the path
> In which the sun-beams gleaming from the west
> Fall on the watry cloud, whose darksome veil
> Involves the orient; and that trickling show'r
> Piercing thro' every crystalline convex
> Of clust'ring dew-drops to their flight oppos'd,
> Recoil at length where concave all behind
> Th' internal surface of each glassy orb
> Repells their forward passage into air;
> That thence direct they seek the radiant goal
> From which their course began; and, as they strike
> In diff'rent lines the gazer's obvious eye,
> Assume a diff'rent lustre, thro' the brede
> Of colours changing from the splendid rose
> To the pale violet's dejected hue.

And for some, including Christopher Smart, the new discoveries of science were a reminder of the Source of Light:

> Who shone supreme, who was himself the light
> Ere yet refraction learn'd her skill to paint,
> And bend athwart the clouds her beauteous bow.

Or, as William Blake quipped:

> That God's Colouring Newton does shew,
> And the devil is Black outline, all of us know.

X

A New Theory

NEWTON AND DESCARTES WOULD APPEAR TO HAVE HAD ALL THE ANSWERS TO the physical and mathematical problems of the rainbow—the shape, the radius, the breadth, and the formation of the colors. Descartes' boast, at the close of Les Météores—that those who understood what had been said no longer would see in the clouds anything of which they would not easily understand the cause—may have been a bit premature; but, bolstered by the complementary work of Newton, it appeared to be justified. Little wonder it is that for ninety-nine years after the Opticks appeared there was nothing of comparable significance in the story of the rainbow. It was generally assumed that the last word had been written. The eighteenth century in many respects was an unspectacular age, satisfied to smooth out the details in the great discoveries made by its predecessor, the "century of genius"; and this was especially true in the history of the theory of the rainbow. In fact, the theory appeared to be in such satisfactory shape that little refinement seemed to be necessary.

The tide of volumes on the rainbow ebbed noticeably after the 1699 Thaumantiados of Volkamer.[1] This was of course due in part to the availability of learned journals to which articles on the rainbow could be submitted. Then, too, the eighteenth century was noted for its multi-volume textbooks; and a Cours de Physique of the time naturally would carry at least a short explanation of the rainbow along Cartesian-Newtonian lines. Newtonianism had been popularized on the Continent by Voltaire, as well as by the numerous editions of the Opticks; and the texts of Rohault had given way to those of a new generation of writers, among whom Pieter van Musschenbroek (1692–1761), a Leyden professor, achieved considerable popularity. Musschenbroek's Essai de Physique of 1739 includes not only the Newtonian explanation (for, says the author, no one had given a better treatment of the rainbow than had Newton), but also the Newtonian prejudices. Musschenbroek praised the work of De Dominis, although he thought he had been anticipated "by some ancients." (Would that we knew who these were!) And he pointed out that Kepler and Harriot had somewhat the same thought. Then, in an evaluation which is grossly unfair, Musschenbroek wrote that "Descartes did nothing but follow the path of Antonio de Dominis, being satisfied to determine some

269

angles a little more neatly." [2] In Germany somewhat earlier (1716) Christian von Wolf had gone so far as to say that Descartes had merely republished the explanation by De Dominis; [3] but later in Italy the prominent Newtonian, Ruggero Boscovich (1711–1787), pointed out the unfairness of such statements. In 1747 Boscovich published extensive notes, partly historical and partly expository, as a supplement to a poem, *De Iride et Aurora Boreali Carmina*, by Carlo Noceti, a fellow Jesuit. These notes included explanations of the bow in the manner of Descartes and Newton.

A defense of Descartes and full accounts of the explanation of the rainbow are found in two of the best known historical works of the eighteenth century, Étienne Montucla's *Histoire des Mathématiques* (1758), and Joseph Priestley's *The History and Present State of Discoveries Relating to Vision, Light and Colours* (1772). Montucla's account is characterized by a particularly full analysis of the work of Halley and Bernoulli on multiple rainbows.[4] At just about the time of this work a still more extensive treatment of multiple rainbows by Simon Kotelnikow appeared in the *Novi Commentarii* of the St. Petersburg Academy, showing that the Cartesian-Newtonian theory was familiar throughout Europe. Kotelnikov first derived a formula, analogous to those of Halley and Bernoulli; and from this he computed the radii, both for red and for violet rays, of the rainbows of the first eight orders. From the differences in the radii for red and violet rays, he found the width of these eight rainbow bands, thus extending the calculations which Newton had carried out for the first two bows. His values showed that the fifth and sixth bows are opposite the sun; and that the width of the bands increases steadily with the order. Inasmuch as data on these multiple bows are not readily found in books, the table of Kotelnikov is given below in full.[5] It will be noted

	red	violet	width	order of colors
1.	42° 1′ 52″	40° 15′ 28″	1° 46′ 24″	r/v
2.	50° 58′ 44″	54° 9′ 36″	3° 10′ 52″	v/r
3.	138° 35′ 12″	142° 51′ 4″	4° 27′ 52″	v/r
4.	136° 6′ 16″	130° 25′ 50″	5° 40′ 26″	r/v
5.	51° 37′ 4″	44° 38′ 24″	6° 58′ 40″	r/v
6.	32° 28′ 9″	40° 46′ 44″	8° 18′ 35″	v/r
7.	116° 12′ 20″	125° 33′ 14″	9° 20′ 44″	v/r
8.	158° 7′ 53″	169° 34′ 16″	11° 26′ 23″	v/r

in this table that the radius is given in terms of the angular distance from the point opposite to the sun (the antisolar point), so that rainbows of the "second kind" have radii between 90° and 180°. The precision implied in this table is, of course, specious, for Kotelnikov's data did not entitle him to give the angles to the nearest second.

In some quarters during the middle of the eighteenth century there apparently was a popular belief that the nearer the rain producing the bow is to the observer, the smaller will be the circle of the bow which is seen. Kotelnikov pointed out that this view, just the opposite of that which Morton had taught (incorrectly) at Harvard, is a fallacy. He also called attention to the fact that, for arcs of the second kind (i.e., bows seen in the direction of the sun), the higher the sun is, the higher is the arc—just the reverse of what the ancients had noted for rainbows of the first kind. Some of the confusion in rainbow terminology is apparent in his criticism of Musschenbroek, whom he understood to say that the tertiary bow appears opposite the sun after three reflections. Kotelnikov apparently preferred to call "tertiary," not the bow corresponding to three internal reflections, but five, inasmuch as this is next in order, following upon the secondary, in a direction opposite to the sun.[6] He believed that bows of type two had not been seen in nature, although he held that under favorable conditions, with a suitably dark background, they might be visible. By a strange coincidence, two observations of the rainbow corresponding to three reflections were reported by the scientist Bergmann in Sweden on September 3rd and 5th, 1759, at virtually the same time that Kotelnikov was presenting his paper to the Russian Academy.[7] Although Bergmann's reported observations were questioned, there is no doubt that the third rainbow arc has, very occasionally, been seen in nature. In particular, a clear observation of this elusive bow was made on the afternoon of September 4, 1878 by Heilermann, on looking out the west window of the train while riding north to Cologne.[8] The quaternary rainbow arc has not been known to appear in nature; but the scientist Mascart believed that he sometimes had seen the fifth bow. The quintic bow very nearly coincides with the secondary, but with colors arranged in the opposite order, and this contributes to the difficulty of observation.[9] The sixth lies within the arc of the primary bow, and has a color order similar to that of the secondary; but it never has been observed out of doors. We shall see that in the laboratory, under well controlled conditions, more than a dozen-and-a-half rainbows have been observed,[10] all confirming the Cartesian theory.

Confidence in the Cartesian-Newtonian theory of the rainbow remained high throughout most of the eighteenth century. In 1772 two candidates for the doctorate in theology at Tübingen, L. T. von Spittler and C. M. T. Breunlin, publicly defended a *Dissertatio Physica de Iride* along the usual lines. They proposed, in clarifying the explanation in Segner's *Physica*, to answer three principal questions: how the seven colors arise, why the rainbow is the size that it is, and what accounts for the width of the bow. The exposition is entirely in Cartesian and Newtonian terms, with reference being made also to the calculations of Jean Bernoulli. The computations are based on Newton's indices of refraction for red and violet rays; and, with exaggerated confidence in the precision which the data warranted, the authors carry out the work to a greater number of figures. The radii of the primary bows for red and violet are given as 42° 1′ 44″ and 40° 16′ 10″ respectively, differing slightly from those of Kotelnikov. Similar computations are completed for the secondary rainbow also, the non-colored interval between the inner and outer bows being placed at 8° 56′ 54″. The authors raised again the question which disturbed "Biblical physicists" as to whether or not the iris existed before the flood. Their affirmative answer is based upon the assertion that even if there were no rain prior to the flood, the rainbow would have been observed in mists.

For at least two thousand years men had been more interested in the theory of the rainbow than in careful observations; but toward the end of the seventeenth century, and especially in the eighteenth century, with the theory assured, attention turned to reports of unusual forms of the rainbow. The *Philosophical Transactions* for the very first year, 1666, carried an account of a strange bow seen near the river Chartres on August 10, 1665. Rising from between the feet of the principal bow was a portion of another arch which, rising higher than the principal bow, intersected the latter. On other occasions anomalous bows were observed to intersect both the primary and secondary arcs. Such bows were observed in 1698 by Halley and in 1743 by Anders Celsius (1701–1744), founder of the centigrade scale of temperatures. Some-

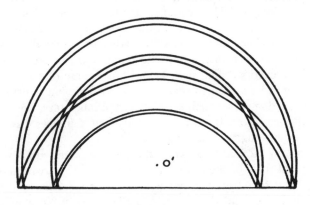

Fig. 55.—One form of rainbow seen by reflection. O is the center of the ordinary bows, O′ of the reflected bows.

times, when the sun is low, the intruder arc lies between the two ordinary bows, a combination observed in 1685 by one Wolferdus Senguerdius; again it may form with the primary bow a figure like a lune or crescent. It was easily recognized that these adventitious rainbows were not in conflict with the Cartesian theory. They invariably were observed near bodies of water (and hence sometimes are called "water rainbows"), and they are explained as "reflection rainbows." By means of simple geometry one can readily determine the altitude h of the reflection bow in terms of the angular altitudes h' and h'' of the sun and the primary rainbow respectively. Let the rays from the sun S be reflected from the surface of a body of water WW' at point R, and let them follow the usual path (but with the deviation in contrary sense) through the drop of rain, ABC, finally reaching the eye of the observer at point O (Fig. 56). Let P be the intersection of the lines RA and CO. The angle CPA

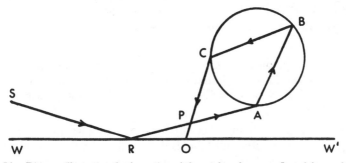

Fig. 56.—Diagram illustrating the formation of the rainbow by rays reflected from a horizontal surface.

is known, from the observation of Roger Bacon, to be 42°; and hence angle OPR is also 42°, because vertical angles are equal. If angle WRS, the altitude of the sun, is h', then angle ARW' is also h', because of the law of reflection. But angle POW', the altitude h of the reflection bow, is an exterior angle of triangle ROP, and hence it is equal to the sum of the two remote interior angles. That is, $h = h' + 42°$, or, because $h' + h'' = 42°$, $h = 2h' + h''$. In other words, the difference between the highest points of the reflection bow and the primary bow, $h - h''$, is twice the altitude h' of the sun. When the sun is at the horizon, the reflection bow coincides with the primary arc; and as the sun rises, or, in other words, as the primary bow sinks, the reflection rainbow rises until it reaches an altitude of 84°, and becomes a complete circle of radius 42°, as the primary bow sinks below the horizon.[11] If the surface of the reflecting body of water is disturbed gently, the reflection bow may become simply two colored vertical bands tangent to the extremities of the ordinary bow, being the envelope of arcs of varying heights having centers on a vertical line. Reflection rainbows corresponding to the secondary arc are of course also possible, the relationship being analogous to that given above in connection with the primary rainbow.

One should be careful not to confuse "reflection rainbows" with the reflection of the primary bow. The reflection bow may, in fact, be visible when the sun is too high in the sky to produce a primary bow visible to the observer. The rainbow is not a "thing" in the sense that it has objective existence. One may not speak of its image in the usual sense; and yet the bow can be produced also by rays which are reflected from the surface of a body of water *after* having passed through the raindrops, just as it is by rays reflected at such a surface *before* they traverse the drops.

Rainbows by reflection raised again the query as to whether the ordinary rainbow itself can be reflected. Here all the old arguments about the subjectivity or objectivity of the bow crop up once more. A century ago Augustus De Morgan, in his charming *Budget of Paradoxes*, raised the following problem:

A few years ago an artist exhibited a picture with a rainbow and its apparent reflection. . . . Some started the idea that there could be no reflection of a rainbow; they were right; they inferred that the artist had made a mistake; they were wrong.[12]

Artists have indeed made mistakes in portraying rainbows. In one of Rubens' paintings the bow, mostly dull blue, is darker than the sky, and the scene is lighted from the side of the bow; in reality the bow is brighter than the background, and the light comes from the sun directly opposite the bow. But De Morgan's artist was not in error. The answer to the paradox is easy if one phrases the question carefully. If one asks whether "it" can be reflected, the answer is, "Certainly not"; there is no material object to be reflected. If one inquires whether rays producing the bow for a given individual can be reflected after leaving the drop and before reaching the eye, the answer is, "Of course." It is indeed possible, therefore, to see a rainbow in the sky and a rainbow by reflection (a "mirror-rainbow") in a smooth surface of water at the same time. Nevertheless, the latter is *not* the image of the former in the usual sense; the drops which, for any observer, produce the bow in the sky are *not* those which cause the bow seen in the water. This is made clear in the accompanying diagram (Fig. 57). Let the eye of the observer be at point O and let the bow be seen directly on the cloud at S. Were the bow an *object* viewed in the surface of the water WW', the image would be seen along the line OP, where angle SPW is equal, by the law of reflection, to the angle OPW'. Actually, the reflected bow is not seen along this line, but along the line OP',

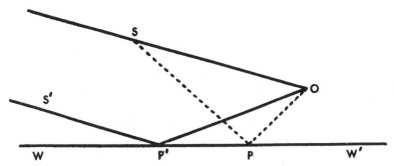

FIG. 57.—The direction in which a rainbow is seen by reflection.

where the angle OP'W' is equal, by the law of reflection, to the angle S'P'W, and S'P' is parallel to SO. A reflection bow almost always is traceable to the agency of a body of water, but occasionally light reflected from a cloud can serve as the source of illumination. An appreciation of the wide variety in anomalous rainbow forms should lead to more understanding judgments on the ancient and medieval failures in the story of the rainbow.

The rainbow invariably is thought of as formed in descending raindrops; but the Cartesian-Newtonian theory applies also to what are known as dew-bows or horizontal rainbows. The French call these arcs-en-terre or arcs-en-mer, because of the places where they are seen; Swiss and German writers often refer to them as Herbstiris, because of the season of the year during which they most frequently appear. Just above the quiet surface of a soot-covered lake, a sheet of ice, or a dewy meadow, there may be a layer of tiny droplets which reflect and refract sunlight in precisely the same manner as do falling drops of rain. Cones of rays with vertex at the eye and having the very same angles as those in the ordinary rainbow, produce colored arcs. The rays have been reflected from a smooth surface either before (as is more usual) or after traversing the drops. Now the shape of the rainbow is determined not only by the cone of rays but also by the position of the background upon which the eye sees the rays projected.

Most frequently, in the case of rainbows, the background is a nearly vertical sheet of rain or cloud, and the arc appears to be practically circular. It must be remembered, however, that only those sections cut off from a right circular cone by a plane are circular for which the plane is perpendicular to the axis of the cone. If the plane is tilted in any direction from the position of perpendicularity, the section will not be a circle but a more general figure called a conic. The theory of conic sections had been developed in antiquity by a number of geometers, including Euclid and Apollonius, with no thought of applicability to science. Kepler had seen that these curves immeasurably simplified astronomy, but long before his time it was appreciated that the explanation of the rainbow depended upon the study of sections of a cone. Ordinarily the ends of the rainbow appear to be equidistant, but it had been noted that sometimes one end is much nearer than the other. This fact is easily understood if one recalls that the background of the bow need not be a plane perpendicular to the axis of the cone; it need not be a plane at all. In the case of the dew-bow the section is made by a plane—a horizontal plane. Menaechmus, tutor of Alexander the Great, had discovered the conics, and he had noted that they are of three types. To these Apollonius later gave the names ellipse, parabola, and hyperbola, according as the cutting plane intersects all of the elements (or lines) of the single-napped conical surface, all but one of the elements, or only some of the elements. Inasmuch as the axis of the cone of effective rays producing the rainbow usually is inclined to the horizontal, and the sheet of raindrops causing the bow is approximately vertical, the rainbow drops generally lie not on a circle but in an arc of an ellipse with major axis vertical. In the case of dew-bows the horizontal plane generally cuts the visual cone of rays in a hyperbola. The observer may, in fact, see several adjacent hyperbolic rainbow arcs, corresponding to the primary and secondary bows and one or more bows by reflection.[13] That there can be elliptic, parabolic, and hyperbolic rainbows formed artificially in water sprays became clear from the ancient and modern study of the conic sections, and in 1772 this was pointed out by von Spittler and Breunlin.[14]

The eighteenth century was very fond of unusual rainbow forms, reporting these with considerable regularity in the periodical literature. Sometimes striking modifications in the colors were noted.[15] When the bow is formed by rays from the rising or setting sun, all of the colors tend to fade out except the red. With the sun on, or even somewhat below, the horizon, the result may be an all-red rainbow. It was understood that such a "twilight rainbow" was due to the fact that the atmosphere somehow had screened out all but the red rays before the sunlight had entered the raindrops. Only that portion of the sun's light remained which had, according to the Newtonian theory, the smallest degree of refrangibility. Just why nature allowed these to remain was not quite clear; but there was another unusual type of rainbow which presented more ominous problems.

Theodoric in the fourteenth century had called attention to white rainbows produced in a fine mist. Such colorless bows are observed especially in northern latitudes on mountains and along seashores. Scotch mist is a medium eminently suited to the reproduction of these "fog bows." The brilliantly white bow has margins somewhat tinted, the outer with yellow and orange-red, the inner with bluish violet. Observations of fog bows were made during the seventeenth century by Mariotte, Menzelius, and others; and with improved scientific communication in the eighteenth century, numerous further instances of white bows were reported. One account by Antonio de Ulloa, who in 1748 reported on a white rainbow observed on Mount Pambamarca in Peru while he was on a scientific expedition, achieved such publicity that a white bow sometimes is called "circle of Ulloa." There seemed to be no satisfactory explanation for the absence of color. It seemed somehow to be related to the size of the drops; and yet under the Cartesian theory, drop size should play no role whatever. One suggested explanation for the white bow was that the minuteness of the drop produces a feebleness of the light, the result being a lack of color; but this was too Aristotelian. The geometry of a small circle is not different from that of a larger; and Descartes had categorically asserted that whether the drops "are larger or smaller, the appearance of the bow is not changed in any way." But the experiences of the eighteenth century showed that the bow does indeed change in many ways. Rainbows are almost as different from each other as are trees. Frequently no blue is seen, even when the sun is not near the horizon; and sometimes there is little or no red. Nor is the intensity of the colors uniform throughout the entire periphery. Why should the colors be more brilliant toward the ends than in the center? The Cartesian-Newtonian theory included no provision with respect to the altitude of the drops. Even the radius of the rainbow, which, according to Descartes, should be invariable for a given index of refraction, was found to vary in some unaccounted-for manner. The radius of the primary rainbow often was found to be smaller, that of the secondary larger, than the value calculated according to the Cartesian rules. And the width, which Newton had explained with such confidence, was subject to unexplained variations. But more embarrassing than all of these vagaries were the increasingly frequent reports of bows seen within the primary arc.

Witelo had remarked on these in the thirteenth century, followed by Theodoric in the fourteenth; and Mariotte and Lahire in the seventeenth and early eighteenth centuries had described with precision three or four arcs adjacent to the inside of the primary rainbow. Were these true rainbows? The term had not been specifically defined, but it was tacitly assumed that a rainbow was an arc of spectral colors formed by light from the sun (or, more rarely, the moon) which has been refracted and reflected within drops of rain or mist. Under this definition the supernumerary arcs were indeed rainbows, yet they did not fit into the classical theory. An account of these arcs by Benjamin Langwith (1684?–1743), Rector of Petworth, in the *Philosophical Transactions* for 1722 was cited by others so frequently that their discovery came to be ascribed to him. In letters to the editor he wrote unusually perceptive descriptions of four rainbows, all different, which he had recently observed. One of these was as follows:

The first series of Colours was as usual, only the Purple had a far greater Mixture of red in it, than I have ever seen in the prismatick Purple: Under this was a colour'd Arch, in which the green was so predominant, that I could not distinguish either the yellow or the blue: Still lower was an Arch of purple, like the former, highly saturate with red, under which I cou'd not distinguish any more Colours.[16]

Langwith correctly suspected that there was more to the rainbow than had been realized.

I begin now to imagine, that the Rainbow seldom appears very lively without something of this Nature, and that the suppos'd exact Agreement between the Colours of the Rainbow and those of the Prism, is the reason that it has been so little observed.

His description of the fourth instance included the color sequence for the primary bow and three spurious arcs as follows:

    I. Red, Orange, Yellow, Green, Light Blue, Deep Blue, Purple.
    II. Light Green, Dark Green, Purple.
    III. Green, Purple.
    IV. Green, Faint Vanishing Purple.

Langwith called attention in particular to a curious fact not previously reported:

I have never observ'd these inner Orders of Colours in the lower Parts of the Rainbow, tho' they have often been incomparably more vivid than the upper Parts, under which the Colours have appear'd. I have taken notice of this so very often, that I can hardly look upon it to be accidental, and if it should prove true in general, it will bring the disquisition into a narrow compass; for it will shew that this Effect depends upon some Property, which the Drops retain, whilst they are in the upper part of the Air, but lose as they come lower." [17]

Fearing that his curious observations would not be credited, Langwith felt it incumbent upon him to point out that he had had five other witnesses, one of whom was a clergyman!

Langwith's conjecture that the properties of drops varied with altitude was a penetrating one, but he failed to carry it far enough. Henry Pemberton (1694–1771), an Oxford professor of medicine who helped to prepare the second edition of Newton's *Principia*, tried unsuccessfully to account for the vagaries of the rainbow in terms of irregular reflection at the rear of the drops; but one of his ideas, presented also in the *Philosophical Transactions* for 1722, was particularly prescient.

The precise Distances between the principal Arch of each respective Colour and these fainter correspondent Arches depend on the Magnitude of the Drops of Rain. In particular, the smallest Drops will make the secondary Arches of each Species at the greatest Distance from their respective principal, and from each other. Whence, as the Drops of Rain increase in falling, these Arches near the Horizon by their great Nearness to their respective principal Arches become invisible.[18]

From 1722 on, the existence of the superfluous bows had to be seriously reckoned with. They acquired a name—supernumerary arcs or, less frequently, spurious rainbows. (In France they are called "supplementary rainbows"; and in Germany they are known as "secondary rainbows," although they are not, of course, to be confused with what in English is known as the secondary rainbow.) It was clear that the supernumerary arcs were not multiple rainbows in the sense of arising from multiplied reflections within the drops. They were much too clear to be thus accounted for. Then, too, the colors of all of these arcs were arranged in the same order as in the primary bow, and their radii did not correspond to those of the "geometric rainbows" which Halley had calculated. Pemberton in the Philosophical Transactions for 1723 compared them, prophetically, to Newton's rings; but this did not help too much because the phenomenon of the rings itself was not really understood. Toward the middle of the century observations of the supernumerary bows were reported with great regularity: in 1742 by Pierre Bouguer (1698–1758), who would dismiss them lightly as subjective phenomena, by Daval in 1748, Musschenbroek in 1751 and 1755, and Le Gentil in 1756. The problem of these arcs became a part of the textbook literature, for Musschenbroek's Cours de Physique included an attempted explanation. (The author did not claim it as his own, saying simply that "physicists" thus explained them.) As the width of the rainbow band before Newton had been somewhat vaguely ascribed to the width of the sun, so now the supernumerary bows were attributed to the fact that rays emerging from neighboring drops are not exactly parallel, but converge slightly. An accompanying diagram (Fig. 58) is supposed to clarify what was intended, but, lacking any quantitative basis, it is not convincing.[19]

Boscovich had been among those who had noted three extraordinary rainbows lying within the primary and with unusually bright colors. Spittler and Breunlin, calling attention to Boscovich's observation, dismissed it with the facile Cartesian-Huygenian explanation that the interior bands were due to an admixture of sulphurous particles which refract rays differently from water. They boasted that, given the sizes of the bows, it would not be difficult to calculate the refractive property of the material in which they were generated. (Fortunately for them, they did not promise to isolate and identify the drops producing the adventitious bows!) A more sophisticated explanation was hinted at in 1787 by another relatively obscure writer. Somehow Andrea Com-

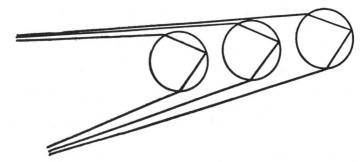

Fig. 58.—Musschenbroek's idea of the formation of the supernumerary rainbow arcs.

paretti (1745–1802) had come to realize the importance of the phenomenon of diffraction, largely overlooked since the time of Grimaldi, and he published a book on the subject: *Observationes Opticae de Luce Inflexa et Coloribus*. In this volume, after describing a rainbow with supernumerary arcs, the author asserted that the principle of inflection (his name for diffraction) suffices to "construct a complete theory of the rainbow." [20] His report that the Royal Society of Sciences of Montpellier had proposed such a project betrays that the late eighteenth century did not feel entirely comfortable about the theory of the rainbow. It was becoming increasingly evident that Descartes and Newton had accounted only for its broader features. But although Comparetti recognized the shortcomings in the classical explanation, he did not clarify his ideas by furnishing a "complete theory."

Attempts to account for the supernumerary rainbows took many forms. One suggestion was that there were rings like waves around the upper half of the drops [21] or that the drops were subject to oscillations. (It is indeed true that thunder affects the colors of the rainbow, a phenomenon not well understood even now; but this does not explain the spurious rainbows.) Another type of theory fixed on the shape of the raindrops: some believed that the drops are spheroidal or elliptical, others that they are hollow, like bladders or vesicles. In some quarters, resort was even had to the stultifying thesis, reminding one of medieval controversies on the reality of the rainbow colors, that the inner bands were optical illusions. [22] All such ad hoc assumptions failed to meet with approval, although some of them persisted well into the following century.

Typical of the mixture of optimism and uncertainty which are apparent in late eighteenth century writing on the rainbow is an article, "Arc-en-ciel" in the *Encyclopédie Méthodique*, an ambitious undertaking in 190 volumes which appeared at intervals from before the Revolution until after the fall of Napoleon. [23] The writer of the article held that we do not know why a stone falls, but we do know the cause of the colors of the rainbow, although the latter is more surprising; and he concludes from this that the study of nature is fitting to make us proud on the one side and to humble us on the other. But the article goes on to give, by implication, reason for humility even in connection with the rainbow. After historical remarks, an explanation of the Cartesian-Newtonian theory, and a description of unusual bows, the article turns to the supernumerary arcs.

How to explain, in the Newtonian theory of the rainbow, the 4 or 5 colored arcs which appear within the interior arc and which are concentric and contiguous to it? This is the most delicate point in the theory of the rainbow and no physicist seems to have tried to explain the production of these arcs.[24]

In this statement there is rhetorical exaggeration, for some attention had been given by physicists to the troublesome arcs. The author of the article himself presents "some conjectures" on the spurious bows, ascribing these to Flaugergnes. The conjectures remind one of the suggestions of Musschenbroek, but they are considerably elaborated. Flaugergnes divided into three categories the rays from the sun which had been refracted upon entering the drop: those which come to a focus (a) before, (b) upon, or (c) after reaching the rear surface of the drop. He argued that, after the rays have been reflected at the back of the drop and have again been refracted upon leaving the front surface, rays in category (a) will diverge, those in (b) will be parallel to each other. Rays of type (a) have no effect on the eye, and those of type (b), according to the theory of Descartes, produce the principal bows. It is some of the rays of the last type, category (c), he thought, which cause the spurious bows. If the arc between the incident rays on the rear concave surface is less than half of the arc between the corresponding incident rays on the convex surface, he said, the rays in category (c) will converge after leaving the drop, if greater, the emergent rays in (c) will diverge. The degree of convergence depends on the ratio of the arcs; and those which converge too soon will diverge before reaching the eye. Some will converge more slowly and reach the eye while still converging; and these, Flaugergnes held, will still be effective. But convergence depends also on the distance from which the rays emerge; and the weaker and narrower supernumerary arcs come from drops farther away. That is, he seems to have believed (incorrectly) that a single drop contributes to a single spurious bow. He was, however, right in saying that there is an overlapping of the colors of successive supernumerary bows; and he was quite correct also in his assertion that, at least potentially, the rainbows of second, third, and higher orders are accompanied by supernumerary arcs, thus extending the conjecture Theodoric had made in the fourteenth century in connection with the secondary bow.

The author of the unsigned rainbow article in the *Encyclopédie Méthodique* included reference to an aspect of the story of the rainbow which had been generally neglected. One might call this the physiology of the rainbow, to distinguish it from the geometrical theory of Descartes and the physical theory of Newton. The article raised the question of why the *arc-en-terre* is almost never seen, despite the fact that the sun shines on the dewdrops of the prairie far more frequently than on falling raindrops; and the response is an anticipation of the principle of the modern motion picture. The eye has to a high degree the faculty of conserving the sensation caused by an exterior object; and the impression made by one drop, too slight by itself to produce a sensation, remains while another drop takes its place and makes another superimposed impression. The author of the article conjectured that in this way perhaps five or six drops make virtually simultaneous impressions, which does not happen in the case of the immobile dewdrops. But the first substantial studies in the physiology of color, as well as the first credible explanation of the supernumerary rainbows, came in the first part of the next century, in the work of Thomas Young (1773–1829).

It is curious to note that Huygens, who had championed a wave of light theory, had not seen any significant periodicity, while Newton, who proposed the opposing corpuscular theory, actually had measured the frequency of vibration of his "fits." (The values are fantastically high. Red light, for example, impinges upon the eye almost 400,000,000,000,000 times in a second, violet more than 750,000,000,000,000 times.) But the true significance of periodicity was not appreciated until, as the nineteenth century opened, Young pondered the difficulties in the traditional explanations of the rainbow. The result was not only a new era in the story of the rainbow, but also one of the most important chapters in the theory of light.

Young, like Leonardo da Vinci, was a remarkably versatile scholar. His epitaph reports that "he first penetrated the obscurity which had veiled for ages the hieroglyphics of Egypt"; but his important work on the Rosetta Stone and the deciphering of the ancient hieroglyphics (anticipating that of Champollion) shows only one facet of a brilliant career.[25] His work in medicine and science later led the physiologist Helmholtz to say of him, "His was one of the most profound minds that the world has ever seen." And it was the profundity and flexibility of his thought that led Young to the solution of the problem of the supernumerary rainbows. Descartes and his successors had assumed, naturally, that the conjunction of many rays in the neighborhood of his effective ray necessarily would augment the quantity of light; but Young made the important discovery that light plus light does not necessarily mean more light. This had been anticipated in part by Grimaldi's observation of the partial quenching of light when two beams under certain circumstances fall on the same spot,[26] a phenomenon not understood at the time. Young showed that this is bound up with the very nature of light. If light were corpuscular, its motion would be the algebraic sum of the motions of its particles. Twice as many light corpuscles travelling in the same direction would mean twice as much light. Young knew that this was not true of sound, which is a wave motion—a propagation of form rather than substance. Aristotle had known that sound is a disturbance carried by the air, and the Roman architect Vitruvius had compared waves of sound to waves of water. Newton himself had been aware of the principle of interference as applied to lunar and solar effects in the theory of the tides; [27] and he had called attention to analogies between light and sound. He also had repeated some of Grimaldi's shadow experiments, and he had admitted that, in the wave theory, the phenomena of the rings could be accounted for on the assumption that "the Vibrations which make *Blew* and *Violet*, are supposed shorter than those which make *Red* and *Yellow*." [28] But, obsessed by the fact that Huygens never had adequately explained the rectilinear propagation of light, Newton had rejected the better half of the undulatory hypothesis and had never thought of an optical principle of interference. And Huygens, the exponent of the wave theory, had denied periodicity to his pulses precisely *because* he had wished to avoid the interference which he believed did not take place. Thus it was that Young could honestly say about his great discovery:

There was nothing that could have led to it in any author with whom I am acquainted, except some imperfect hints . . . in the works of the great Dr. Hooke . . . and except for the Newtonian explanation of the tides.

And as Young recounts the way in which he was led to invent the interference of waves, it was not through the influence of Huygens, the undulatorian, but under the stimulus of Newton, the corpuscularian:

It was in May 1801 that I discovered, by reflection on the beautiful experiments of Newton, a law which appears to me to account for a greater variety of interesting phenomena than any other optical principle that has yet been made known.[29]

Presumably the "beautiful experiments" to which he referred are those on the rings of color in thin plates. It is odd that the similarity of these rings to the rainbow, which Pemberton had noted in 1723, had not led to a persistent search for a causal relationship. Many of the eighteenth-century observations on the supernumerary arcs call to mind remarks Newton had made on the rings. The similarity in the order of the colors, the regularity in the spacing, the narrowing breadths of the successive bands, the diminishing intensity in the illumination: these apply so well to the diagram of the rings taken from Newton's *Opticks* that Young's simultaneous solution of the two problems occasions little surprise. But the solution required a mind sufficiently imaginative to reject the generally accepted concept of light as made up of a stream of corpuscles. The wave theory had not entirely disappeared during the eighteenth century, but it had few proponents. One of these was Leonhard Euler (1707–1783), who was none too successful in explaining Newton's rings in terms of a periodic vibration.[30] Benjamin Franklin (1706–1790) was another who upheld the undulatory hypothesis (and who, incidentally, admitted to having observed a lunar rainbow [31]); but neither he nor Euler answered satisfactorily the objections Newton had raised. In 1797 Ambrogio Fusinieri leaned toward the idea of light as an "intestine motion"; [32] but not until Young made his discovery of 1801 was the wave theory really reborn. Young, a physician, had made a study of voice, including the production and propagation of sound, and this made him conscious of vibration theories. But the example which Young used to illustrate the manner in which two beams of light would interfere destructively was that of equal water waves moving with equal speed on a lake and arriving at a certain channel exactly out of phase; and later he compared optical interference with the Newtonian theory of the tides.

From the time of Young's publication of his discovery in the *Philosophical Transactions* for 1802 the corpuscular doctrine was thrown on the defensive, for the interference of light waves afforded a far simpler and more plausible explanation of the ring phenomenon than the Newtonian theory of "fits." Young described his principle as follows:

The law is, that wherever two points of the same light arrive at the eye by different routes, either exactly or very nearly in the same direction, the light becomes most intense when the difference of the routes is any multiple of a certain length, and least intense in the intermediate state of the interfering portions; and this length is different for light of different colours.[33]

Descartes had thought color was determined by a speed of rotation of particles, Boyle had regarded color as determined by the structure of atoms in the surface of a medium, and Newton had thought of color as an indication of the size of particles; but Young subscribed to the theory that each color corresponded to waves of light of a specific length and periodicity. And while Newton had believed that the action producing the rings took place at one of the two surfaces, Young saw that it was the result of two reflections, one at each of the two surfaces. Rays reflected from one of the surfaces had to travel slightly farther, in reaching the eye of the observer, than those reflected from the other surface; and hence the conditions expressed in his "law" were satisfied. Where the difference in distance traversed is an integral multiple of the wave length, the two reflected beams reinforce each other; when the difference in distance is such that the two reflected beams are out of phase, destructive interference takes place. At first the dark spot in the center of the rings puzzled Young; but this difficulty was cleared up by assuming that reflection introduced a change of phase equal to half a wave length.

The discovery of optical interference unlocked one of nature's best-kept secrets—the cause of the supernumerary rainbows. Young saw that for each angle of incidence upon a raindrop greater than that of the Cartesian effective ray, there is another of smaller angle such that the two rays emerge from the drop in parallel or nearly parallel paths. As a matter of fact, it can be shown that these two rays are reflected at the same point on the rear surface of the drop. These two rays, being deviated more than the Cartesian ray, will appear *inside* the primary bow. (For the secondary rainbow they appear *outside* the bow.) From the accompanying diagram (Fig. 59) it is clear that the rays on traversing the drop will have followed paths which are not quite equal in distance, and so they are subject to Young's "law." For certain pairs of paths, the two red rays will augment one another, while the blues will tend to cancel each other; for other paths the opposite will take place. If the difference in

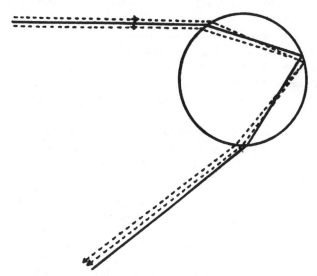

Fig. 59.—Diagram illustrating Young's explanation of the supernumerary rainbow arcs.

the lengths of the two paths is an integral multiple of the wave length of a given color, the rays will be reinforced; if it is an odd multiple of half a wave length, the rays will extinguish each other. The differences in distance increase steadily as the pairs of rays depart more and more from the Cartesian ray, so that there will be not one, but several positions where reinforcement takes place, and also other intermediate positions where the rays annihilate each other. One should expect, therefore, not a single primary rainbow, but the formation of a whole series of bows, as in Newton's rings; and this is what Young realized actually does take place.

There are potentially infinitely many bands of each hue in the primary bow, the bands becoming fainter and narrower as the radii diminish. The spacing of the bands depends on the variations in the length of the path, and these are determined by the size of the drops. Ordinarily there is considerable overlapping in the bands of various colors, especially when the drops are not uniform in size; and this accounts in part for the fact that most people see only a single primary rainbow. If the drops are unusually minute (as in a fine mist), the interference bands may become so intermingled that the result is a synthesis of all colors, that is, a white bow. For large drops of rain, only one brightly colored primary rainbow usually is seen (for reasons which Pemberton had correctly divined); but Young found that supernumerary bows are clearly visible when the raindrops are uniformly sufficiently small, and he noted that there was a regularity in their spacing which corresponds to that in Newton's rings.

In the Bakerian Lecture read by Young on November 24, 1803 he advanced the phenomena of Newton's rings and the multiple rainbow arcs as arguments supporting his doctrine of interference. Beginning with the rings, and with Grimaldi's "crested fringes," Young asserted that the bows described by Langwith "admit also a very easy and complete explanation from the same principles." He pointed out that the arcs depend on the size of the drops, "according to the differences of time occupied in the passage of the two portions, which thus proceed in parallel directions to the spectator's eye, after having been differently refracted and reflected within the drop." Then Young gave the only quantitative statements to be found in his account of the arcs:

This difference in the length of path varies at first, nearly as the square of the angular distance from the primitive rainbow: and, if the first additional red be at a distance of 2° from the red of the rainbow, so as to interfere a little with the primitive violet, the fourth additional red will be at a distance nearly 2° more; and the intermediate colours will occupy a space nearly equal to the original rainbow. In order to produce this effect, the drops must be about 1/76 of an inch, or .013, in diameter: it would be sufficient if they were between 1/70 and 1/80.[34]

Young, unfortunately, did not indicate the nature of the calculations by which he arrived at his estimate for the size of the drops (which he sometimes gave less precisely as between 1/50th and 1/100th of an inch).

Young's theory of the rainbow shed an intriguing new light on the old problem of multiple rainbows, for it showed that, for each of the potentially infinitely many orders of bows, infinitely many arcs are possible. Such a magnificent conception of the double infinitude of bows should stir the soul of every poet who mistakenly believes that science has taken the romance out of the story of the rainbow. For Young, however, the problems of the rainbow had relevance of a different order; he saw in them a means of convincing sceptics of the superiority of the undulatory theory of light. From his consideration of rainbow arcs he went on to say,

Those who are attached to the Newtonian theory of light, or to the hypotheses of modern opticians, founded on views still less enlarged, would do well to endeavour to imagine any thing like an explanation of these experiments, derived from their own doctrines; and, if they fail in the attempt, to refrain at least from idle declamation against a system which is founded on the accuracy of its application to all these facts, and to a thousand others of a similar nature.[35]

The interference theory of the rainbow explained a number of the puzzling qualitative vagaries which had been observed. It made clear why the bow is brighter near the earth and why the supernumerary arcs seem to appear near the highest part of the bow. Raindrops tend to increase in size as they fall, and the results of Young showed that where the drops are uniformly larger, there will the bow be brighter, but unaccompanied by supernumerary arcs. It had been noted also that the finest rainbows occur principally during the warmer half of the year, and especially following the retreat of an afternoon thunderstorm. The reasons for this are obvious, in the light of Young's theory, when one realizes that such atmospheric conditions are most likely to produce large raindrops.[36] But there still remained at least one vagary of the rainbow which stubbornly refused to be reconciled with any theory which had yet been proposed. Descartes had thought that he had definitively answered the ancient question as to whether the radius of the rainbow was constant or variable. He had believed that, for a given index of refraction, the radius of the primary bow was always the same. But eighteenth-century observations had shown this belief to be untenable. The radius does indeed vary considerably. For a white bow it may be as many as half a dozen degrees smaller than the traditional figure of about 42°. Young was unable to give his rainbow theory the quantitative elaboration which was needed to account for such differences, and the explanation awaited a new generation of mathematical physicists.

The interference theory of the rainbow was published in the *Philosophical Transactions*[37] for 1804, but it was to be many years before Young's work received the recognition it deserved. A series of prejudiced articles in the

Edinburgh Review heaped scorn upon the man who had revived the wave theory of light and explained the supernumerary rainbows. Young had made a point of showing how his ideas were but a natural outgrowth of the Newtonian queries; but he was accused nevertheless of disrespect for the authority of the author of the *Opticks*. The undulatory theory found few recruits in England during the early nineteenth century; and on the Continent his explanation of the spurious rainbows was pretty much disregarded. Moreover, the revival in the wave theory of light was very nearly snuffed out by an optical discovery made in 1808 by Étienne Malus (1775–1812). Ever since the seventeenth century it was known that light which had traversed certain types of transparent crystals, such as Iceland spar, was modified so that it no longer behaved like ordinary direct sunlight. Huygens had assumed that the waves had acquired a peculiar disposition; Newton had believed that the crystal somehow had acted like a sieve to sort out certain types of light particles. Newton's further experimentation showed that light rays act as though they were made up of components with lateral asymmetry, as though they had "sides." He found that, after refraction through Iceland spar, one type of component, possessing a peculiar orientation, remained, while the component oriented at right angles to it had been screened out. Huygens had no explanation to offer, because he could not see how waves could have the polarity which seemed to be called for; but he was not seriously worried because the phenomenon was thought to be quite unusual. Malus, however, showed that the polarization of light is an everyday occurrence. He had been one of the group of scientists who had accompanied Bonaparte on the Egyptian campaign, and on his return to Paris he had been spurred to compete for a prize offered by the Académie des Sciences for a theory of anomalous refraction. One evening he was looking through a doubly refracting calcite crystal at the reflection of the setting sun seen in a window of the Luxembourg Palace. He noticed that as the crystal was rotated, sometimes one, sometimes the other, of the two images disappeared. He realized that the light from the window must be polarized; but why? At first he thought that the atmosphere was the polarizing agent; but further experimentation that night led him to the correct conclusion that it was the reflection which had polarized the light. Polarization of light, far from being an isolated peculiarity, is the rule rather than the exception, for both refraction and reflection. To the unaided eye, reflected light appears to be indistinguishable from that seen directly; but Malus discovered that the two do indeed differ, and his calcite crystal gave him a means of determining precisely the angle of orientation of the light polarized by the reflection. He found that the degree of polarization of the light depended on the angle of incidence, and that at a particular angle, which he called the polarizing angle for the medium in question, virtually all of the reflected light is polarized. Sir David Brewster (1781–1868) in England was able to show in 1815 that the tangent of the polarizing angle of incidence is equal to the refractive index of the medium. That the blue light from the sky is produced by reflection from minute particles in the air was confirmed by his demonstration that this light is polarized. Brewster also applied the phenomenon of polarization to the rainbow, verifying in 1812 the observation of Jean-Baptiste Biot (1774–1862) that the light from the two rainbow arcs is almost entirely polarized in the planes which pass through the eye and the radii of the arcs.[38] This observation constituted a beautiful confirmation of the Cartesian theory of the rainbow, for it showed that the arcs are formed by reflections at just about the angle of polarization for water—i.e., 53°.

The polarization of light from the rainbow can be verified without complicated apparatus. A nicol prism is the most appropriate device for observing this; but it can be detected also by looking at the image of the rainbow in a mirror formed by a sheet of glass backed with black. As the orientation of the plane of the mirror is varied from a horizontal to a vertical position, the intensity of the reflected light will be found to vary greatly. Calculation shows that for the primary rainbow the greatest intensity of the light for a given direction is more than twenty times the intensity in a direction at right angles to this, indicating virtually complete polarization. For the secondary bow the polarization is also pronounced, the ratio of maximum and minimum intensities being about eight to one.[39]

Brewster could well say that "observation agrees so well with the results of calculation that there remains no doubt of the truth of the Cartesian explanation." [40] But while the polarization confirmed Descartes' geometrical explanation, it played havoc with Young's interference theory. The years from 1808 to 1816 were just about the darkest in the entire history of the wave theory. How could a light wave have "sides," if one can not conceive of the polarization of sound? Even Young was somewhat worried by the turn of events. He wrote to Malus in 1811, "Your experiments show the insufficiency of a theory which I have adopted, but they do not prove it false." [41]

The undulatory theory seemed to have reached an impasse at just the time that it had won an ardent young advocate in the brilliant mathematical physicist, Augustin Fresnel (1788–1827). Like Malus, Fresnel was a Polytechnician; but whereas the one was a Bonapartist, the other declared allegiance to the Bourbons. During the Hundred Days, when Fresnel had been ousted from his position, he quietly began the great experiments and calculations which made the years from 1815 to 1825 one of the greatest decades in mathematical optics. Fresnel is not noted for any direct contribution to the theory of the rainbow, but through his optical discoveries his indirect influence was substantial. Rectilinear propagation had been an awkward point in the optical theories of Huygens and Young; but the mathematical analysis of Fresnel accounted both for this phenomenon and for diffraction.

Independently of Young, Fresnel next worked out the principles of interference, and he set out to see how this was affected by polarization. The results he obtained showed that two rays polarized in the same plane interfere as do rays of ordinary light, but that rays polarized at right angles to each other do not have any effect upon each other. How could this be reconciled with an undulatory theory? Pondering this difficulty, Fresnel had a brilliant idea—one that came to Young at almost the same time during the years 1814 to 1816. Proponents of the wave theory up to this time had pictured light as a disturbance in a fluid. As in the transmission of sound, it was believed that the expansions and contractions took place in the direction in which the disturbance is carried. That is, the waves were assumed to be *longitudinal*. Fresnel and Young finally saw that polarization made it necessary to abandon this preconception and to conceive instead of *transverse* vibrations in which the displacements take place at right angles to the direction of transmission. If, for example, a stone is dropped into a placid pool, the disturbance travels out radially, but the water itself moves up and down. Light waves, said Young and Fresnel, were to be pictured as particles of ether dancing *up and down* like corks in water—not oscillating *back and forth* like molecules of air in the transmission of sound. Again, if one end of a string is fastened to a wall and the free end is whipped from side to side, the wave-like motion travels along the rope while the elements of the rope are in reality moving from side to side. Newton had toyed with this conception when he had asked, "Are not rays of light bent backwards and forwards with a motion like that of an eel?"

But it was Young and Fresnel who seized upon this idea of eel-like transverse vibrations to explain polarization and thus to give the undulatory hypothesis a new lease on life. They assumed that light is a bundle of transverse waves in planes variously oriented, and that in reflection and refraction some of the planes of vibration are screened out to leave a beam which is wholly or partially polarized. This brilliant idea saved the wave theory from the incubus presented by the non-interference of polarized light, for transverse vibrations in different planes could scarcely be expected to affect each other. This was the first satisfactory explanation of polarization, but it was bought at a price which was found too high by some contemporaries otherwise favorably disposed toward the undulatory hypothesis. Transverse vibrations were regarded as incompatible with the fluid state, and hence Fresnel was forced to assume that the luminiferous ether behaves like an elastic solid. Mathematically the arguments of Fresnel were convincing, but scientists found it difficult to believe that the heavenly bodies are moving resistlessly through a solid ether with an elasticity greater than that of steel. How could such an ether be reconciled with Young's assumption that it passes through ordinary matter as freely as the wind passes through a grove of trees? So contrary to preconceived notions was such an hypothesis of transverse undulations that it was not until almost a dozen years after Fresnel's death (in 1827) that the Newtonian emission theory faded from the field to leave the vibratory theory of Young and Fresnel triumphant.

The almost simultaneous discovery by Young and Fresnel that transverse vibrations would save the fortunes of the wave theory might have resulted in a priority dispute; but the principals afforded the trouble-makers no opportunity. Each magnanimously accorded to the other full credit for independence of thought. This situation contrasts pleasantly with the jealous fears of plagiarism which proponents of Young and Champollion had aired on the question of the decipherment of Egyptian hieroglyphics.

The single-mindedness of Fresnel made him the effective founder of the wave theory which Young had revived; but the many-sidedness of Young carried him into fields in which Fresnel played no part. The place of Young in the story of the rainbow is not fully told without reference to his work on the physiology of color. Young realized that the physics of the rainbow was a thing apart from the perception. He confirmed the subtlety of the bow when he photographed the rays in Newton's rings. The rainbow comprises infra-red and ultra-violet rays which are unperceived by human eyes but which, whether or not they are visible to other forms of life, form part of the total phenomenon. As polarization had indicated, there is indeed more to the rainbow than meets the eye!

Too often it had been assumed that the last word on color had been said by Newton. While Young was working on the wave-theory, the poet Wolfgang Goethe (1749–1832) in Germany was opposing the Newtonian optics with vitriolic obstinacy, branding it "cunning lawyers' tricks, and sophistical distortions of nature." (Schelling likewise contemptuously referred to Newton's *Opticks* as "the greatest illustration of a whole structure of fallacies.")

Goethe was too poor a mathematician to cope with the intricacies of physical optics, and his substitute for the spectrum of Newton was an unsatisfactory sort of Aristotelian combination of light and shade. On the other hand, he does seem to have realized that a quantitative study of the spectrum does not in any sense explain the sensation or the perception of color. The physical and chemical sides of color had been explored, but not the psychological. Goethe collected materials for a history of color theory, but his work failed of definitive status because the author lacked exhaustive and specialized knowledge. Even today there is no truly comprehensive treatise on the subject.[42] Had Goethe not undertaken to demolish the Newtonian physical theory, his own contributions to the chemistry and physiology of color—the study of complementary colors and the persistence of after-images—might have been more significant. In 1791 he had published his Beiträge zur Optik, but this had been pretty generally ignored, partly because of its polemic overtones. Goethe thereupon became sensitive on the subject and it became his greatest love. He wrote,

As for what I have done as a poet I take no pride in whatever . . . but in my century I am the only person who knows the truth in the difficult science of colors, . . . of that I say I am not a little proud.[48]

But he was better at synthetic thinking than at analytic investigation, and his failure in experimental techniques too often misled him. Toward the end of his life he took an interest in the rainbow, but he did not add anything of significance. Only a few weeks before his death he wrote to a friend that words, lines, and letters are not enough in the study of the bow. One must also think and do. He recommended experimenting with a glass sphere such as tailors were accustomed to use to concentrate lamplight on their work. Little did he realize that in this he was repeating advice which Theodoric had given half a millennium before.

Goethe understood the importance of physiological optics (to the exclusion, one might almost say, of physical optics), but it is Young who is known as the founder of this branch of sicence. He explained the accommodation of the eye through the curvature of the lens, including the nature of astigmatism; but more important in the story of the rainbow is the physiology of color. Here Young proposed the hypothesis, afterward developed by Helmholtz, that color perception depends on the presence in the retina of three kinds of nerve fibers which respond respectively to red, green, and violet light. Color perception must not, of course, be confused with the problem of pigmentation. Painters and dyers long had used a three-color doctrine even before the physicists and physiologists took it over. Newton's color triangle, in which red, yellow, and blue are the basic elements, had been used to study color production, rather than color perception. To know how the colors of the rainbow are produced spectrally is of little help to the artist, for he must mix not wave-lengths but absorptive capacities. The physiologist is concerned with something else again: the sensitivity of nerve fibers to stimulation by varying hues. What we see as red is not the result of stimulation by the red spectral rays only; it is a reaction to a composite of light of various wave-lengths.

Hooke had held that all colors were produced by a mixture of the two hues, yellow and blue, together with white and black. Young in 1801 proposed instead that in the retina there are three kinds of nerve fibres (or else that each fibre consists of three distinct parts), sensitive respectively to red, yellow, and blue in the ratio of 8:7:6. The following year he substituted the colors red, green, and violet, and a sensibility ratio of 7:6:5. He did not hold that each of the three colors would excite but one of the sets of fibres. Green, for example, might excite the fibres appropriate to red or violet, but in a smaller proportion compared with the excitation of the "green" fibres.[44] His theory is based on the assumption that color perception is given, as it were, by the three-parameter equation $P = aX + bY + cZ$, where a, b, and c are constants and X, Y, and Z are the sensitivities of the retina to the frequencies of colors red, green, and violet. The trichromatic theory of Young, modified by restoring blue rather than violet as basic, has been further developed by Helmholtz and Maxwell, but there are today other rival theories. The physiology of color perception still is far from satisfactorily explained. We really have very little idea how color vision works—how combinations of stimuli from light of varying wave-lengths are seen as single hues. It is not even known yet whether each cone in the retina is able to respond to every one of the three basic color sensations, or whether there are three entirely different types of cones, one for each of the three colors. How much better understood is the physical formation of the rainbow than is its physiological perception.

The interference theory of the rainbow was included in the lectures which Young delivered at the Royal Institution between 1801 and 1803. Young gave up his professorship there in favor of his medical practice, but his rainbow theory appeared in print again in 1807 with the publication of his *Course of Lectures on Natural Philosophy*. Nevertheless, the new theory found few adherents for the next thirty years. This neglect is explained in part by the reluctance of scientists to accept the new ether with its fantastic properties. Then, too, neither Young nor Fresnel gave for the formation of the rainbow the adequate mathematical exposition which was needed. Young made a number of quantitative statements on the sizes of drops as compared with the widths of the supernumerary bands, but he gave no insight into the reasoning by which he had associated these. Finally, the new explanation was slow in gaining disciples for the same reason that the ideas of Theodoric, Descartes, and Newton were tardily adopted. It is not enough for one scientist to present a new theory in a single context. He must follow it up with additional evidence, or else he somehow must bestir others to bring the new views under review and discussion. This Young failed to do, and the interference theory of the rainbow consequently went virtually unnoticed until half a dozen years after the death in 1829 of its author.

XI

The Mathematician's Rainbow

YOUNG HAS BEEN CALLED THE LAST MAN TO KNOW EVERYTHING.[1] THIS OBVI-
ously over-simplified statement is to be taken to mean that during the nineteenth century the world of learning rapidly was becoming much too broad
for any polymath to master more than a fragment of it. Even within a single
well-circumscribed compartment of man's knowledge, such as mathematics,
complete coverage became an impossibility; and the great contemporary of
Young, Carl Friedrich Gauss (1777–1855), is regarded as the last mathematician to be conversant with substantially his whole field. Specialization was
inevitable; and the story of the rainbow was swept deep into the esoteric realm
of advanced mathematics. Young, who always was on the lookout for unifying
principles, had carried the theory of the rainbow about as far as one could
without a concentrated quantitative study of just what goes on within a drop
of rain when rays of the sun strike it. So intricate were the operations of nature
found to be that mathematicians were compelled to make use of their most
powerful analytical weapons to subdue the intractable phenomena, and physicists were urged to correspondingly greater ingenuity in devising experimental
verification of theoretical predictions.

The era of specialization did not dawn suddenly, and the definitive explanation of the rainbow (to the extent that any theory in the cumulative advance of science may be regarded as definitive) was to a large extent the work
of three men not one of whom was primarily a mathematician, although two
of them had won prizes in mathematics. One of the three, George Biddle Airy
(1801–1892), was an astronomer; another, William Hallowes Miller (1801–
1880), was a mineralogist; and the third, Richard Potter (1799–1886), was a
chemist and physicist with medical training. All of them had two things in
common, longevity and Cambridge training. Because of the crucial nature of
his contribution, the modern theory of the rainbow usually is ascribed to Airy
alone, but it might more appropriately be referred to as the "Cambridge
theory." It had its origin in Newton's Cambridge lectures; it was given a new
direction by Young, a Cambridge graduate; it received definitive formulation
through the efforts of the above-mentioned triumvirate, all of whom held
Cambridge degrees and two of whom taught there; it was refined by other

294

Cambridge scholars and professors, notably Sir George Gabriel Stokes (1819–1903) and Sir Joseph Larmor (1857–1942); and it achieved further mathematical elegance through John William Strutt (1842–1909), better known as Lord or Baron Rayleigh, a Nobel prize-winner, who was a Cambridge student, professor, and chancellor. Were a monopoly possible in the story of the rainbow, it would certainly go to Cambridge University; but there will be need to mention contributions from other places as far away as the University of Tokyo.

Young's work on the rainbow for some reason lacked popular appeal, and it remained little known even in England. As an undergraduate at Cambridge, Potter rediscovered the interference theory of the supernumerary bows; and it was only through speaking with the very learned William Whewell, an historian of science, that he learned that he had been anticipated.[2] But though he admired the discoveries of Young, Potter stubbornly refused to accept the wave theory of light. The principle of interference fitted neatly into the undulatory theory; but it had not completely destroyed the rival doctrine, for the Newtonian adherents had found ways of adapting the corpuscular doctrine to the new phenomenon through the assumption of the emanation of successive rings of a subtle fluid. When the rings coincided, it was assumed, they somehow destroyed each other.

The proponents of corpuscles or emanations found their position occasionally quite awkward; but they took great comfort in the fact that the wave theory had been unable to iron out all difficulties which had been encountered. "On these grounds," Brewster wrote, "I have not yet ventured to kneel at the new shrine, and I must even acknowledge myself subject to that national weakness which urges me to venerate, and even to support, the falling temple in which Newton once worshipped." [3] Where the older man hesitated, one should not severely criticize the young Potter for not seizing upon the new doctrines. Potter had been impressed by the assertion of Sir John Herschel that if as much talent had been spent on the corpuscular theory as on the undulatory, a satisfactory corpuscularian explanation of interference would have been found.[4]

But if the prescience of Potter with respect to the nature of light was not comparable to that of Young, nevertheless his application of interference to the explanation of the rainbow went beyond that of his predecessor. On December 14, 1835 he read to the Cambridge Philosophical Society a paper entitled, "Mathematical Considerations on the Problem of the Rainbow, Shewing it to Belong to Physical Optics," [5] in which he made clear the direction in which study had to proceed. He began by pointing out the shortcomings in the Cartesian-Newtonian theory—its failure to explain how the size of the drops influences the colors of the bow, and its inadequacy in accounting for the supernumerary arcs. He held, as had Young, that only the principle of interference could explain the phenomena. Potter, therefore, went into a detailed mathematical study of the collective paths of rays through a raindrop. To study a single path by itself, as Descartes had done, would not suffice, for it is the *interaction* of rays which is important here. Two important ideas were intimately involved; and although each one had been developed in the seventeenth century, it was only from 1835 on that they were systematically applied to the rainbow. The first of these is the Huygenian concept of a wave-front, a surface perpendicular to the rays of light—that is, an orthogonal trajectory of a one-parameter family of rays. A good approximation to this idea can be had by picturing the light rays as columns of marching soldiers, the cross-ranks in this case corresponding to the wave-fronts. Such a picture, in fact, comes close to the emanations which Potter and other corpuscularians visualized instead of Young's "waves." The second mathematical concept which was needed was that of a caustic surface—the envelope of a family of reflected or refracted rays. A good example of a caustic curve, which is a plane section of a caustic surface, can easily be seen by noting the bright arcs formed by reflected light rays on the bottom of a teacup. A caustic by reflection can be formed readily by placing a bright metal ring on a sheet of paper in the sun's light (Fig. 60). In this case the caustic is a curve known as an epicycloid, or the curve traced out by a point on the circumference of one circle while it

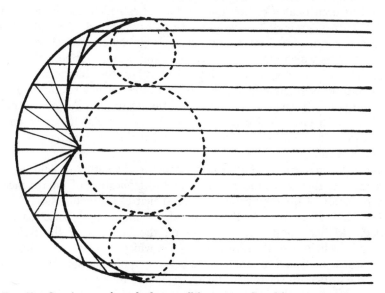

FIG. 60.—Caustic curve formed when parallel rays are reflected by a semicircular mirror.

rolls smoothly about another circle, the radius of the rolling circle being one-quarter that of the ring mirror, and the radius of the fixed circle being half that of the ring.[6] If the equations of the lines representing light rays are known, the equations of the orthogonal trajectories (wave-fronts) and envelopes (caustics) can be found by the methods of advanced calculus. Caustic curves associated with the reflections and refractions of the rainbow are more complicated than the epicycloid, but they are produced by light rays in much the same manner.

Potter's account of the formation of the rainbow, considerably modified below in order to elude the heavy mathematical detail which necessarily accompanied it, is substantially as follows. Let the circle in Figure 61 represent

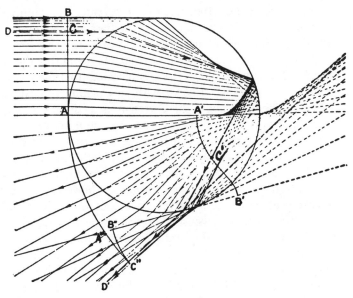

Fig. 61.—The paths of light rays through a raindrop, indicating some of the associated wavefronts and caustics.

a cross-section through the center of a raindrop, and let the solid lines represent rays from the sun which strike the upper half of this circle, are refracted into the drop, are reflected at the rear surface, and are refracted out again through the lower portion of the drop. The dotted lines are the backward extensions of the refracted rays, and the heavy barred line represents the Cartesian ray. From the diagram it is clear that, if the incident rays are taken as parallel, the wave-front (or "section of the luminous surface," as Potter calls it) will be a straight line, ACB. (If one thinks of the plane figure as rotated about the axis AA', the result will be an indication of what actually is taking place in three dimensions; and the wave-front indicated by ACB in reality will be a plane rather than a line. Throughout the present discussion the language used will be that of two dimensions; but it is to be understood that this is only done to simplify the exposition, for the phenomenon actually is three-dimensional.)

Following the refraction at the concave surface of the drop, the wave-front is no longer rectilinear, but curvilinear. In fact, some of the rays intersect others in the family even before they strike the rear surface, as is indicated in the diagram. Descartes had traced the path through the drop of one ray at a time, and so he failed to call attention to this intersection. Theodoric, too, had overlooked it, but Kamal al-Din had been aware of it, and it was rediscovered by geometers in the latter part of the seventeenth century. These rays, after the first refraction, form a caustic by refraction which can be seen in Figure 61. Potter did not call attention to this, for he was primarily concerned with the rays which emerge from the drop. He found that the orthogonal trajectory A'C'B' (wave-front) of the rays reflected from the rear concave surface of the drop is an s-shaped curve, with an equation approximately of the form $y = kx^3$. The beam above the Cartesian ray is divergent, and hence the wave-front A'C' or A''C'' is convex (forward); the beam below the Cartesian ray at first is convergent, so that here the wave-front C'B' is concave. The point C' on the Cartesian ray is known as a point of inflection, for it is the point at which the sense of concavity changes. After the lower convergent rays have passed through their center of convergence, however, they too form a family of divergent rays with a convex wave-front C''B''. The combined wave-front A''C''B'', therefore, consists of two convex portions, mutually tangent but with unequal radii of curvature, which form a cusp at the point C'' near, but slightly below, the Cartesian ray. The totality of the vertices of all these cusps, known as a cusp-locus, is a curve along which there will be many almost-parallel rays of light, and hence this is a caustic.

This caustic curve (the curve Bb in Figure 62, taken from Potter's paper) approaches more and more closely to the line representing the Cartesian ray, the Cartesian ray being known as the asymptote of the caustic. For most ordinary purposes the curve and its asymptote are so close to each other that they may be regarded as coincident, and the Cartesian rainbow band therefore can be thought of as a caustic formed in much the same way as is the bright curve on the bottom of a cup. This, of course, confirms the fundamental property which Descartes had derived through an entirely different type of argument. But Potter carried the explanation further. The light causing the cup caustic has not been refracted, but that producing the rainbow has; and hence the

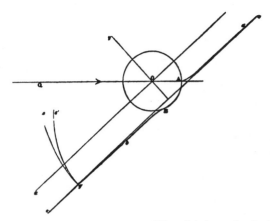

FIG. 62.—Potter's diagram showing caustic curves (Bb and Aa) associated with the rainbow.

rays of each color have their own caustic, each one corresponding to a colored band in the rainbow as explained by Newton in different terms. Potter called attention to the fact that close to the caustic the nearly-parallel rays do not necessarily reinforce each other, for they will exhibit the interference phenomena which Young had pointed out. If the two sets of waves, indicated by the two arcs forming the cusp, are in phase along the effective ray, they will at first be more and more out of phase as one moves away from this ray (that is, as the difference in the lengths of the paths increases). Finally a distance from the effective ray will be reached at which the crest of the one wave will coincide with the trough of the other, and here the rays will extinguish each other. (The language used here is that of the wave theory; that used by Potter did not commit him to this theory.) As one continues to depart still further from the effective ray (Fig. 63), the difference in path-length now will come closer to a full wave-length until, when crest again coincides with crest, reinforcement reaches another relative maximum. Beyond this the combined effect will tend again to a relative minimum; and so the cycle continues, resulting in alternate light and dark bands within the circle of the primary rainbow. That is, the intensity of illumination does not fall off monotonically as one departs from the effective ray, as Descartes believed; the decline in intensity is oscillatory.

If, now, one recalls that wave-length varies with color, it becomes clear that the positions of relative maximum and minimum reinforcement for red rays are not quite the same as for violet. Each bright band is in actuality determined by a series of juxtaposed maxima, the result being a band of rainbow colors. The rainbow, in other words, is really a superposition of indefinitely many bows of each color. The bright colors, however, are limited by their own

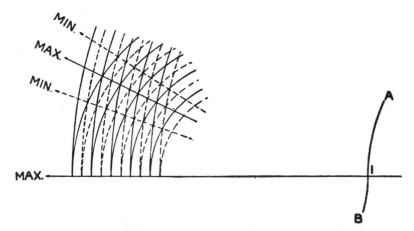

Fɪɢ. 63.—Diagram illustrating the superposition of light waves in the formation of supernumerary rainbow arcs.

self-destruction to a narrow band (the principal rainbow) and sometimes a few partial arcs (spurious bows). The positions of the bands depend, of course, on the size of the drops. Potter calculated that for drops with a diameter of about 1/72 of an inch, the second red maximum will fall 1° 46′ within the first red maximum, and will thus coincide with the first violet arc. This was in fairly good agreement with the results of Young, who had found that for drops with a diameter of 1/76 of an inch, the second red band should be 2° inside the first red band and should just about coincide with the first violet arc. Potter asserted, with justification, that from observations of the rainbow one can calculate the size of the drops producing it. This is a far cry from the Cartesian belief that all rainbows, at least for a given index of refraction, are alike. But there remained in the rainbow family a black sheep—the white rainbow—which perversely refused to fit into the Young-Potter theory. When Theodoric, more than six hundred years earlier, had noted the white fog-bow, he had denied that it belonged in the family circle of the ordinary bows; but observations of the eighteenth century had indicated that its status as a proper rainbow was legitimate. It was confirmed that the refractions and reflection which produce it are basically not unlike those which are involved in other rainbows. But why, for exceedingly minute drops should the resulting bow be so radically different? What made the white bow such a white elephant was not so much its color. That could somehow be attributed to the overlapping of colored bands. But the white bow was much too small to fit in with the theoretical results. Bouguer, noted for experiments on the inverse square law in photometry, had measured the radius of a white rainbow and had found it to be a full 8° less than that of the usual colored bow. How could such a deviation from normality be accounted for without endangering the fundamental theory? A clue had been available since 1665 when Grimaldi announced that light is only approximately rectilinear, for it can be deviated from a rectilinear path not only by reflection and refraction, but also by what he called inflection (now known as diffraction). Light is indeed inflected or bent around corners and into the geometrical boundaries outlining shadows; but the bending is slight in comparison with the bending of sound waves because the waves of light are exceedingly short in comparison with those of sound. (Relatively small objects produce light shadows, but only comparatively large objects cause sound "shadows.") Could it be that diffraction would answer the difficulty with the theory of the rainbow? Comparetti had asserted this in the eighteenth century; but the theory of diffraction had not then been worked out. It was Fresnel who, during the years immediately following his conversion to the undulatory faith, first computed the intensity of illumination of the diffraction bands near the edges of shadows. His quantitative analysis, which won the prize offered by the Académie des Sciences for an essay on diffraction, led him to the evaluation of the expressions $\int_0^\infty \sin x^2 dx$ and $\int_0^\infty \cos x^2 dx$. These "Fresnel integrals" [7] can not be evaluated in terms of the usual elementary functions; they call for the application of difficult numerical methods of integration or else the use of infinite series. Thus it was largely Fresnel's treatment of diffraction which carried the study of optics from the Huygenian elementary *geometrical* stage over into what may be called the modern advanced *analytic* stage. And what Fresnel had done for optics in general, Airy did for the rainbow in particular. No longer could one be satisfied with relatively simple and easily understood geometrical diagrams. Even Potter's application of the calculus to the wave-fronts and caustics was inadequate as an explication. What was essential was a precise analytic expression for the intensity of illumination at each and every point of the area brightened by the bow. Extraordinary mathematical powers were called for, and these were possessed by Airy, who had carried off the prizes in mathematics during his student days at Cambridge. The crucial date in the modern diffraction theory of the rainbow therefore is 1836, the year in which Airy gave a formula for intensity of illumination as a function of the angular deviation of a ray from the Cartesian least-deviated ray. For this reason the modern theory usually is known as "Airy's theory," or sometimes as the "complete theory." His computations are too forbidding to include here, for they led him to an integral which can not be evaluated in terms of the elementary functions. He found that the intensity of light is given by the square of an integral which since has come to be known as "Airy's rainbow integral." This he wrote as $\int_w \cos \frac{\pi}{2}(w^3 - mw)$ $\left\{\text{from } w = 0 \text{ to } w = \frac{1}{0}\right\}$, where modern mathematicians use the more compact form $\int_0^\infty \cos \frac{\pi}{2}(w^3 - mw)\,dw$. The parameter m determines the angular departure of the ray from the Cartesian ray, and Airy tabulated the values of his integral (and of its square) to seven decimal places for intervals of 0.2 from $m = -4$ to $m = +4$. The methods of computation, in which use is made of mechanical quadratures or approximate integration, are tedious and involved; but an understanding of the implications of his work can be had by noting the graphical representation of the intensities (Fig. 64) which he appended to his famous paper of 1836. This paper, which he entitled "On the Intensity of Light in the Neighbourhood of a Caustic," was published two years later in 1838, in the same volume of the *Transactions of the Cambridge Philosophical Society* as that which carried Potter's explanation of the rainbow.[8] The graph which he included (Fig. 64) illustrates how his results (indicated by the heavy solid curve) compared with those of Descartes (the dotted curve) and Young (the lighter solid curve); and the differences are so striking that it has often been charged, with considerable justification, that the theory of Descartes is not only incomplete, but that it is downright erroneous. Before Airy presented his results it had been assumed that the intensity of illumination was greatest at the angle for which the deviation is least, that is, along the Cartesian ray or the caustic curve of which it is the asymptote (the vertical axis in Figure 64). The calculations of Airy showed, however, that this is not so; the region of greatest brightness lies appreciably *within* the radius computed on the basis of the geometrical theory. As he expressed it,

One of the most important points to be remarked is, that the maximum illumination does not take place at the Geometrical Caustic, or where m = 0, but where m = + 1.08, that is, on the external side of the convexity of the caustic, or on the luminous side of the geometric position of the rainbow, that is, (for the primary bow), within it. The following rule . . . will suffice, in practice, to determine the geometrical position. When the first spurious bow is visible, measure the distance of its maximum intensity from that of the brilliant bow; then the geometrical bow is exterior to the brilliant bow by 11/24 of this distance.[9]

That is, whereas Young and Potter had shown that the *colors* and *spacing* of the arcs in the rainbow depend on the size of the drops, Airy showed that the *radius* of the primary rainbow itself varies with the magnitude of the raindrops.

A glance at Airy's graph forcibly brings out another respect in which his theory differs from the earlier explanations. According to Descartes, Young, and Potter, there should be no light whatever returned to the eye at an angle greater than that of the least deviated ray, following a single internal reflection within the drops. Airy's calculations (and their representation in the graph of Figure 64) show that this assumption also is erroneous. As he phrased it,

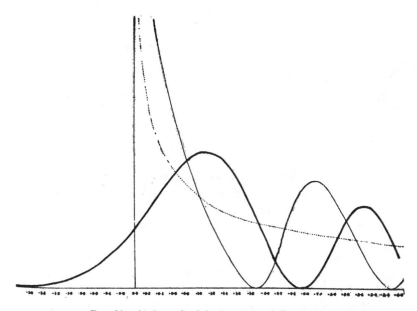

Fig. 64.—Airy's graph of the intensities of illumination.

"We have found above, on the complete theory that there is sensible light for negative values of m." [10] Although the intensity of illumination falls off rapidly for angles greater than that of the Cartesian ray, there nevertheless is an appreciable amount of light outside of the boundary set by the geometrical theory—as indicated by the portion of the graph (Fig. 64) to the left of the vertical axis. This bending of light into what should be a region of complete shadow shows that diffraction must be taken into account in any "complete" theory of the rainbow. Moreover, the diffraction theory shows that "Alexander's dark band" is not so thoroughly devoid of light as previously had been believed.

Airy was an extremely hardworking man who published over five hundred scientific articles, but the rainbow was far from occupying the center of his attention. In 1836, the year of his classic paper on the rainbow, he became Astronomer Royal, a post he held for forty-six years and which he made the real business of his life.[11] Only once again did he make a contribution to the story of the rainbow: a "Supplement to a paper on the Intensity of Light in the Neighbourhood of a Caustic," published in the *Cambridge Philosophical Transactions* for 1849.[12] In this he described an alternative method for computing values of his rainbow integral, based on a suggestion communicated to him by the mathematician Augustus De Morgan (1819–1904). Making use of infinite series instead of mechanical quadratures, he extended his table of intensities from $m = \pm 4$ to $m = \pm 5.6$, but this still was sufficient to give only the first two maxima. More effective was another series device used in 1850 by Stokes to put the integral into a form from which the value can be calculated "with extreme facility" when m is fairly large. Replacing sines and cosines by their equivalents in terms of complex exponential quantities (using equations known as Euler's identities), Stokes calculated intensities of illumination sufficient to place the first fifty maxima,[13] thus going far beyond the range of Airy.

Except for devices which might facilitate the evaluation of his integral, Airy's concern for the problems of the rainbow seems to have vanished. But the rainbow seldom has lacked devotees, and among these was the Cambridge mineralogist, W. H. Miller, a man who did much to make Airy's theory scientifically acceptable.

Airy (and later Stokes) had left his analysis of the rainbow in the mathematician's proverbial ivory tower. Theoretically it looked impeccable; but every scientific theory sooner or later must meet the acid test of agreement with experience. It was on March 22, 1841 that Miller presented to the Cambridge Philosophical Society a paper, entitled "On Spurious Rainbows," in which he summarized the results of his exhaustive comparison of calculations and observations. Neither the theoretical basis nor the experimental procedure which he used was original; but his work was important because it effectively bridged the gap between the two. Miller candidly admitted that his computations hinged upon those of the Astronomer Royal, although he extended these to include the secondary rainbow as well as the primary. His experimental method was derived from that which had been used at Paris by the physicist Jacques Babinet (1794–1872). The theory of the rainbow is exhibited most easily by studying a cross-section of a raindrop, thus reducing the problem from one of three dimensions to another of two (as was done in most of our diagrams above); and Babinet and Miller had recourse to an experimental equivalent of this device. For the light of the sun they substituted a horizontal beam of light admitted through a vertical slit, and instead of a spherical raindrop they used a very thin vertical cylinder. The intersection of the horizontal beam and the vertical cylinder is tantamount to the circular cross-section of a raindrop by a plane through the center of the drop. The result of the refractions and reflections will, of course, not be complete rainbows, but only bright arcs representing the feet of the rainbows; but these rainbow fragments are entirely adequate for determinations of the radii of the bows and of their supernumeraries. Moreover, experimental refinements make extreme precision of measurement and observation possible. Using cylinders of glass, Babinet observed rainbows of the first fourteen orders; and Babinet reported that he saw sixteen supernumeraries for the primary arc, nine for the second arc, adding that "one sees these also for the third arc." [14] Whereas in nature the third bow is seen with extreme rarity, in the laboratory not only the tertiary arc but also its accompanying spurious arcs can be observed. Believing that "the formula for the rainbow has not been given in any work," Babinet worked out expressions for multiple rainbows similar to those which Halley and Bernoulli had given long before. With respect to the supernumerary arcs Babinet expressed the opinion, before Airy's work had been published, that "the explanation of Young, adopted and published long ago by Arago, leaves nothing to be desired." [15] Miller, on the other hand, performed his own experiments with the express purpose of verifying the results of Airy. Using vertical cylindrical streams of water with diameters of from 0.0206 in. to 0.0135 in., and monochromatic light with index of refraction varying from $n = 1.3318$ to $n = 1.33453$, he was able in some cases to see as many as thirty bars corresponding to spurious arcs inside the primary bow, and up to twenty-five outside the secondary. But the aspect of his work which pleased him most was the "remarkable agreement" he noted between the theory of Airy and his own observations. With a diameter of 0.0206 in. and an index of refraction of 1.3318, for example, he observed the radii of the primary and secondary rainbows to be 41° 51'.4 and 51° 25' respectively. According to the geometric theory the values should have been 42° 15' and 50° 34', he calculated, and from the integral of Airy he computed radii of 41° 45'.4 and 51° 27'.5 respectively.[16] Other cases showed a correspondingly closer agreement between observation and Airy's theory which stood in marked contrast to the inaccuracy of the older geometric explanation.

The work of Airy and Miller on the rainbow was reported in scientific journals on the continent [17] and met with favorable reception, for in essentials the laboratory observations of Miller did not differ from the conditions under which rainbows are formed in nature. Such scepticism of Airy's theory as may have remained was dealt a further and more direct blow by careful observations of natural rainbows made by the German astronomer, J. G. Galle (1812–

1910) in 1843. The mean of his observations gave a radius of 41° 32′.8. He calculated that under the explanation of Airy the radius should be 41° 27′, and according to the theory of Descartes 41° 53′.6, again giving a very decided edge to the "complete" or diffraction theory.[18] Nevertheless, there remained a few who were not entirely convinced of the correctness of Airy's theory; and one of these, strangely enough, was Potter. In a paper of 1838 entitled "On the Radii and Distance of the Primary and Secondary Rainbows, as Found by Observation, and on a Comparison of Their Values with Those Given by Theory," [19] Potter claimed that observations of the bows agreed neither with the Cartesian nor the Airy theory. He asserted that the angular distance between the bows is observed to be 8° 30′, whereas according to the geometrical theory it should be 7° 25′ and according to Airy's work 7° 45′. He concluded that the discrepancy "leaves a strong presumption against the theory, and induces a corresponding argument in favour of the corpuscular theory, with which this fact is in accordance." [20] Again in 1855 he published a paper, with the heading "On the Interference of Light Near a Caustic and the Phenomena of the Rainbow," in which he asserted that Airy's theory "fails in so many cases when examined with impartial eyes." [21] Potter questioned in particular the prediction made by Airy of the existence of light outside the Cartesian ray. He admitted that Miller had claimed to have found results agreeing with this crucial prediction of the diffraction theory, but Potter questioned the accuracy of Miller's interpretation. Having tried some new experiments which he believed confirmed his stand that there is no light outside the rainbow caustic curve, Potter suggested that Miller's observation of such light may have been due to the fact that the velocity of the thin vertical stream of water may have increased slightly the refractive index. Potter, however, found no support for his suggestion, and one can not help wondering whether or not personal factors were responsible for his opposition to the work of Airy. In 1833 the future Astronomer Royal had criticized rather bluntly some mathematical calculations Potter had published in a paper on interference and the corpuscular theory of light.[22] Airy had predicted that Potter, if he continued his experimental work, would become an "undulationist"; but the latter replied, rather testily, that the probability of his turning undulationist became less day by day. Making allowance for stubborn adherence to a moribund theory and for the operation of personal factors (Airy and Miller won appointments to Cambridge professorships, but Potter had to be satisfied with an appointment at the University of London), it is possible that Potter's opposition to the modern theory stemmed, at least in part, from some inadequacy in his mathematical background. The theory of the rainbow had become so intricate that even the calculus, with which Potter was familiar, failed to suffice. The story of the rainbow had become inextricably interwoven with the development of mathematics.

Airy had shown how to account for the variability in the radius of the rainbow, but his table of intensities had been quite circumscribed, including only the first two relative maxima of illumination—that is, the principal bow and the first supernumerary arc. Miller had extended the table to include other values, but no one had undertaken the arduous and almost thankless task of computing values of the rainbow integral for large numbers of drop sizes and apparent radii. The problem of the white rainbow had not, consequently, been definitively disposed of. The view that the white bow is caused by non-spherical droplets had been espoused by a number of scientists, including Giambatista Venturi (1746–1822),[23] the man who in 1814 had rediscovered Theodoric's *De Iride*; and in France the older vesicular theory continued to boast at least one ardent exponent. The last stronghold of the vesicular theory was August Bravais (1811–1863), a French crystallographer, who in two papers of 1845 defended the thesis that the droplets producing the "cercle d'Ulloa" are hollow spherical shells. The ratio of the inner and outer radii, he held, is fixed for drops of a given size, and this ratio increases with the size of the drops. He believed that if the ratio is about 1.38 or 1.40, the white bow with a radius of 33° to 35° is formed; and as the ratio gradually increases, the size of the bow increases, until, for a ratio of 1.555 or greater, the white bow is replaced by the ordinary colored rainbow with a radius of 41° 38′. Bravais gave a geometrical explanation (Fig. 65) in which he argued that the effective rays for the white bow are those which are tangentially incident to the outer surface (shades of Kepler's theory once more), and also those which, after the first refraction and reflection, are tangent to the interior surface.[24] A few years later he wrote another article, tracing historical developments in the story of the rainbow from Theodoric to Airy, and although he accepted the theory of Airy to account for the ordinary rainbow with its spurious bands, he again advanced his peculiar hypothesis for the white bow.[25] Nevertheless, his contemporaries, including even Potter, judged the vesicular theory unfavorably, holding it to be quite improbable.

By the middle of the nineteenth century the modern theory of the rainbow had triumphed over virtually all obstacles but one—indifference. An article in the *Encyclopaedia Britannica* for 1858 concluded, on the basis of the work of Airy and Stokes, that "at last we begin to believe that we understand this matter [the formation of the rainbow] completely." [26] Inescapable difficulties in computation, however, had greatly hampered the adoption of Airy's theory by authors of textbooks and popular expositors. In 1857 F. Raillard presented to the Académie des Sciences a paper, "Explication nouvelle et complète de l'arc-en-ciel," in which he called attention to the fact that Airy's theory

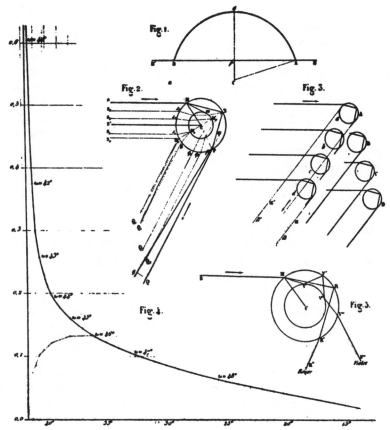

FIG. 65.—Diagrams illustrating the hypothesis of Bravais that the white rainbow is formed by hollow drops.

"accords perfectly with the observed facts," including the formation of the white rainbow. He presented tables of values comparing the radii of rainbows according to Descartes and as given by Airy. "All of these," he reported, "show clearly how illusory was the theory of effective rays." And he added that they show also the falsity of the hypothesis of vesicular state. He closed with the words, "I dare to hope that these two hypotheses will from now on be banished from the teaching of science which they have fettered." [27]

But almost a decade later Raillard complained that the Cartesian-Newtonian explanation continued to hold sway in physics books, whereas only the theory of Airy sufficed to explain the phenomenon of the rainbow with all of its "variations and accessories." He called particular attention to four respects in which only the new theory accorded with experience, as shown by Miller and Galle: 1) the intensity of the geometric arc is only 0.442 of the intensity of the first maximum; 2) the deviation of the first maximum from the position of least deviation is variable, depending on the size of the drops; 3) the illumination of the least-refracted rays extends above the Cartesian arc; and 4) the new theory "explains perfectly" the white rainbow.[28] In Germany, too, one heard the same complaint—that because of the difficulty of the mathematics, most textbooks of physics did not include the theory of Airy. (From India in 1852 one heard the complaint that even the simpler calculations of the interference theory were not available in books.[29]) To counteract this situation, expository papers on the complete theory appeared with mounting frequency in scientific periodicals of the later nineteenth century. One of the more ambitious of these was an account in Wiedemann's *Annalen der Physik* for 1883 of new experimental results carried out by Carl Pulfrich (b. 1858). Using glass cylinders instead of Miller's columns of water, Pulfrich verified the new explanation for multiple rainbows, concluding that "Airy's theory holds for all bows of higher order." [30] Experimental work on rainbows of unusually high order had been given from 1863 to 1868 by Félix Billet (1808–1882), a professor at Dijon; but he had not related this to the new theory. Through vertical cylindrical threads of water Billet passed in turn slits of monochromatic light of five different colors. The scientific identification of spectral colors had been facilitated by Fraunhofer's discovery of the dark absorption lines in the solar spectrum, some of the most prominent of which he designated by letters of the alphabet. The colors Billet used were thus designated by the letters B (the red lithium line), D (the yellow sodium line), E (the green magnesium line), F (a bluish hydrogen line), and H (a potassium line). For each of these five colors he noted the positions of the feet corresponding to the rainbows of the first nineteen orders, and he represented these graphically on a chart which he called a "rose" (Fig. 66). This chart depicts vividly the relative positions of the feet of the bows (for orders seventeen, eighteen, and nineteen only one of the feet is illustrated), and it indicates clearly how the width of the bands increases as the number of internal reflections becomes greater. The rainbows of orders eighteen and nineteen, for example, appear in roughly the same direction as the primary bow (number 1 in the lower left corner of Figure 66), but they are more than ten times as broad as the first. The widths given in Billet's rose are not quite so great as they are in nature, for there are colors at the extreme ends of the spectrum which he did not use in his experiments. The third and fourth bows, for example, really should overlap as they are actually seen, but Billet's chart does not indicate their full extent.[31] With such precision were his observations carried out that he was able to make an experimental study of the supernumerary arcs of the first eleven rainbows.[32] From such elaborate studies one looks back with bemusement at the ancient rejection of the possibility of even a tertiary arc!

It is easy to fall into the error, after a contribution to the story of the rainbow, of believing that the ultimate explanation finally has been achieved. One must never assume that the last word has been written, even though the author may have been Airy. While the diffraction theory of the rainbow was making its way in the scientific world, profound changes were taking place in the theory of light. At Cambridge the mathematician Stokes had not only facilitated the computation of values of Airy's intensity integral, but he and George Green (1793–1841) were working out the mathematical details of the undulatory hypothesis, showing that the familiar properties of light—rectilinear propagation, reflection, refraction, interference, polarization, diffraction, and others—are consistent with the assumption of a transverse vibration. One of the prominent difficulties in the wave theory was that of explaining dispersion, the very basis of the formation of the rainbow. Why should lights of different wave lengths have different velocities, as inequalities in refrangibility imply? The corpuscular theory had attributed dispersion to differing attractions between the corpuscles and the medium. The great mathematician Augustin Cauchy (1789–1857), however, had shown by a complex analysis that longer waves should be expected to travel more rapidly, and hence to be least refracted; and Airy, too, had reconciled dispersion with the vibratory theory.[33] Finally, the last remnants of the corpuscularian faction were swept away in 1850 by the crucial experiments of Fizeau and Foucault which showed that the velocity of light is indeed greater in air than in water, contrary to the conclusion which Descartes and Newton had had to draw. As Foucault put it, the theory of corpuscles was "incompatible with the truth of the facts." [34] But even as the wave theory swept the field, strangely disquieting new observations on light were being made. Young, Fresnel, Cauchy, and Airy (and later also Green and Stokes) had tacitly assumed that the luminiferous ether behaves like an elastic solid. Not long afterward Michael Faraday (1791–1867) found it necessary to postulate the existence of a medium with very similar properties in order to account for the transmission of electrical and magnetic forces. Faraday discovered not only that electricity and magnetism are closely related, but also that they affect the polarization of light. He announced in 1845 that he had succeeded in "magnetizing and electrifying a ray of light, and in illuminating a magnetic line of force." [35] In 1862 James Clerk Maxwell (1831–1879) deduced the fact that the elasticity of the magnetic medium is the same as that of the luminiferous ether. In the face of the coincidence of the speeds of propagation of light and of electromagnetic effects, Maxwell concluded that "We can scarcely avoid the inference that light consists in the transverse undulations of the same medium which

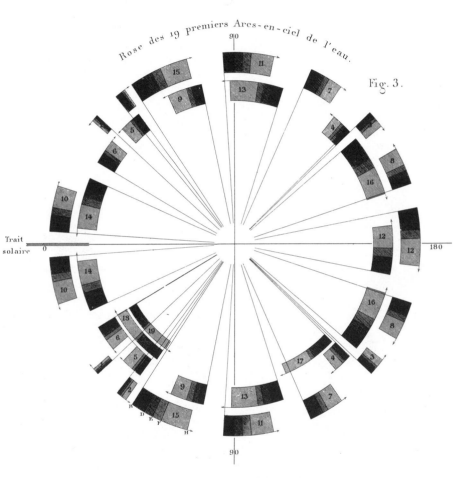

FIG. 66.—Billet's "Rose," showing the position, breadth, and color arrangement (sense of deviation) for each of the rainbows of the first nineteen orders.

is the cause of electric and magnetic phenomena." [36] In 1873 he published his great treatise on *Electricity and Magnetism* in which, through the so-called "Maxwell equations," he gave a coherent mathematical explanation of the simpler optical phenomena (including reflection and refraction) in terms of the electromagnetic theory of light. His equations are analogous to those for mechanical vibrations in an elastic solid ether, but he interpreted them in terms of electric and magnetic displacements. That is, light waves are a combination of two transverse vibrations, one electrical and one magnetic, polarized in planes perpendicular to each other. But is this really the way in which the colors of the rainbow are carried to the eye? Do such waves actually exist? Can they be produced and detected? Maxwell had put them down on paper; it remained for Heinrich Hertz (1857–1894) to put them into the air—to give physical meaning to the mathematical link between light and magnetism. In 1888 Hertz produced and studied the electromagnetic waves which now are known as radio waves. Their properties are in most respects (including the phenomena of reflection, refraction, interference, and diffraction) similar to light waves—except for size, for they are measured in meters rather than microns (thousandths of a millimeter). The story of the rainbow finally had reached our age—the radio age. But the difficulties are not yet over. The Maxwell-Hertzian waves in an electromagnetic field cannot be reconciled with two other essential elements in the theory of the rainbow—dispersion and diffraction. The interaction of matter and energy, such as takes place when rays of light traverse a raindrop, continues so to baffle analysis that it is not possible to say that the nature of light is satisfactorily understood. The structure of vibrating atoms is vastly more complicated than the few simple motions proposed by Huygens, Newton, Young, Fresnel, and Maxwell. In 1887, the year before Hertz discovered his waves, Stokes wrote of the wave theory as "a thing at the present day resting on evidence quite overwhelming." [37] Today the situation has changed so radically as to occasion the facetious remark that physicists accept the wave theory three days a week and hold to the corpuscular doctrine three days a week, and every seventh day they humbly admit their lack of understanding. If light consists of waves, how does one account for the fact that its energy is not continuous but quantized? On the other hand, if light is bundles of energy, how is one to explain interference phenomena, one of the great triumphs of the vibratory theory?

Maxwell, professor of physics at Cambridge, was interested in a wide range of physical phenomena, but he did not contribute directly to the story of the rainbow. Indirectly, through his work on the nature of light and color, his ideas were of considerable significance. He revived, in modified form, the theory of color which Young had proposed, for his experiments had indicated that there are three primary colors. Young had proposed red, green, and violet

as the basic hues, but Maxwell found that red, green, and blue are more satisfactory. He made such extensive use of the well-known color triangle, originally suggested by Newton and Young, that frequently it is known as "Maxwell's color triangle." This triangle, which is used to illustrate the relationships between complementary colors, is formed by placing at the vertices of
an equilateral triangle the three fundamental colors red, green, and blue, and
by choosing for every other point of the triangle a combination of the primary
colors in such a proportion that, if three weights at the vertices of the triangle
were in this same proportion, the center of gravity would lie at the point in
question. The center of gravity of the triangle, corresponding to equal proportions of the three basic hues, will then be white.

The mathematical and experimental versatility of Maxwell was rivaled only
by that of his successor as Cavendish professor at Cambridge, Baron Rayleigh, a student of the atmosphere and of abnormal psychology. Lord Rayleigh made no discovery which captured the popular imagination (unless the
discovery of argon can be thus described), but he added analytic refinement
to innumerable branches of physics. It long had been known, for example,
that the blue of the sky was due to the scattering of light by particles in the
air; but whereas it had been thought that the particles responsible were dust,
Rayleigh in 1870 demonstrated mathematically that they had to be much
smaller, and that they were in fact molecules of air.[38] For particles approaching
in size the wave-length of light, he showed that the "Rayleigh scattering" of
light varies inversely as the fourth power, and that hence far more blue than
red rays are returned to the eye.[39] In a somewhat analogous role in the story
of the rainbow, Lord Rayleigh extended the mathematical analysis of the bow.
Airy had derived his intensity integral on the hypothesis of the wave theory
of light, but this is not entirely adequate when the minuteness of the drops
concerned is such that they become of the order of magnitude of a wave-
length of light—that is, somewhat less than a millionth of a meter. In this case
the application of the electromagnetic theory of light leads to improved representation of the phemonena, and this was where Rayleigh made his contribution to the story of the rainbow. In 1879 and 1881 he began a project of
extending the application of Maxwell's electromagnetic theory to various optical phenomena, including the passage of light through an infinitesimal sphere.[40]
This was side-tracked by competing interests in his many-sided career, but
almost thirty years later another paper on the passage of light rays through
infinitesimally small spheres was published posthumously.[41] In this he showed
that when the circumference of the sphere is equal to the wave-length of the
light, the behavior is not markedly different from what it would be for an
infinitely small sphere. When the ratio of circumference to wave-length is
more than two, however, certain characteristics change markedly. Polarization,

for example, is just the reverse of what it would be for an infinitesimal sphere. This showed that the theory of Airy could not be extended downward to ever smaller spheres by a simple extrapolation procedure. As a matter of fact, two experimenters, both of whom published their findings in the same year, 1888, had found discrepancies between results calculated by means of the so-called "complete theory" and their own observations. Boitel, in an article on the supernumerary arcs,[42] recalled that Airy had calculated the intensities for only the first two bows or maxima, but that Stokes had given a quicker method of calculation which had enabled him to find the positions of many more bands. Boitel then derived, from the formula for the wave front, $y = kx^3$, the expression

$$\tan \theta = \frac{m}{(54)^{\frac{1}{3}}} \frac{(\lambda^2 \sin I)^{\frac{1}{3}}}{\cos I} \cdot \frac{1}{R^{\frac{2}{3}}}$$

giving the angular deviation θ (between the emergent ray in question and the lease deviated ray) in terms of the angle of inclination I of the ray, the wavelength λ of the light, the radius R of the drop, and the quantity which Airy had designated by m. By a strange coincidence Larmor had derived independently and at about the same time an essentially similar formula [43] for the angular separation of the supernumerary bands from the position of the geometric bow. But whereas Larmor noted the fairly close agreement between the calculated and observed positions of the first twenty-three bands, Boitel, comparing his computed deviations with the experimental results of Miller and Pulfrich, found "some disaccord." He concluded that "the theory of Airy seems to need to be completed." Moreover, through very carefully conducted experiments of his own, Boitel again obtained results not entirely consistent with those from the Airy integral, and he suggested that "the theory of Airy therefore is but a first approximation." How familiar a situation this is in the history of science, and especially in the story of the rainbow! What one generation hails as a "complete theory," the next relegates to the status of a "first approximation."

Eleuthère Mascart (1837–1908) in 1888 also reported on a formula, derived from the theory of Airy by the celebrated mathematician Henri Poincaré (1854–1912), for the angular deviation of the supernumerary fringes. Moreover, with experimental refinements, he was able to see two hundred of these, produced with a fine thread of glass, and to measure "exactly" the first hundred of these. Comparing the observed positions of the arcs with those calculated by formula, he too found discrepancies.[44] He concluded that "there still are several points to elucidate" in the theory of the rainbow. And this again is a characteristic of the development of science: no matter how refined an explanation may be, there always remains some point calling for further investigation. Mascart was especially fond of the study of the rainbow, spending considerable pains in computing values of the intensity integral by alternative devices; and sometimes the nineteenth-century diffraction explanation of the rainbow is known as the Airy-Stokes-Mascart theory. He published other papers on the rainbow, including one in 1892 on the white bow,[45] in which he estimated that achromatism sets in for drops of diameter 72.5 microns, and that for diameters of only 30 microns all color is gone except for an outside border tinged lightly with red. Drops in mist and fog he reported as varying from 6 to 100 microns, for beyond this upper limit the fog tends to be resolved into rain. Because of the fascination which the rainbow held for him, Mascart devoted an unusually substantial section of his multi-volume Traité d'Optique[46] to the explanation of the bow.

The two most important centuries in the story of the rainbow without any doubt were the seventeenth and the nineteenth; but whereas the contributions of Descartes and Newton before long had taken their place in elementary textbooks of physics, those of Young and Airy were not readily welcomed in the less advanced expository works. The inherent difficulty of the treatment of the bow from the point of view of interference and diffraction constituted a handicap not easily overcome; but toward the turn of the century the explanation of the rainbow found an indefatigable champion who made it a point to spread knowledge of its theory at every opportunity and on all levels. This was Josef Pernter (1848–1908), professor and dean of the faculty of philosophy at the University of Vienna. He complained repeatedly that the Cartesian theory still dominated instruction, and he sought to introduce the complete theory on the secondary school level.[47] He contributed numerous expository and research papers to learned periodicals, especially of the Akademie der Wissenschaften at Vienna,[48] seeking to stir up interest in what to him evidently was the most fascinating of studies. And in his ambitious book on meteorological optics,[49] completed after his death by Felix M. Exner, he devoted some seventy-five pages to the rainbow (and again as much to the halo). Pernter took particular pains to point out the inadequacy of the Cartesian explanation and to clarify the theory of Airy. As he pointed out, the phrase "Cartesian effective ray" is misleading. It is not the only effective ray; and, indeed, Airy had shown that among all of the effective rays, it is not even the most efficacious. The phrase "least deviated" is the correct designation for Descartes' ray. Then, too, Pernter objected to the name "supernumerary arcs" and "interference bows" for the bands inside the principal bow. They are, he said, no more supernumerary than is the primary bow, nor is the primary bow any less an interference rainbow. He preferred to have them known as "secondary bows," but this conflicts with the customary use in English, and hence his terminology has not been followed. Pernter's most useful contribution to the rainbow was a painstaking experimental verification of the theory of Airy. He extended the calculations of intensities of illumination through the rainbow integral, and he constructed tables and charts showing how the rainbow changes in appearance with changes in the size of the raindrops. For very large drops several millimeters in diameter, as in a deep tropical rain, there may be insufficient uniformity in the size to produce a rainbow (although in general uniformly large drops produce bright rainbows). If the colors of the principal rainbow arc are clear and bright, with an intense and wide violet band, and if the supernumerary arcs include rose, green, and blue, the drops causing the bow are about 1 or 2 mm. in diameter. If the red in the principal bow is weak, if there is no interval between the principal arc and the supernumeraries, and if the latter arcs include only green and violet, the diameter of the drops is about 0.5 mm.[50] If the primary bow has no red, and if there is a narrow interval between the principal arc and the supernumeraries, the drops are about 0.3 mm. in diameter; and if yellow appears in the first supernumerary arc, the drops are not more than 0.2 mm. across. If there are several spurious arcs not well separated, and if they show only a bright violet-rose and a weak blue-green, the drop-diameter is close to 0.1 mm. When the first spurious arc is clearly separated from the principal bow and includes some white, the size of the drops is from 0.08 to 0.1 mm.; and if a white streak appears in the principal arc, the drops are only 0.06 mm. Finally, a white bow with a tinge of orange along the outer edge and a little blue along the inner border signifies drops of the order of 0.05 mm. Figure 67, taken from Pernter's work, illustrates schematically some of these variations. The white bow in particular, he noted, is the best foundation for Airy's theory; but he cautioned that one should not confuse the true white bow with the pale lunar rainbow, the achromatism of which is due merely to the weakness of the source of light. The white rainbow is not generally seen during a rainfall because a falling drop has a radius about ten times that appropriate for the production of a white bow. Within the white bow, the width of which is about double that of the ordinary rainbow, one or two supernumerary bows may be seen; but the order of colors is not the same as that of the ordinary supernumerary arcs seen by Witelo and Theodoric. As the size of the drops approaches more closely to the minute wave-length of light, the radius of the red arc becomes smaller than that of the blue, and therefore the red band lies within the blue—just the reverse of the usual primary and supernumerary arrangement. In this case the size of the drops can be computed from the formula $r = \dfrac{.012}{(41.7 - \theta)^{3/2}}$, where r is the radius of the drop in inches and θ is the apparent radius in decimal degrees of the dark ring between the primary bow and the first supernumerary arc.[51] The fog or mist producing the white bow may be very tenuous indeed;

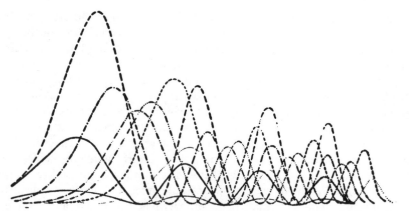

FIG. 67.—Pernter's diagram illustrating the overlapping of maxima and minima of various colors.

and such a bow is sometimes seen in temperatures as low as 0° Fahrenheit, indicating that the tiny droplets can be greatly supercooled without freezing. On rare occasions a double fog-bow had been observed.[52] Pernter reiterated that the geometrical theory cannot account for any of these variations, "and inasmuch as it holds only for drops of infinite radius, it must be completely abandoned." While Pernter's extreme position is of course correct, in a strict sense, the Cartesian theory has remained the backbone of the schoolboy's explanation of the rainbow because it affords so eminently clear and direct a first approximation.

The missionary zeal with which Pernter spread information on the Cambridge diffraction theory of the rainbow could not entirely overcome the inertia of expository writers, and the explanation of Descartes remains to this day solidly intrenched in the more elementary textbooks. His heroic efforts, nevertheless, were not in vain, for he had successors in research and exposition both at home and abroad. The voluminous *Treatise on Physics* of O. D. Chwolson (b. 1852), which appeared in Russian, German, French, and English during the first decade of the twentieth century, gave full recognition to Pernter's support of Airy's theory and referred to Descartes' explanation as "completely false." [53] And at Leipzig in 1907, the year before Pernter died, the young Willy Möbius (b. 1879) earned his doctorate with a thesis entitled *Zur Theorie des Regenbogens und Ihrer Experimentellen Prüfung*. This constituted a thoroughgoing examination, on an advanced level, of the mathematical and experimental work of Airy, Miller, Stokes, Pulfrich, Mascart, Pernter, and others. It had been inspired in part by Christian Wiener (1826–1896), a scholar who had included a highly technical analytic treatment of the modern rainbow theory as an incidental portion of a monumental study of the brightness of the sky.[54] A few years later there was offered at Leipzig a prize for the best essay on the "Theory of the Rainbow," and in 1912 the winning essay, by Möbius, was published.[55] The chief function of this work was to fill a gap which had been left between the calculations of Rayleigh and those of Mascart. The former, using the electromagnetic theory, computed the intensities of illumination for drops for which the circumference is not greater than twice the wave length of light; Mascart, employing the Huygens-Fresnel wave theory, had extended the tables of Airy downward to include spherical drops so small that the diameter was about ten times the wave length of light. Between these two limits there was a gap which was closed by Möbius' computation of the intensities corresponding to drops from one to ten times the wave-length of light. Light wave-lengths are a fraction of a micron, and the droplets Möbius was concerned about had radii of from roughly a ten-thousandth to a hundred-thousandth of an inch. Inasmuch as raindrops ordinarily have a radius in the order of a fiftieth of an inch, it can be seen that Möbius' calculations applied only in the case of an exceedingly fine mist.

In the theory of the rainbow it is assumed that raindrops are spherical in shape, despite the fact that popular fancy persists in picturing them with the familiar "tear-drop" deformation. High-speed photography has shown that very large drops, far from being elongated, are if anything somewhat flattened, like a hamburger bun. The usual small-sized drops are almost perfectly spherical, due to the fact that the surface tension tends to reduce the surface of a free mass of liquid to the smallest possible surface area. (A cloud droplet of one micron radius has an internal pressure of more than two atmospheres.) Another wide-spread erroneous notion assumes that the speed of falling raindrops increases greatly as they approach the earth, whereas in actuality they reach a virtually terminal uniform speed of from 15 to 25 ft. per sec. after falling but a few yards.[56]

Was the theory of the rainbow now complete? The answer in science is always, "Of course not." Just so long as curiosity remains, there will be unsolved problems. Throughout the computations of the rainbow integral the light was assumed to come from a point source. Under the experimental conditions of the nineteenth century this assumption was virtually realized; but for the natural rainbow the source is the sun, and this has a very distinct breadth. In far away Tokyo two scientists tackled the problem of determining the intensity of illumination in a rainbow produced by a circular source of light. These Japanese mathematicians, Keiichi Aichi (b. 1880) and Aikischi Tanakadate (b. 1856), had been struck by the fact that, according to the theory of Airy, one should anticipate numerous supernumerary arcs, whereas in nature the bow generally is accompanied by only a very limited number. They suspected, from some approximations which Pernter had given, that this discrepancy might be accounted for by the fact that the sun is not a point source of light. Through an elaborate chain of further calculation and the application of the Maxwell color triangle, they successfully showed that, because the source is finite rather than infinitesimal in area, the supernumerary arcs of the natural rainbow lose most of their color, especially for large drops. They found, in other words, that if the sun were smaller, the rainbow colors would be clearer; or, to put it differently, rainbows on Mars could be better than those seen here on earth. Aichi and Tanakadate demonstrated that the degree of luminous intensity depends on the breadth of the source, as well as on the magnitude of the drops, and that at certain points the positions

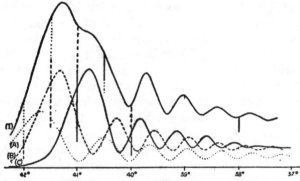

FIG. 68.—Diagram of Aichi and Tanakadate showing the overlapping of maxima and minima for light of various colors and for drops of different sizes.

of the maxima and minima may even be interchanged. Here Maxwell's theory of the compounding of colors accounts for the predominance of white in supernumerary bows produced by a source, such as the sun, with an angular diameter of about half a degree. Airy's theory was found to hold good only when the source is very narrow and the drops small. As the luminous source widens, and more particularly for relatively large drops, the color distribution varies considerably. Fig. 68, taken from the work of Aichi and Tanakadate, shows how the intermingling of the maxima tends to extinguish the boundaries of the supernumerary arcs as calculated from the theory of Airy for a point source, thus effectively limiting the number which are ordinarily seen. When the sun shines through a thin mist or haze, the effect is to magnify several fold the effective angular diameter of the source of light, and the increased overlapping of the colors tends to accentuate still further the whiteness in the rainbow.[57]

Benjamin Franklin, once asked by a "practical man" what was the use of the curious properties of electricity which he had discovered, replied, "What is the use of a baby?" The rainbow at the present time is perhaps no more "useful" than electricity was two centuries ago, and man continues to study it for the sense of satisfaction which understanding brings. But lest the pragmatist of the mid-twentieth-century should lightly dismiss the bow as completely without applicability to the workaday world, it may be well to point out that studies published in 1948 contain a suggestion that the rainbow may yet find a place in modern technology. Aircraft icing investigations have necessitated a more complete knowledge of the structure of clouds, especially of the free-water content, and the size of drops here becomes a matter of very immediate concern. In this connection two optical phenomena have been considered as an indirect means of determining drop size—the corona and the rainbow. The former seems to present greater practical difficulties, and hence the National Advisory Committee for Aeronautics in 1944 focussed upon the rainbow as a means of calculating how large the drops of water in a cloud are. From 1945 to 1947 a camera "rainbow recorder" was used, both in natural clouds and in experimentally-controlled fog chambers, to find the difference in viewing angle between the first two rainbow maxima—that is, the angle between the principal bow and the first supernumerary arc. Allowance had to be made for many factors, but the results nevertheless were encouraging— the rainbow-calculated drop diameters differed by as little as from 2% to 5% from those computed by other means. As reported by W. V. R. Malkus and his associates,

The fundamental soundness of the rainbow recorder theory has been proved by reasonably good agreement with several comparisons and checks against independent methods of measuring free-water content and drop diameter of fog.[58]

They point to the possibility of wider applicability of their methods, suggesting that rainbow recorders may be used in the study of particle size in carburation problems and in chemical problems dealing with liquid aerosols. The pursuit of such suggestions holds out the prospect of further changes in the theory of the rainbow, for relations between theory and practice generally are mutual. In fact, the Malkus group found, as had earlier observers, that Airy's theory "proves inadequate in the range of drop sizes from 10 to 50 microns of most common occurrence in clouds." They therefore substituted for the approximate wave-front equation previously used a more precise parametric

form: $$\begin{cases} x/a = \cos(4q - g) + G\cos(4q - 2g) \\ y/a = \sin(4q - g) + G\sin(4q - 2g) \end{cases}$$

where $\sin g = m \sin q$ and $G = 4m(1 - \cos q) - (1 - \cos g)$ and m is the index of refraction, g the angle of incidence, q the angle of refraction, and

320   The Rainbow

a the radius of the spherical drop. Following a chain of involved computations, they found that the intensity of light I_s for points on this curve near the cusp at a distance s from the x-axis is given by

$$I_s/I_B = \frac{R(1-R)^2 \sin 2g}{2(y/a)\, d(s/a)\, dg},$$

where
$$R = \frac{1}{2}\left[\frac{\sin^2(g-q)}{\sin^2(g+q)} + \frac{\tan^2(g-q)}{\tan^2(g+q)}\right]$$

and I_B is the intensity of the incident beam. For points not near the cusp, correspondingly more elaborate formulas, found by integrating over the surface, are obtained. That these new formulations are the last word on the rainbow is, of course, far from likely.

The twentieth century so far has not contributed to the story of the rainbow on so spectacular a scale as did the seventeenth and nineteenth centuries; but then it is perhaps too early to write it off as a prosaic century comparable to the eighteenth. There have been fundamental changes in ideas as to the nature of light, and these may be an augury of alterations to come in the theory of the rainbow. Through the work of H. A. Lorentz (1853–1928), Max Planck (1858–1947), Albert Einstein (1879–1955), and others, electricity and light have been found to have a degree of atomicity (in the electron and the photon) not anticipated by electromagnetic wave theory. For over a century physicists have been wrestling with the phenomena of reflection, refraction, interference, and diffraction, seeking to reconcile them with the properties of the medium postulated as the carrier of light. One looks in a mirror or through a magnifying glass, and accepts the reflection and refraction of light as a matter of course—as simple and familiar phenomena. Familiar they may be; but they are far from simple. Cauchy, Green, and Stokes were among the great mathematicians who studied these "simple" phenomena; and as a preliminary to their study they had to use powerful new tools of analysis: directional derivatives, potential theory, Laplacian operators, and line and surface integrals. Few look upon these as the modern counterparts of the ancient Greek myths concerning Iris; but such they are. How amazed would Anaxagoras be if he could know of the mathematical apparatus required to study even his simple reflection theory of the rainbow. Aesthetes were grieved at Newton's simple materialistic explanation of color. But if they could know of the intricate maze of creative mathematics required for the study of the rainbow, they would realize how misguided they are—or else how thoroughly avenged! Within a raindrop the interaction of light energy with matter is so intimate that one is led directly to quantum mechanics and the theory of relativity. It is assumed that it is the energy of moving electrons that is propagated through the ether as light; but under the theory of relativity the whole concept of energy has changed. The old dichotomy between matter and energy has been abandoned in favor of their equivalence through the famous equation $e = mc^2$, a relationship confirmed from the observed bending rays of light in a gravitational field. The corpuscular and wave theories of light seem to have far more in common than Newton and Huygens had ever dreamed, for in 1924 Louis de Broglie (b. 1892) established a synthesis of views in which light is made up of tiny entities dragging trains of waves with them.[59]

Descartes once boasted that he had put an end to the mystery of the rainbow; poets later lamented that Newton had despoiled the bow of its inner secrets. Scientists now glory in the ultimate refinements of Airy's explanation, and the story of the rainbow seems almost to have come to an end. But the curiosity of the true scientist is never fully satisfied. The answer to one question serves but to raise further problems. The story of the rainbow had passed from Iris to Mathesis through a mythological stage, a reflection stage, a refraction stage, a geometrical stage, a dispersion stage, an interference stage, and a diffraction stage. But although much is known about the production of the rainbow, little has been learned about its perception. It has long been known that the position and distance one attributes to the rainbow depend on psychological as well as physical factors. The background against which terrestrial and celestial objects are viewed is a very significant factor, as the phenomenon of the "horizontal moon" shows. But the psycho-mechanics of vision are not really understood, nor is the eye the perfect optical instrument one could wish for. It is known, for example, that if one estimates the altitude of a celestial object by eye and then determines it by means of a non-subjective instrument, the former estimate is always the greater.[60] That is, the sky, even in the absence of clouds, appears to be a somewhat flattened vault. But whether this is because of some physical factor, such as the luminosity of the background, or for purely entoptic reasons, is not known. Nor can one boast that the perception of the rainbow colors is satisfactorily understood. The Young-Helmholtz trichromatic theory has very respectable rivals,[61] and the whole question of what goes on between the eye and the brain when one sees a rainbow is pretty much in a state of flux.

Will the twenty-first century point to our age as the psychological stage in the story of the rainbow? Perhaps some of us will live to know. And what will remain then? Will mankind tire of the endless search for understanding concerning the nature of reflection, refraction, interference, diffraction, and the physiological processes which operate when one views the rainbow? Not unless human nature should somehow be fundamentally changed. As long as man by nature desires to know and yearns for beauty, just so long will Iris continue to inspire both exact science and romantic literature. For poets and fabulists the rainbow has served as a ubiquitous source of inspiration, but mathematics also has given to the bow a "beauty bare" which only the deeply initiated can fully appreciate.

Notes

I

1. William Wordsworth, *My Heart Leaps Up*.
2. *Genesis*, IX, 13.
3. *Iliad*, V, 350.
4. *Paradiso*, XII, 10-18.
5. James Patrick, "Rainbow," A *Dictionary of the Bible* (edited by James Hastings, New York, 1903), IV, 196.
6. Judah David Eisenstein, "Rainbow," *The Jewish Encyclopedia* (New York, 1905), X, 311-12.
7. G. L. Walls, "The Evolution of Color Vision," *Journal of Applied Physics*, XIV (1943), 161-65.
8. See K. T. A. Halbertsma, A *History of the Theory of Color* (Amsterdam, 1949); also Dora Ilse, "The Colour Vision of Insects," *Proceedings of the Royal Philosophical Society of Glasgow*, LXV (1940–41), 68-82.
9. Commandant Rouch, "La météorologie dans l'Iliade," *Académie de Marine. Communications et mémoires*, XIV (1935), 81-91.
10. Hesiod, *Theogony*, p. 265.
11. Apollonius Rhodius, *Argonautica*, II, 286. In Hesiod's account, however, it was Hermes who made the brothers turn back.
12. Virgil, *Aeneid*, V, 606.
13. Wolfgang Menzel, *Mythologische Forschungen und Sammlungen* (Stuttgart and Tübingen, 1842), pp. 235-76.
14. James A. and Vincent A. FitzSimon, *The Gods of Old and the Story That They Tell* (London, 1899), pp. 119-21.
15. Article on "Iris" in A *Dictionary of Greek and Roman Biography and Mythology* (edited by William Smith, 3 vols., London, 1876). A more extensive account of the role of Iris in mythology is found in W. H. Roscher, *Ausführliches Lexikon der griechischen und römischen Mythologie* (Leipzig, 1890–97), vol. II, part I, cols. 320-57.
16. Aristophanes, *The Birds*, p. 575.

17. Shakespeare, Henry VI, second part, iii, 2, 407.
18. Aydin Sayili, "The Aristotelian Explanation of the Rainbow," Isis, XXX (1939), 65-83, especially pp. 82-83.
19. Hesiod, Theogony, p. 777.
20. Frazer, The Golden Bough (third edition, New York, 1935), III, 258.
21. Ezekiel, I, 28.
22. Eisenstein, op. cit.
23. Patrick, loc. cit.
24. Eisenstein, op. cit.
25. Charles Daremberg and Edmund Saglio, "Iris," Dictionnaire des antiquités grecques et romaines (Paris, 1900), vol. III, part 1.
26. Jacob Grimm, Teutonic Mythology (translated from the fourth edition, London, 1880-88), II, 731-34.
27. Gerhard Bähr, "El arco iris y la vía láctea en Guipúzcoa (vocables, etimologías y difusión, creencias populares)," Revista International de los Estudios Vascos, XXII (1931), 397-414.
28. For these and similar allusions see Canon John Arnott Mac Culloch, The Mythology of All Races (13 vols., Boston, 1930-32), II, 278, 329, and passim.
29. Ch. Renel, "L'arc-en-ciel dans la tradition religieuse de l'antiquité," Revue de l'Histoire des Religions, XLVI (1902), 58-80.
30. Supplement to volume X, pp. 18-19.
31. Hoffmann-Krayer and Hanns Bachtold-Stäubli, "Regenbogen," Handwörterbuch des deutschen Aberglaubens (Berlin and Leipzig, 1935-36), vol. VII, cols. 586-97.
32. C. M. Blaas, "Die Regenbogenschüsselchen," Berichte und Mittheilungen aus Alterthums-Vereines zu Wien, XXI (1882), 149-51.
33. Franz Streber, "Ueber die sogenannten Regenbogenschüsselchen," Abhandlungen der Philosoph.-Philologischen Classe der Königlich Bayerischen Akademie der Wissenschaften, Munich, IX (1860), 165-276, 547-730.
34. Samuel Merian, Die französischen Namen des Regenbogens (Halle, 1914).
35. George Sarton, "A History of Science." Ancient Science Through the Golden Age of Greece (Cambridge: Harvard University Press, 1952), pp. ix, 194.

II

1. Gustav Hellmann, "Die Anfänge der Meteorologie," Meteorologische Zeitschrift, XX (1908), 481-91
2. Ibid.

3. Menzel, *loc. cit.*

4. Otto Gilbert, *Die meteorologischen Theorien des griechischen Altertums* (Leipzig, 1907), p. 606.

5. Cicero, *De natura deorum*, III, 20, 51.

6. Gilbert, *loc. cit.*

7. I. B. Cohen, "Roemer and the First Determination of the Velocity of Light (1676)," *Isis*, XXXI (1940), 327-79.

8. Gilbert, *loc. cit.*

9. *Timaeus*, 46. The translation is that of B. Jowett in *The Dialogues of Plato* (second edition, 5 vols., Oxford, 1875), III, 628-29.

10. *Theaetetus*, 155D.

11. *Cratylus*, 408B.

12. For the little that is known of "Philipus Platonis Sodalis" see Pauly-Wissowa, *Realencyclopädie*. For the suggestion that Philip may have held a theory of the rainbow based on refraction, see my paper, "Refraction and the Rainbow in Antiquity," *Isis*, XLVII (1956), 383-86.

13. Wilhelm Capelle, "Aus der Vorgeschichte einer Fachwissenschaft," *Archiv für Kulturgeschichte*, X (1912), 1-24. This is a résumé of ideas expressed by the author in more technical papers in *Philologus*, LXXI (1912), 414-48 and *Hermes*, XLVII (1912), 514-35.

14. An account of Aristotle's work on reflection was given in my paper on "Aristotelian References to the Law of Reflection," *Isis*, XXXVI (1946), 92-95.

15. *Physica*, II, 2, 194a. Translations here are taken from *The Works of Aristotle* (edited by W. D. Ross and J. A. Smith, 11 vols., Oxford, 1908–31).

16. *De sensu*, II, 438a.

17. *Meteorologica*, III, 2, 372a. A treatise on optics by Aristotle is cited by the ancients but appears to be lost. See Albert Rochas, *La science des philosophes* (second edition, Paris, 1912), p. 27, footnote. Cf. also Gino Loria, *Le scienze esatte nell' Antica Grecia* (second edition, Milano, 1914), p. 568. However, see also A. Sayili, *op. cit.*, especially p. 75.

18. *Meteorologica*, III, 2, 372a-372b.

19. *Meteorologica*, III, 4, 373-374. Aristotle frequently lapses into a Platonic terminology and speaks of the reflection of *sight* rather than *light* (See *De caelo*, II, 8, 290.), but in general he rejected the doctrine of visual rays for his theory of light as an activity in a pellucid medium. (See *De sensu*, II, 437b.) As far as the law of reflection and the rainbow are concerned, the question of the mechanics of vision has no special significance.

20. For analyses of this viewpoint see Sayili, *op. cit.*, and Fr. Poske, "Die Erklärung des Regenbogens bei Aristoteles," *Zeitschrift für Mathematik und Physik*, XXVIII (1883), 134-38. See also Aristotle, *Météorologie* (translated by J. Barthélemy Saint-Hilaire, Paris, 1863) for extensive notes on Aristotle's text.

21. Sayili, *op. cit.*

22. *Meteorologica*, III, 5, 377b.

23. See, for example, *Problemata*, XI, 45, 904a, 58, 905.

24. *Problemata*, XI, 23, 901b. Cf. 51, 904b. The word refraction is, of course, to be taken as meaning reflection. See editor's note to *Problemata*, XI, 45, 904a, or the English translation by W. S. Hett (2 vols., Cambridge, Mass., 1936–37), pp. 268-69. The *Problemata* probably is not by Aristotle himself, but it is thoroughly Peripatetic and may be regarded as derived largely from the master's teaching.

25. *Analytica posteriora*, II, 15, 98a.

26. There is no reference to it, for example, in D. C. Miller, *Anecdotal History of the Science of Sound* (New York, 1935). This achievement would tend to indicate that the science of sound in antiquity was not exclusively concerned with "a semi-mystical arithmetic of music." (H. T. Pledge, *Science Since 1500*, London, 1939, p. 16.) Incidentally, there is in Book XIX of the *Problemata* an extensive treatment of harmonics.

27. Carl Sondhauss, "Ueber die Refraction des Schalles," *Annalen der Physik und Chemie*, LXXXV (1852), 378-84.

28. *Problemata*, XVI, 13, 915b. Cf. 3-4, 913b. See also Josephus Blancanus, *Aristotelis loca mathematica ex universis ipsius operibus collecta, et explicata* (Bononiae, 1615), pp. 243-46. Here Biancani more than three hundred years ago drew the obvious conclusion that, the explanation of the rainbow notwithstanding, Aristotle knew the law of reflection.

29. I. B. Hart, *Makers of Science* (London, 1923), p. 50.

30. *Problemata*, XI, 49, 904b; 58, 905; XV, 5, 911; 9, 912; XXV, 9, 939.

31. In *De anima*, II, 8, 419b-420a one finds a further illustration of this consistency. We read, "An echo occurs when a mass of air . . . rebounds . . . like a ball from a wall. . . . What happens here must be analogous to what happens in the case of light." The dispersion of light rays by small or uneven objects is then compared with the dissipation of sound which falls upon a rough surface. As reflection of light from "water, bronze, and other smooth bodies" results in the formation of images, so does an echo occur when the surface upon which sound impinges is quite smooth.

32. The original diagrams of Aristotle have not survived. The figures used here are based upon those furnished by later compilators. See, for example, Charles Graux and Albert Martin, "Figures tirées d'un manuscrit des Météorologiques d'Aristote," Revue de Philologie, de Littérature et d'Histoire Anciennes, new series, XXIV (1900), 5-18.
33. Meteorologica, 373a.
34. The editor of the Loeb Classical Library edition of Aristotle's Meteorologica (Cambridge, Mass., 1952), p. 241, mistakenly adds in a note, "The size of the circle does not in fact vary." The radius of the bow is indeed subject to a variation of half a dozen degrees; but this variation is not related, as Aristotle thought, to the altitude of the sun.
35. In this respect Aydin Sayili (op cit., p. 77), in his otherwise thoroughly admirable account, fails to give Aristotle full credit.
36. De sensu, 439, and De anima, 418.
37. Gilbert, op. cit., p. 606. See also Halbertsma, op. cit.
38. Pierre Gassendi, Opera omnia (Florentiae, 1727), III, Caput VI.
39. Meteorologica, 374b.
40. De caelo, 268a.
41. Meteorologica, 372a.
42. Ibid., 375a.
43. Ibid.
44. Meteorologica, 373b.
45. See Vasco Ronchi, Storia della luce (second edition, Bologna, 1952).
46. Terminology in this respect is not entirely uniform. In Germany the phrase "secondary rainbow" has a different meaning, for it refers instead to what now are known as "supernumerary arcs." See chapter X.
47. Meteorologica, 375b.
48. Cf. Sayili, op. cit., pp. 73-74.
49. Meteorologica, 375b.
50. E.-M. Antoniadi, "La météorologie en grèce antique," Bulletin de la Société Astronomique de France, XLV (1931), 373-83, says that Anaximenes first published the observation that the moon produces rainbows; but I have not been able to verify this report.
51. Meteorologica, 372a.
52. Meteorologica, 375a.
53. W. J. Humphreys, "Why We Seldom See a Lunar Rainbow," Science, new series, LXXXVIII (1938), 496-98.
54. Wilhelm Capelle, "Zur Geschichte der meteorologischen Litteratur," Hermes, XLVIII (1913), 321-58.
55. Gilbert, op. cit., pp. 604-18.

56. J. L. Ideler, *Meteorologia veterum Graecorum et Romanorum* (Berolini, 1832), p. 191.
57. Karl Reinhardt, *Poseidonios* (München, 1921), pp. 162-66.
58. Virgil, *Aeneid*, V, 88 and IV, 700.
59. John Clarke, *Physical Science in the Time of Nero, Being a Translation of the Quaestiones Naturales of Seneca* (London, 1910), pp. 16-33.
60. Rochas, *op. cit.*, p. 25.
61. Florian Cajori, *A History of Physics* (New York, 1906), p. 16. Antoniadi, *op. cit.*, reports that astronomical refraction was known earlier to Aristotle.
62. George Sarton, *Introduction to the History of Science* (3 vols. in 5, Baltimore, 1927–47), I, 274. Cf. Otto Neugebauer, *The Exact Sciences in Antiquity* (Princeton, 1952), p. 146.
63. Cortes Pla, *El enigma de la luz* (Buenos Aires, 1949), p. 68.
64. Albert Lejeune, *Euclide et Ptolémée. Deux stades de l'optique géométrique grecque* (Louvain, 1948). Université de Louvain. *Recueil de Travaux d'Histoire et de Philologie*, series 3, fascicule 31. The opinions of Posidonius and Ptolemy on the number of colors show that it is incorrect to hold, as does Sayili, *op. cit.*, p. 67, that the Aristotelian theory of three colors went unchallenged until the time of Theodoric in the fourteenth century.
65. G. Théry, O. P., "Autour du décret de 1210: II.—Alexandre d'Aphrodise. Aperçu sur l'influence de sa noétique," *Bibliothèque Thomiste*, VII (1926); Paul Moraux, "Alexandre d'Aphrodise. Exégète de la noétique d'Aristote," *Bibliothèque de la Faculté de Philosophie et Lettres de l'Université de Liège*, fascicule 99 (1942). The bibliography of the latter work shows how much more fully Alexander's philosophy has been treated than his science.
66. *Alexandri Aphrodisiensis maximi peripatetici, in quatuor libros Meteorologicorum Aristotelis, commentatio lucidissima, Alexandro Piccolomineo interprete* (Venetiis, 1548), folio 35v. An earlier edition of this work appeared, also at Venice, in 1540. A Greek version, edited by Michael Hayduck, was published at Berlin in 1899.
67. *Loc. cit.*
68. F. C. E. Thurot, "Observations critiques sur les Meteorologica d'Aristote," *Revue Archéologique*, new series, XX (1869), 415-20; XXI (1870), 87-93, 249-55, 339-46, 396-407.
69. *In quatuor libros meteorologicorum*, folio 32v.
70. *Ibid.*, folios 33v-34r.
71. *Ibid.*, folio 34r.

III

1. Quoted from John Tyndall, *Six Lectures on Light* (New York, 1895), p. 13.
2. *Recognitions*, VIII, 42, cited on the basis of Immanuel Hoffmann, "Die Anschauungen der Kirchenväter über Meteorologie," *Münchener Geographische Studien*, XXII (1907), 1-96. See especially pp. 80-82.
3. Hoffmann, *loc. cit.*
4. Raymond Vancourt, "Les derniers commentateurs alexandrins d'Aristote," *Facultés Catholiques de Lille, Mémoires et Travaux*, LII (1941), 1-66.
5. A third Greek commentary by John Philoponus covers only Book I of the *Meteorologica* and hence does not touch on the rainbow. Excerpts from the commentaries of Alexander, Olympiodorus, and Philoponus are found in Aristotle, *Meteorologicorum libri IV* (edited by J. L. Ideler, 2 vols., Lipsiae, 1834-36).
6. For the little that is known of the life of Olympiodorus see the article "Olympiodorus" by R. Beutler in Pauly-Wissowa, *Realencyclopädie*, XVIII (1), 207-27.
7. *Olympiodori Philosophi Alexandrini in Meteora Aristotelis Commentarii ... Ioanne Baptista Camotio Philosopho interprete* (Venetiis, 1551), folios 57-74. I have used the copy of this rare book in the library of the University of Virginia. The British Museum has also an edition of the Camotius translation of Venice, 1567. The Greek is readily available in an edition by Wilhelm Stüve, Berlin, 1900.
8. *Ibid.*, folio 60.
9. *Ibid.*, folio 63r.
10. For a charming account of some of these see M. Minnaert, *The Nature of Light and Colour in the Open Air* (translated by H. M. Kremer-Priest, London, 1940), reprinted by Dover Publications (New York, 1954) as *Light and Color in the Open Air*.
11. Olympiodorus, *op. cit.*, folio 64.
12. See the meteorological excerpts from Isidore's *De natura rerum* in Gustav Hellmann, *Neudrucke von Schriften und Karten über Meteorologie und Erdmagnetismus* (15 nos., Berlin, 1893-1904), XV, 15. Cf. Hoffmann, *loc. cit.*
13. Hellmann, *op. cit.*, XV, 7-10.
14. Hellmann, *op. cit.*, XV, 11-19.
15. Eilhard Wiedemann, "Arabische Studien über den Regenbogen," *Archiv für die Geschichte der Naturwissenschaften und der Technik*, IV (1912-13), 453-60.

16. This has been translated into English. See *Book of Treasures*. *Encyclo-paedia of Philosophical and Natural Sciences As Taught in Baghdad About 817* (Syriac text edited and translated by A. Mingana, Cambridge, 1935).

17. See *Book of Treasures*, pp. 208-10.

18. See Sarton, *Introduction to the History of Science*, I, 660-61, for references concerning this organization.

19. See "Meteorologie der 'Läuteren Brüder' " in Hellmann, *Neudrucke*, XV, 23-41, especially pp. 38-39.

20. For a summary of this work, with references, see Sarton, *Introduction to the History of Science*, I, 709-13.

21. See *Isis*, VI, 138.

22. Eilhard Wiedemann, "Ibn Sina's Anschauung vom Sehvorgang," *Archiv für die Geschichte der Naturwissenschaften und der Technik*, IV (1912–13), 239-41. See also *Journal of the History of Medicine*, I (1946), 330-34.

23. M. Horten and E. Wiedemann, "Avicenna's Lehre vom Regenbogen nach seinem Werk al Schifa," *Meteorologische Zeitschrift*, XXX (1913), 533-44.

24. K. Lokotsch, *Avicenna als Mathematiker* (Erfurt, 1912).

25. Horten and Wiedemann, *op. cit.*, p. 542.

26. For references covering his work see Sarton, *Introduction to the History of Science*, I, 721-23.

27. Michael Jan de Goeje, "Notice biographique d'Ibn al-Haitham," *Archives Néerlandaises des Sciences Exactes et Naturelles*, series 2, VI (1901), 668-70.

28. The *Treasury of Optics* has not been translated into English, but the Latin edition, *Opticae thesaurus* (edited by Risner, Basileae, 1572), is available in many libraries.

29. Eilhard Wiedemann, "Theorie des Regenbogens von Ibn al Haitam," *Sitzungsberichte der Physikalish-Medizinischen Societät in Erlangen*, XLVI (1914), 39-56. See also another article by Wiedemann, "Über die Brechung des Lichtes in Kugeln nach Ibn al Haitam und Kamal al Din al Farisi," in the same journal, XLII (1910), 15-58, and his paper, "Zu Ibn al Haitams Optik," in *Archiv für die Geschichte der Naturwissenschaften und der Technik*, III (1910–12), 1-53, especially pp. 44-45.

30. Joseph Würschmidt, "Die Theorie des Regenbogens und des Halo bei Ibn al Haitam und bei Dietrich von Freiberg," *Meteorologische Zeitschrift*, XXXI (1914), 484-87.

31. See Wiedemann, "Arabische Studien über den Regenbogen," cited above in note 15.
32. Aristote. Stagyrite Meteorum libri quatuor: cum Aver. cordubensis exactiss. commentariis (Lugduni, 1530), folios 47-53. See also Christoph Gottlieb Volkamer, Thaumantiados thaumasia, sive iridis admiranda (Noribergae, 1699), pp. 51, 103.

IV

1. Translated from the text of De imagine mundi, Book I, Chapter 58, as given in Patrologia latina (edited by J. P. Migne, 221 vols., Paris, 1844–55), CLXXII, column 137.
2. See the meteorological excerpts from the Dragmaticon in Hellmann, Neudrucke, XV, 42-68, especially p. 47.
3. Les méthéores d'Aristote (A translation of the thirteenth century by Mahieu Le Vilain, published for the first time by Rolf Edgren, Uppsala, 1945).
4. Ibid., pp. xxvii, 136, 154.
5. For a list of his works see Sarton, Introduction to the History of Science, II (2), 829-31.
6. J. C. Russell, "Hereford and Arabic Science in England About 1175–1200," Isis, XVIII (1932), 14-25.
7. Much valuable information in this connection is found in A. C. Crombie, Robert Grosseteste and the Origins of Experimental Science 1100–1700 (Oxford, 1953).
8. I have not seen this. I have made use here of D. A. Callus, "Introduction of Aristotelian Learning to Oxford," Proceedings of the British Academy (London), XXIX (1943), 229-81, and August Pelzer, "Une source inconnue de Roger Bacon, Alfred de Sareshel, Commentateur des Météorologiques d'Aristote," Archivum Franciscanum Historicum, XII (1919), 44-67.
9. George Lacombe, "Alfredus Anglicus in Metheora," Beiträge zur Geschichte der Philosophie und Theologie des Mittelalters, Supplementband, III (1), 463-71. Cf. Callus, op. cit., pp. 236-50.
10. Pelzer, op. cit., pp. 63-64.
11. C. Baeumker, "Die Stellung des Alfredus van Sareshel (Alfredus Anglicus) und seiner Schrift De Motu Cordis in der Wissenschaft des Beginnen den XIII Jahrhunderts," Sitzungsberichte der Königlich Bayerischen Akademie der Wissenschaften, Philos.-philolog. und Hist. Klasse, München, 1913, Abh. IX.
12. The best account of this work is in Crombie, Robert Grosseteste. See also

Robert Grosseteste, Scholar and Bishop. Essays in Commemoration of the Seventh Centenary of His Death (edited by D. A. Callus, Oxford, 1953); F. S. Stevenson, Robert Grosseteste, Bishop of Lincoln (London, 1899); and S. Harrison Thomson, The Writings of Robert Grosseteste, Bishop of Lincoln (Cambridge, 1940). For Grosseteste's relations to the Franciscans, see A. G. Little, "The Franciscan School at Oxford in the Thirteenth Century," Archivum Franciscanum Historicum, XIX (1926), 803-74. For further references see the bibliography in Crombie's Robert Grosseteste.

13. Crombie, Robert Grosseteste, pp. 10, 61. The novelty and modernity of Grosseteste's work are perhaps here exaggerated, but this valuation is a welcome antidote to the prevalent lack of appreciation of the medieval period.

14. Crombie, Robert Grosseteste, p. 51.

15. This is published in Beiträge zur Geschichte der Philosophie des Mittelalters, IX (1912), 72-78. See also pp. 59-65. For a list of early manuscript copies see Thomson, op. cit. Passages in English translation appear in Crombie, Robert Grosseteste, pp. 119-26.

16. De iride (Beiträge, IX), pp. 74-75.

17. Ibid., p. 75.

18. Ibid., pp. 75-77.

19. Commentary on the Posterior Analytics, quoted from Crombie, Robert Grosseteste, pp. 113-14.

20. Crombie, op. cit., p. 115.

21. Crombie (op. cit., p. 125, note 2) suggests instead that Grosseteste here may have been influenced by Averroës.

22. Crombie, op. cit., p. 116.

23. See Grosseteste, De fractionibus et reflexionibus radiorum (published at Nuremberg in 1503 as Libellus de phisicis lineis angulis et figuris per quas omnes acciones naturales complentur). This is found also in Bibliotheca Mathematica (3), I (1900), 55-59, II (1901), 443-44.

24. See Thomson, op. cit.

25. This is found in Albertus Magnus, Opera (edited by Peter Jammy, 21 vols., Lugduni, 1651), vol. II.

26. Opera, II, 125-26.

27. Opera, II, 130-31.

28. Eisenstein, op. cit., pp. 311-12.

29. Opera, II, 131-33.

30. Crombie, op. cit., pp. 198-200.

31. Opera, II, 133.

32. Crombie, op. cit., pp. 198-99.

33. See excerpts on meteorology in Hellmann, *Neudrucke*, XV, 124f.
34. *Ibid.*, pp. 201-20; also Bartholomew Anglicus, *Medieval Lore* (Gleanings from his *Encyclopedia*, edited by Robert Steele, London, 1893), pp. 23-24. Cf. also Carle Salter, "Medieval Meteorology," *Symons's Meteorological Magazine*, XLIV (1909), 141-44.
35. Sarton, *Introduction to the History of Science*, II (2), 915. See also pp. 914-21.
36. A. C. Crombie, *Augustine to Galileo. The History of Science* A. D. *400–1650* (London, 1952), p. 42.
37. *S. Tho. super meteo. Habes solertissime lector in hoc codice Aristotelis stagirite in libros Meteororum cum duplici interpretatione antiqua et Francisci vatabli: Expositore divo Thoma Aquinate: cuius lucidissima commentaria: nunc primum in lucem exeunt cum indice copiosissimo* (Venetiis, 1537), folios 44-54; or see *Sancti Thomae Aquinatis . . . Opera omnia* (iussu impensaque Leonis XIII. P. M. edita, 16 vols., Romae 1882–1948), vol. III, p. LXXV-CX, especially pp. XC and C; or see *Sancti Thomae Aquinatis . . . Opera omnia* (16 vols., Romae, 1882–1948), vol. III, Book III, pp. LXXV-CX.
38. Edmund Hoppe, *Histoire de la physique* (translated by Henri Besson, Paris, 1928), p. 318.
39. *The Opus Majus of Roger Bacon* (a translation by Robert Belle Burke, 2 vols., Philadelphia, 1928), I, 234-35. See also I, 46, 234; II, 608.
40. *Ibid.*, I, 235.
41. *Ibid.*, II, 592.
42. *Ibid.*, II, 601.
43. *Ibid.*, II, 609.
44. *Ibid.*, II, 610.
45. *Ibid.*, II, 611-12.
46. Sebastian Vogl, "Roger Bacons Lehre von der sinnlichen Spezies und vom Sehvorgange," in A. G. Little, *Roger Bacon Essays* (Oxford, 1914), 205-27. See also the other essays in this well-known collection, as well as Vogl, *Die Physik Roger Bacos* (Erlangen, 1906).
47. Crombie, *Robert Grosseteste*, p. 161.
48. See Clemens Baeumker, "Witelo, ein Philosoph und Naturforscher des XIII Jahrhunderts," *Beiträge zur Geschichte der Philosophie des Mittelalters*, III (2), 1908; also Aleksander Birkenmajer, "Études sur Witelo," in *Bulletin International de l'Académie Polonaise des Sciences et des Lettres, Classe d'Histoire et de Philosophie* (Cracow), 1918–20, 1922, 1926. For further references on Witelo see Sarton, *Introduction to the History of Science*, II (2), 1027-28.
49. Witelo, *Opticae* (Basileae, 1572). This is bound with Alhazen, *Opticae*

thesaurus, of the same place and date. The two treatises are separately paginated. The section on the rainbow appears in Book 10, pp. 457-74 of Witelo's *Opticae.*

50. *Opticae*, p. 457.
51. J. C. Poggendorff, *Geschichte der Physik* (Leipzig, 1879), p. 92.
52. For example, by Crombie, *Robert Grosseteste*, p. 219. Here one finds a convenient and full account of Witelo's work on refraction.
53. Crombie, *Robert Grosseteste*, p. 275, does essentially this.
54. *Opticae*, p. 474.
55. *Ibid.*, p. 471.
56. *Ibid.*, pp. 464-65.
57. See the *Summa philosophiae* of the Pseudo-Grosseteste in *Beiträge zur Geschichte der Philosophie des Mittelalters*, IX (1912), 275-643, especially pp. 617-19. Cf. Pierre Duhem, *Système du monde* (8 vols., Paris, 1913–58), III, 460-71; Crombie, *Robert Grosseteste*, p. 164. For other aspects of the *Summa* see C. K. McKeon, *A Study of the Summa philosophiae of the Pseudo-Grosseteste* (New York, 1948).
58. For biographical notes and bibliographical references, see Sarton, *Introduction to the History of Science*, II (2), 1028-30. For an account of Peckham's ecclesiastical and administrative activities, see Decima L. Douie, *Archbishop Peckham* (Oxford, 1952),
59. See John Peckham, *Perspectiva communis* (Norimbergae, 1542), Book III, propositions 18-21. This book appeared at Leipzig in 1504, in a rare edition at Paris in 1511, and in many other editions from 1482 to 1627; but the editions differ little from each other.
60. *Paradiso*, XII, 10-12.
61. *Ibid.*, XXXIII, 118-20.
62. *Purgatorio*, XXIX, 77-78. The girdle of Delia is, of course, the halo about the moon.

V

1. The fullest account of his life and work is by E. Krebs, "Meister Dietrich (Theodoricus Teutonicus de Vriberg). Sein Leben, seine Werke, seine Wissenschaft," *Beiträge zur Geschichte der Philosophie des Mittelalters*, V (5-6), 1905–06. See also Duhem, *Système du monde*, III, 383-96.
2. A full chapter in Crombie, *Robert Grosseteste*, is devoted to Theodoric's explanation of the rainbow.
3. Theodoric of Freiberg, *De iride et radialibus impressionibus* (Dietrich von Freiberg *Ueber den Regenbogen und die durch Strahlen erzeug-*

ten Eindrücke, edited by J. Würschmidt), in *Beiträge zur Geschichte der Philosophie des Mittelalters*, XII (5, 6), 1914. See p. 37. References below to *De iride* are also to this published edition.

4. Krebs, *op. cit.*, p. 49.
5. *De iride*, pp. 36-37.
6. *De iride*, p. 38.
7. *De iride*, p. 47.
8. *De iride*, p. 61.
9. *De iride*, p. 63.
10. *De iride*, p. 62.
11. *De iride*, p. 92.
12. *De iride*, p. 97.
13. *De iride*, pp. 138-40.
14. *De iride*, p. 134.
15. *De iride*, pp. 149-51.
16. *De iride*, pp. 161, 164.
17. *De iride*, p. 159.
18. *De iride*, p. 167.
19. This is suggested by Krebs, *op. cit.*, p. 49. See also Crombie, *Robert Grosseteste*, p. 252.
20. See especially *De iride*, pp. 126f and 141-42.
21. *De iride*, pp. 129-33.
22. *De iride*, pp. 166-70, especially p. 169.
23. *De iride*, p. 150.
24. *De iride*, p. 195. Cf. Krebs, *op. cit.*, p. 29.
25. Brief modern accounts of the main points in Theodoric's work are found in many places. The earliest is in G. B. Venturi, *Commentarj sopra la storia e le teorie dell'ottica* (Bologna, 1814), especially pp. 149f. A French translation of part of this is found in *Annales de Physique et Chemie*, VI (2), 1817, pp. 141-59. See also Hellmann, *Neudrucke*, XIV. By far the best published summary is that found in Chapter IX of Crombie, *Robert Grosseteste*.
26. For summaries of their work see Sarton, *Introduction to the History of Science*, II (2), 1017-20, III (1), 707-08.
27. A. M. Sayili, "Al Qarafi and his Explanation of the Rainbow," *Isis*, XXXII (1940, published 1947), 14-26. See also Wiedemann's article, "Arabische Studien über den Regenbogen," cited above.
28. Sayili, *op. cit.*, p. 19.
29. For his life and work see Sarton, *Introduction to the History of Science*, II (2), 1001-13.
30. H. J. J. Winter and W. Arafat, "A Statement on Optical Reflection and

'Refraction' Attributed to Nasir ud-Din al-Tusi," *Isis*, XLII (1951), 138-42.
31. Cortes Pla, *El enigma de la luz* (Buenos Aires, 1949).
32. I have here depended on Sarton, *Introduction to the History of Science*, II (2), 1018.
33. See p. 42 of Wiedemann's article, "Theorie des Regenbogens von Ibn al Haitam," cited above.
34. There is no adequate account of this work in English. See Eilhard Wiedemann, "Ueber die Brechung des Lichtes in Kugeln nach Ibn al Haitam und Kamal al Din al Farisi," *Sitzungsberichte der Physikalisch-Medizinische Sozietät in Erlangen*, XLII (1910), 15-58. See also his article on "Theorie des Regenbogens von Ibn al Haitam," in the same journal, XLVI (1914), 39-56. Cf. his articles, "Zur Optik von Kamal al Din," *Archiv für die Geschichte der Naturwissenschaften und der Technik*, III (1910–12), 161-77, and "Ueber das Sehen durch eine Kugel bei den Araben," *Annalen der Physik und Chemie*, new series, XXXIX (1890), 565-76. Relevant comments are found also in J. Würschmidt, "Ueber die Brennkugel," *Monatshefte für den Naturwissenschaftlichen Unterricht*, IV (1911), 98-113.
35. See p. 56 of Wiedemann's article, "Ueber die Brechung des Lichtes in Kugeln nach Ibn al Haitam und Kamal al Din al Farisi," cited in the note above.
36. See p. 49 of Wiedemann's article, "Theorie des Regenbogens von Ibn al Haitam," cited in note 34.
37. Relevant portions of the *Buch der Natur* are reproduced in Hellmann, *Neudrucke*, XV (1904), 221-38. See also Hellmann *Meteorologische Volksbücher* (Berlin, 1891), p. 11.
38. See Pierre Duhem, "Physics, History of," in *Catholic Encyclopedia*, XII, 47-67. See also, Konstantyn Michalski, "La physique nouvelle et les différents courants philosophiques au XIVe siècle," *Bulletin International de l'Académie Polonaise des Sciences et Lettres, Classe d'Histoire et de Philosophie*, 1927, pp. 93-164. Cf. Duhem, *Système du monde*, IV, 128f.
39. M.-D. Chenu, "Aux origines de la science moderne," *Revue des Sciences Philosophiques et Théologiques*, XXIX (1940), 206-17.
40. As, for example, is very critically done by Anneliese Maier, "Die Anfange des physikalischen Denkens im 14. Jahrhundert," *Philosophia Naturalis* (Meisenheim am Glan), I (1950), 7-35.
41. J. R. Weinberg, *Nicolaus of Autrecourt. A Study in Fourteenth Century Thought* (Princeton, 1948), p. 100.
42. A judicious evaluation of his proper place in the history of science is given

by D. B. Durand, "Nicole Oresme and the Medieval Origins of Modern Science," *Speculum*, XVI (1941), 167-85.
43. See Sarton, *Introduction to the History of Science*, III (2), 1486-97.
44. For further details see Pierre Duhem, *Études sur Léonard de Vinci* (3 vols., Paris, 1906-13), I, 159f. See also Duhem, *Origines de la statique* (2 vols., Paris, 1905-06), II 48f, 94f, 326f; Sarton, *Introduction to the History of Science*, III (2), 1539-40; Crombie, *Robert Grosseteste*, pp. 261-68.
45. I have used especially the edition of Themo's *Quaestiones* edited by George Lokert (or Lockhart), published at Paris in 1518. There is a copy at the University of Pennsylvania. I have used also a copy of a manuscript of Themo's *Quaestiones* (Ms Vat. Lat. 2177, 14c, folios 1-92), but references are to the printed edition. I have compared the Themo manuscript with a manuscript copy of Oresme's *Quaestiones* (St. Gall MS 839), finding the material to be very much the same. Neither of the manuscripts has any diagrams, whereas the printed edition of Themo's *Quaestiones* has a number of clearly drawn figures. That the content of these manuscripts is similar to that in Buridan's *Quaestiones* is apparent from the description of the latter given by E. Faral, "Jean Buridan. Notes sur les manuscrits, des éditions et le contenu de ses ouvrages," *Archives d'Histoire Doctrinale et Littéraire du Moyen Age*, XV (1946), 1-53. For the similarity with the work of Albert of Saxony I depend on the authority of Sarton, introduction to the History of Science, III (2), 1428-32, 1539-40.
46. We here are following the material as found in *Quaestiones et decisions physicales insignium virorum: Thimonis in quatuor libros meteorum* (Parisiis, 1518), folios CLXXVII-CCIIII. The *incunabula scientifica et medica* of A. C. Klebs [*Osiris*, IV(1938), 1-359] cites editions at Pavia in 1480 and at Venice in 1496. There was another Paris edition in 1516 and another Venice edition in 1522.
47. Folio CLXXXV, recto and verso.
48. Crombie, *Robert Grosseteste*, pp. 265f, 275, appears to make too much of vague medieval references to differential refraction as anticipations of Newton's discovery.
49. *Quaestiones*, folios CXCII-CXCIII. The Latin of this passage is found also in Crombie, *Robert Grosseteste*, pp. 266-67, note.
50. In the article on "Physics, History of," in *Catholic Encyclopedia*. See especially p. 50. A similar exaggeration is found in Crombie, *Augustine to Galileo*, p. 81: "Theodoric's theory was not forgotten; it was discussed by Themon Judaei later in the 14th century." Cf. also p. 352.

51. The *Quaestiones* of Oresme has also been ascribed to Albert of Saxony. See Lynn Thorndike, "Oresme and Fourteenth Century Commentaries on the Meteorologica," *Isis*, XLV (1954), 145-52.
52. Cf. Duhem, *Système du monde*, III, 383.
53. Crombie, *Robert Grosseteste*, p. 233.
54. *Ibid.*, pp. 260-69.
55. This work is found in Joannes Duns Scotus, *Opera omnia* (new edition, 26 vols., Parisiis, 1891–95), IV, 1-263, especially pp. 200-08. See also Pierre Duhem, "Sur les meteorologicorum libri quatuor faussement attribués a Jean Duns Scot," *Archivum Franciscanum Historicum*, III (1910), 626-32.
56. *Opera*, IV, 208.
57. See Crombie, *Robert Grosseteste*, p. 268, note, for references.
58. Some scepticism of the broad claims often made in this connection is shown by George E. Nunn, "The Imago Mundi and Columbus," *American Historical Review*, XL (1935), 646-61.
59. Sarton, *Introduction to the History of Science*, III (2), 1108.
60. Pierre d'Ailly, *Tractatus Petri de Eliaco . . . super libros metheororum* (Argentinae, 1504), folios XVIII verso-XIX recto. I have used the copy in the De Golyer collection at the University of Oklahoma.
61. *Ibid.*, folio XIX verso.
62. *Ibid.*
63. I have used a microfilm at Columbia University of Gaetan de Thiene, *Meteorologicorum . . . cum commentariis* (Venetiis, 1491). For a recondite study of the relations among medieval versions of the *Meteorologica* see F. H. Fobes, "Medieval Versions of Aristotle's Meteorology," *Classical Philology*, X (1915), 297-314. Cf. also pp. 188-214.

VI

1. Crombie, *Robert Grosseteste*, p. 269.
2. See the listings in Klebs, "Incunabula scientifica et medica," cited above.
3. A. R. Hall, *The Scientific Revolution* (London, 1954), p. 11.
4. Wilberforce Eames, *A List of Editions of the Margarita philosophica, 1503–1599* (New York, 1886). The New York Public Library has copies of a dozen editions of the *Margarita*, including those of 1503, 1504, 1508, 1512, 1517, 1535, 1583, 1599, and 1600.
5. Book IX, Chapter xxii. The edition of 1503 is unpaginated. In the Italian edition of 1599 this material is on pp. 550-51.
6. Meteorological excerpts from the *Margarita philosophica* are given in

Hellmann, Neudrucke, XV (1914), 243-69. See especially pp. 265-67. An account of this material is found also in Vasco Ronchi, Storia della luce (second edition, Bologna, 1952), pp. 49-51.

7. Meteorologia Aristotelis. Eleganti Iacobi Fabri Stapulensis paraphrasi explanata. Commentarioque Ioannis Coclai Norici declarata (Norinbergae, 1512), folios LXXXIIII verso-LXXXVI verso. This book is of interest also for a very early reference (folio 62 verso) to "nova illa America terra."

8. Totius naturalis philosophiae Aristotelis paraphrases per Iacobum Fabrum Stapulensem recognitiae iam . . . et scholijs doctissimi viri Iodoci Clichtouei illustratae (Friburgi Brisgoiae, 1540), folio cxli verso; or folio 272 verso of the edition of 1501.

9. Alexandri Aphrodisiensis maximi peripatetici in quatuor libros Meteorologicorum Aristotelis commentatio lucidissima: quam Latinitate donavit Alexander Piccolomineus . . . Accedit insuper eiusdem Alexandri Piccolominei Tractatus de iride (Venetiis, 1540), folios 59-64, especially 64 recto.

10. Article on "Physics, History of," in Catholic Encyclopedia, XII, 50.

11. De iride, folio 61 recto.

12. See Chapter XLII, "For and Against Aristotle," in Lynn Thorndike, History of Magic and Experimental Science, VI, 363-89.

13. Quoted from W. T. Sedgwick, H. W. Tyler, and R. P. Bigelow, A Short History of Science (New York, 1939), p. 268.

14. The Notebooks of Leonardo da Vinci (edited by Edward MacCurdy, Garden City, 1941-42), p. 284.

15. Ibid., pp. 925-26.

16. This material is found in De subtilitate (Norimbergae, 1550), Book IV, pp. 97-102; or see his Opera (10 vols., Lugduni, 1663), III, 420f.

17. Julius Caesar Scaliger, Exotericarum exercitationum lib. XV. De subtilitate ad Hieronymum Cardanum (Francofurti, 1576), pp. 297-312.

18. See Crombie, Augustine to Galileo, for a good general account of medieval technology.

19. Leonard Digges, A Prognostication of Right Good Effect (London, 1555, reprinted at Oxford, 1926), p. 27.

20. Ibid., p. 6. For other items of folklore and popular notions on the rainbow in the sixteenth century see Thorndike, History of Magic, V, 657, VI, 80, 267, 368, 432.

21. Francisci Vicomercati Mediolanensis in quatuor libros Aristotelis Meteorologicorum Commentarij (Venetiis, 1565), folios 148-49. Rainbow material is found in folios 131-56.

22. Ibid., folios 149-51.
23. Ibid., folio 156.
24. Crombie, Robert Grosseteste, p. 273, suggests an opposite conclusion.
25. See Henry Crew, The Photismi de lumine of Maurolycus. A Chapter in Late Medieval Optics (New York, 1940). This is an English translation of the treatise in question, the Latin version having appeared in several editions under varying titles. I have used also two editions titled Theoremata de lumine (Lugduni, 1613 and 1617), but references below are to the Crew edition. For correction of misstatements by Crew concerning the Photismi de lumine, see Edward Rosen, "The Title of Maurolico's Photismi," American Journal of Physics, XXV (1957), 226-28.
26. Photismi (Crew), p. 103.
27. Photismi (Crew), p. 93.
28. Quite a number of modern historians have misinterpreted his use of the word refraction. See, e.g., Paul F. Schurmann, Historia de la física. Anales de la Universidad. Entrega no. 139, 1936. República Oriental del Uruguay, Montevideo. See p. 160. Cf. A. Mieli, Panorama general de la historia de la ciencia (7 vols. in 4, Buenos Aires, 1950–54), VI, 104.
29. Photismi (Crew), p. 132.
30. Ibid., pp. 92-93.
31. Ibid., p. 99.
32. Ibid., p. 131.
33. Ibid., p. 130.
34. Ibid., p. 103. Cf. p. 131.
35. Johannes Fleischer, De iridibus doctrina Aristotelis et Vitellionis, certa methodo comprehensa, explicata, et tam necessariis demonstrationibus, quam physicis et opticis causis (Witebergae, 1579, i.e., 1571), Chapter VI, pp. 138-74. A good summary of Fleischer's ideas is given by A. G. Kaestner, Geschichte der Mathematik (4 vols., Göttingen, 1796–1800), II, 248-50. A briefer account is available in J. E. Montucla, Histoire des mathématiques (new edition, 4 vols., Paris, 1799–1802), I, 701-02.
36. De iridibus, p. 90.
37. Trattato di M. Francesco de Vieri, Cognominato il Verino Secondo . . . nel quale si contengono i tre primi libri delle metheore. Nuovamente ristampati e da lui ricorretti con l'oggiunta del quarto libro (Fiorenza, 1582), p. 135.
38. Ioannes Demerlierius, Iridis coelestis, et coronae brevis descriptio (Parisiis, 1576), folio ciij verso.

39. See the article "Conimbricenses" by John J. Cassidy in *Catholic Encyclopedia*.
40. *De refractione optices parte: libri novem* (Neapoli, 1593), pp. 42, 54, 64, 223, 224.
41. See Ronchi, *Storia della luce*.
42. *De refractione*, pp. 200, 202.
43. *De mundo nostro sublunari, philosophia nova* (Amstelodami, 1651), Chapters X-XIII of Book IV, pp. 270-76.
44. Schurmann, *op. cit.*, p. 160.
45. For the next few pages the account follows closely the material in our article, "William Gilbert on the Rainbow," *American Journal of Physics*, XX (1952), 416-21.

VII

1. The next few pages follow closely the account given in my article on "Kepler's Explanation of the Rainbow," *American Journal of Physics*, XVIII (1950), 360-66.
2. Johann Kepler, *Opera omnia* (edited by Ch. Frisch, 8 vols., Francofurti a. M. and Erlangae, 1858–70), I, 200.
3. *Opera*, I, 425.
4. *Opera*, II, 119-397.
5. *Opera*, II, 100.
6. For the little that is known of his life see Kepler, *Opera*, II, 37.
7. See Pierre Gassendi, *Opera omnia* (second edition, 6 vols., Fiorentiae, 1727), II, 86-93.
8. *Opera*, II, 67-71.
9. *Opera*, II, 71-72.
10. *Opera*, I, 570.
11. *Opera*, II, 530.
12. See Kepler's *Epistolae* (edited by M. G. Hansch, Lipsiae, 1718), letters 152 and 223.
13. For Kepler and refraction see his *Opera*, IV, 176-226. See also H. Bögehold, "Kepler's Gedanken über das Brechungsgesetz und ihre Einwirkung auf Snell und Descartes," in *Kepler Festschrift* (edited by Stöckl, 1930), pp. 150-67.
14. Ambrosius Rhodius, *Optica* (Witebergae, 1611), pp. 410-18.
15. Charles Singer, in *Studies in the History and Method of Science* (2 vols., Oxford, 1917–21), II, 413, thinks that, "in view of the well-known dishonesty of the notorious Archbishop of Spalato," De Dominis

may have been indebted for ideas on the telescope to Kepler's *Dioptrice*.

16. Antonio de Dominis, *De radiis visus et lucis* (Venetiis, 1611), p. 14. Cf. p. 77. Attention is called to some of the more egregious errors of De Dominis by R. E. Ockenden, "Marco Antonio de Dominis and His Explanation of the Rainbow," *Isis*, XXVI (1936), 40-49. An erroneous criticism of Ockenden's criticism of an error in De Dominis is made by Crombie in his learned *Robert Grosseteste*, p. 273, note 3.

17. *De radiis visus et lucis*, pp. 75-76.

18. *Ibid.*, pp. 64-65.

19. *Ibid.*, p. 65.

20. *Ibid.*, p. 62.

21. *Ibid.*, pp. 66, 71.

22. *Ibid.*, pp. 65-70.

23. It has been generally assumed that De Dominis did perform experiments on a globe of water, but doubt of this was expressed almost two centuries ago by Joseph Priestley, *History and Present State of Discoveries Relating to Vision, Light, and Colours* (London, 1772), p. 51.

24. For references see Ockenden, *op. cit.*, pp. 45-49.

25. E. White, *Modern College Physics* (New York, 1948), p. 401. Cf. Florian Cajori, *A History of Physics* (New York, 1929), p. 96.

26. See Christiaan Huygens, *Oeuvres complètes* (22 vols., La Haye, 1888–1950), XVII, 357.

27. See P. van Geer, "Notice sur la vie et les travaux de Willebrord Snellius," *Archives Néerlandaises des Sciences Exactes et Naturelles*, XVIII (1883), 453-68. Cf. J. A. Vollgraff, "Snellius' Notes on the Reflection and Refraction of Rays," *Osiris*, I (1936), 718-25.

28. Blancanus, *Aristotelis loca mathematica*, pp. 115-30.

29. Bartholomew Keckermann, *Systema compendiosum totius mathematices* (Hanoviae, 1621), pp. 162-63.

30. Galileo Galilei, *Opera* (Edizione Nazionale, 20 vols., Firenze, 1890-1909), VI, 447f. Cf. pp. 296, 373.

31. Francis Bacon, *Works* (edited by James Spedding and Robert Leslie Ellis, new edition, 14 vols., London, 1887–1902), V, 164.

32. *Ibid.*, IV, 295.

33. Cunradus Cellarius, *Partitiones meteorologicae* (Tubinga, 1627), pp. 517-39.

34. L. Dufour, "Esquisse d'une histoire de la météorologie en Belgique," *Institut Royal Météorologique de Belgique, Miscellanées*, fasc. XL, 1950. Dufour gives a date of 1613 for the first edition, but this seems to be an error.

35. Libertus Fromondus, *Meteorologicorum libri sex* (Oxoniae, 1639), p. 439. The material on the rainbow is found on pp. 415-73. I have seen also the editions of 1627 (Antwerp), 1656 (London), and 1670 (London), but there is little to distinguish the various editions.
36. Article XIII of Book VI.
37. *Meteorologicorum*, pp. 450-60.
38. *Ibid.*, pp. 461-63.

VIII

1. See *Oeuvres de Descartes* (edited by Charles Adam and Paul Tannery, 12 vols. and supplement, Paris, 1897–1913), VI, 1-78.
2. *Oeuvres*, VI, 367-485.
3. *Oeuvres*, VI, 79-228.
4. See J. W. Shirley, "An Early Experimental Determination of Snell's Law," *American Journal of Physics*, XIX (1951), 507-08.
5. *Oeuvres*, VI, 93-105, especially pp. 100-01. Cf. Gaston Milhaud, "Descartes et la loi des sinus," *Revue Générale des Sciences*, 1907.
6. For details and references on such work see Cohen's paper cited in note 7 of Chapter II. Cf. my "Early Estimates of the Velocity of Light," *Isis*, XXXIII (1941), 24-40.
7. Pierre Fermat, *Oeuvres* (edited by Paul Tannery and Charles Henry, 4 vols. and supplement, Paris, 1891–1922), II, 354, 464, 472, 483.
8. See the opening pages of his appendix, "Les météores," in his *Oeuvres*, VI, 229-366.
9. The 1624 edition had appeared under the pseudonym Van Etten. (See Moritz Cantor, *Vorlesungen über Geschichte der Mathematik*, vol. II, Leipzig, 1892, p. 701.) I have used the edition of Rouen, 1628.
10. Jean Leurechon, *Récréations mathématiques* (Rouen, 1628), pp. 63-64.
11. *Op. cit.*, p. 65.
12. See E. Gilson's annotated edition of the *Discours* (Paris, 1930), p. 80.
13. E. Gilson, "Météores cartesians et météores scholastiques," *Études de Philosophie Mediévale*, 1921, pp. 247-86.
14. *Ibid.*, pp. 75-77.
15. See Hyman Stock, *The Method of Descartes in the Natural Sciences* (New York, 1931).
16. *Oeuvres*, VI, 325-44.
17. The most important sections on the rainbow are found in translation in *A Source Book in Physics* (edited by W. F. Magie, New York, 1935), pp. 273-78. See especially p. 274.
18. *A Source Book in Physics*, pp. 274-75.

19. *Ibid.*, p. 275.
20. *Oeuvres*, VI, 329.
21. *A Source Book in Physics*, p. 277. Italics are mine.
22. E. Gnau, "Der Regenbogen," *Humboldt*, II (1883), 226-36.
23. *Oeuvres*, VI, 338. The table of Descartes includes other lines and angles which, for clarity of exposition, are not included here. The excerpt in *A Source Book in Physics* unfortunately does not include the tables of Descartes.
24. *Oeuvres*, VI, 339.
25. *Oeuvres*, VI, 340-41.
26. Crombie, in *Robert Grosseteste*, p. 275, mistakenly attributes this secret, the idea of dispersion, to Descartes.
27. *Oeuvres*, VI, 342-44.
28. *Oeuvres*, VI, 366.
29. J. F. Scott, *The Scientific Work of René Descartes* (London, 1952).
30. Gilson, reference in note 13.
31. F. J. Studnicka, "Joannes Marcus Marci a Cronland, sein Leben und gelehrtes Wirken," *Jahresbericht der Königlichen Böhmischen Gesellschaft der Wissenschaften*, 1891. An account of Marci's investigations on the colors of the prismatic spectrum is given by L. Rosenfeld, "Marcus Marcis Untersuchungen über das Prisma und ihr Verhältnis zu Newton's Farbentheorie," *Isis*, XVII (1932), 325-30.
32. Marcus Marci, *Thaumantias, liber de arcu coelesti deque colorum apparentium natura, ortu et causis* (Pragae, 1648), p. 1. I have used the copy of this rare book at the University of Michigan.
33. *Ibid.*, p. 5.
34. *Ibid.*, p. 98.
35. *Ibid.*, pp. 83, 100.
36. *Ibid.*, pp. 109-12.
37. *Ibid.*, p. 128.
38. *Ibid.*, p. 103.
39. See *Thaumantias*, especially pp. 179-81. A good summary of Marci's explanation is given by J. S. T. Gehler in the article "Regenbogen" in *Physikalisches Wörterbuch* (new edition, Leipzig, 1834), VII (2), 1335-40.
40. *Thaumantias*, pp. 206-07.
41. *Ibid.*, pp. 179-90.
42. *Ibid.*, pp. 198, 210, 213.
43. *Ibid.*, pp. 216-17.
44. *Ibid.*, pp. 249f.
45. See Crombie, *Robert Grosseteste*, p. 275, note 6.

46. Huygens, *Oeuvres complètes*, I, 240f, III, 6.
47. Marin Cureau de la Chambre, *Nouvelles observations et coniectures sur l'iris* (Paris, 1650), pp. 107f. Cf. pp. 18, 137.
48. *Ibid.*, pp. 166, 266.
49. *Ibid.*, pp. 331-32.
50. *La lumière* (Paris, 1662), especially p. 43.
51. *La lumière*, pp. 182, 401.
52. *Ibid.*, pp. 311f, 325f.
53. *Nouvelles observations*, p. 235.
54. J. B. Duhamel, *De meteoris* (Parisiis, 1660), Book I, Chapter V, pp. 72-97. This is found also in his *Operum philosophicum* (2 vols., Norimbergae, 1681), I, 329-50.
55. *De meteoris*, pp. 93, 97.
56. Gassendi, *Opera omnia*, II, 86-93.
57. J. B. Riccioli, *Almagestum novum* (Bononiae, 1651), Book II, Chapter xix.
58. Nicolaus Zucchi, *Optica philosophia* (2 vols., Lugduni, 1652–56) I, 73f.
59. J. C. Kohlhans, *Tractatus opticus* (Lipsiae, 1663), p. 112.
60. Giuseppe Antonio Barbari, *L'iride: opera fisicomatematica* (Bologna, 1678), especially pp. xxii, xxviii.
61. See the diagrams at the end of Barbari's *L'iride*.

IX

1. Huygens, *Oeuvres complètes*, X, 398-406, especially p. 405.
2. See Henry Crew, *The Rise of Modern Physics* (Baltimore, 1928), pp. 157-59.
3. Henry Crew, *The Wave Theory of Light. Memoirs by Huygens, Young and Fresnel* (New York, 1900).
4. Huygens, *Oeuvres complètes*, XIII (1), 146-53, 163-68.
5. *Ibid.*, p. 226.
6. *The Works of the Honourable Robert Boyle* (edited by Thomas Birch, 5 vols., London, 1774), V, 372-73.
7. *Ibid.*, II, 21, 49.
8. Benedictus Spinoza, *Opera* (edited by J. van Vloten and J. P. N. Land, 2 vols., Hagae Comitum, 1882–83), II, 507-20. This little work was published with Latin translation in *Opera quae supersunt omnia, Supplementum* (Amstelodami, 1862). The original was republished also in *Nieuw Archief voor Wiskunde*, XI (1884), 49-82.
9. See, e.g., Friedrich Just, *Geschichte der Theorien des Regenbogens* (Marienburg, 1863).

10. Francesco Grimaldi, *Physico-mathesis de lumine, coloribus, et iride* (Bononiae, 1665), pp. 420-72.

11. A full account of his basic experiments is found in Ernst Mach, *The Principles of Physical Optics. An Historical and Philosophical Treatment* (translated by J. S. Anderson and A. F. A. Young, New York, 1953, reprint of the 1926 English edition), pp. 133-36.

12. Newton later gave the year 1666 as the time of his early experiments, but recent studies indicate that they probably date from 1664 or early 1665. See A. R. Hall, "Sir Isaac Newton's Note-Book, 1661–65," *Cambridge Historical Journal*, IX (1947), 239-50.

13. Accounts of Newton's optical discoveries are available in many places, including histories of physics and optics. See Michael Roberts and E. R. Thomas, *Newton and the Origin of Colours* (London, 1934); also Louis Rosenfeld, "La théorie des couleurs de Newton et ses adversaires," *Isis*, IX (1927), 44-65.

14. Sir Isaac Newton, *Opticks: or a Treatise of the Reflections, Refractions, Inflections & Colours of Light* (New York, 1952, based on the fourth edition of London, 1730), p. 175. See also *Lectiones opticae* (London, 1729; bound with *Optices libri tres*, Patavii, 1749), pp. 56-58, 105-10.

15. Stephen Peter Rigaud, *Correspondence of Scientific Men of the Seventeenth Century* (2 vols., Oxford, 1841), II, 315.

16. Sir Isaac Newton, "A New Theory of Light and Colours," *Philosophical Transactions of the Royal Society of London*, VI (1672), 3075-87, especially p. 3083. This paper has been often reprinted. See, for example, *Popular Science Monthly*, LXI (1902), 461-71, and *Isis*, XIV (1930), 326-41.

17. Quoted from Cajori, *History of Physics*, p. 86.

18. Edme Mariotte, *Oeuvres* (new edition, 2 vols., La Haye, 1740), I, 244-68.

19. Athanasius Kircher, *Arca noë* (Amstelodami, 1675), pp. 173, 176-78.

20. Kircher, *Ars magna lucis et umbrae* (Amstelodami, 1671), p. 55.

21. Christian Seyfried, *Q. D. B. V. iridem diluvii* (Jenae, 1696).

22. The thesis was presented under the praesidium of Johann Christoph Sturm, and frequently it is catalogued incorrectly under Sturm instead of Volkamer. There are several copies, to be found under Sturm, at the New York Public Library; but the book is nevertheless relatively rare.

23. Christoph Gottlieb Volkamer, *Thaumantiados thaumasia, sive iridis admiranda . . . exposita sub praesidio Joh. Christophori Sturmii . . . a Christophoro Theophilo Volcamero* (Noribergae, 1699), p. 123.

24. *Ibid.*, p. 87. Cf. also p. 20.

25. Jacob Hermann, "Méthode géométrique et générale de déterminer le diamétre de l'arc-en-ciel," Nouvelles de la République des Lettres, XXXII (1704), 658-71.

26. For Newton's calculations see his Lectiones opticae, pp. 58, 108; for those of Halley see "De iride, sive de arcu coelesti," Philosophical Transactions of the Royal Society of London, XXII (1700), 714-25; for those of Bernoulli see "Pour la détermination des iris ou arc-en-ciels de toutes les classes," Opera omnia (4 vols., Lausannae and Genevae, 1742), IV, 197-203.

27. Opticks (4th edition), p. 178.

28. Opticks, Book I, Part II, Prop. IX.

29. W. Le Conte Stevens, "Theory of the Rainbow," Monthly Weather Review, United States Weather Bureau, XXXIV (1906), 170-73.

30. Opticks (4th edition), p. 178.

31. Ibid., Book II, Part III, Prop. XII.

32. Ibid., Prop. XIII.

33. Queries 17 to 23 appeared in the Latin edition of 1706, queries 24-31 appeared in the next English edition of 1717.

34. Opticks (4th edition), pp. 362-63.

35. Ibid., Query 29, p. 370.

36. Quoted from Cajori, History of Physics, p. 56. The reference here refers specifically to the difference between the Newtonian action at a distance through a void and the Cartesian theory of vortices in a plenum.

37. John T. McCarthy, "Physics in American Colleges Before 1750," American Physics Teacher, VII (1939), 100-04.

38. The work was first published in 1940 from a manuscript of a copy by a student. See Charles Morton's Compendium Physicae. Publications of the Colonial Society of Massachusetts. Volume XXXIII. Collections. Boston, 1940.

39. Ibid., p. 111. Spelling in student notes evidently was as unconventional then as now.

40. Ibid., p. 109. This and other passages cited below are from Chapter 16, pp. 108-12.

41. Compendium physicae, p. 160.

42. See Samuel Eliot Morison's "Biographical Sketch of Charles Morton," in Compendium physicae, p. xxvi.

43. For details see G. L. Kittredge, "Cotton Mather's Election into the Royal Society," Publications of the Colonial Society of Massachusetts, XIV (1913), 81-114. For details on his life see Lamb's Biographical Dictionary of the United States and Dictionary of American Biography.

44. Cotton Mather, Thoughts for the Day of Rain (Boston, 1712). There

are copies of this work at the Boston Public Library and at Yale University.

45. "Gospel of the Rainbow" (in the volume cited in the note above), p. 16.
46. Cotton Mather, *The Christian Philosopher* (London, 1721), p. 58.
47. O. T. Beall, "Cotton Mather, the First Significant Figure in American Medicine," *Bulletin of the History of Medicine*, XXVI (1952).
48. A good account is found in Justin Winsor, "The Literature of Witchcraft in New England," *Proceedings of the American Antiquarian Society*, new series, X (1895), 351-73.
49. Egbert C. Smyth, "The 'New Philosophy' against which Students at Yale College were Warned in 1714," *Proceedings of the American Antiquarian Society*, new series, XI (1896-97), 251-52.
50. For details on his life see the *Dictionary of American Biography*.
51. This material on the rainbow is found in *Puritan Sage. Collected Writings of Jonathan Edwards* (edited by Vergilius Ferm, New York, 1953), pp. 8-10. It is included also in E. C. Smyth, "Some Early Writings of Jonathan Edwards. A. D. 1714-1726," *Proceedings of the American Antiquarian Society*, new series, X (1895), 212-47, and in appendix to vol. I of *The Works of President Edwards* (New York, 1829-30).
52. Rufus Suter, "An American Pascal: Jonathan Edwards," *Scientific Monthly*, LXVIII (1949), 338-42.
53. For this and other verses quoted below see Marjorie Hope Nicolson, *Newton Demands the Muse: Newton's "Opticks" and the Eighteenth Century Poets* (Princeton, 1946).

X

1. For references to some of the works on the rainbow for the period from 1540 to 1777, see F. W. A. Murhard, *Litteratur der mathematischen Wissenschaften* (5 vols. in 2, Leipzig, 1797-1805), V, 115-23. There are a few books cited which I have been unable to locate, but there are a greater number which I have examined which are not included in Murhard's list.
2. Pieter van Musschenbroek, *Essai de physique* (translated by Pierre Massuet, 2 vols. in 1, Leyden, 1739), I, 809.
3. See Ockenden's paper, cited above, in *Isis*, XXVI (1936), p. 48. This includes further references in this matter.
4. J. E. Montucla, *Histoire des mathématiques* (2nd edition, 4 vols., Paris, 1799-1802), II, 541-46.
5. Simon Kotelnikow, "Phaenomenorum iridis seu arcus coelestis disquisitio," *Akademiya Nauk. Mémoires*, series 2, VII (1758-59), 252-76

(published 1761). Radii for the first five bows are included in the articles "Arc-en-ciel" in Le Grande Encyclopédie and Encyclopédie Méthodique.

6. Kotelnikow's account (p. 275) refers to this as arising after five reflections and six refractions, but this obviously is a slip of the pen or an error on the part of the printer.

7. F. W. G. Radicke, Handbuch der Optik (2 vols., Berlin, 1839), II, 305; or see Siegmund Günther, Handbuch der Geophysik (2nd edition, 2 vols., Stuttgart, 1897–99), II (2), 119-25; or A. Bravais, "Notice sur l'arc-en-ciel," Annuaire Météorologique de la France pour 1849, pp. 311-34.

8. See his article, "Ueber den dritten Regenbogen," Zeitschrift für mathematischen und naturwissenschaftlichen Unterricht, XI (1880), 72-73. Cf. Günther, loc. cit.

9. J. Cabannes, "L'explication scientifique de l'arc-en-ciel," La Science Moderne, VIII (1931), 217-26.

10. O. D. Chowlson, Traité de physique (translated from the Russian, revised edition, Paris, 1906), II, 553.

11. For a fascinating account of a variety of unusual phenomena in meteorological optics, see the book of Minnaert cited in note 10 of Chapter III.

12. Augustus De Morgan, A Budget of Paradoxes (2nd edition, 2 vols., Chicago and London, 1915), II, 334.

13. For further details see W. J. Humphreys, Journal of the Franklin Institute, CCVII (1929), 661; or Chancey Juday, "Horizontal Rainbows on Lake Mendota," Monthly Weather Review, United States Weather Bureau, XLIV (1916), 65-67; or Minnaert, op. cit., pp. 187-88; or Kokichi Otobe, "Equations of Horizontal Rainbows," Monthly Weather Review, United States Weather Bureau, XL (1917), XLV (1917), 151-53.

14. L. T. von Spittler and C. M. T. Breunlin, Dissertatio physica de iride (Tubingae, 1772).

15. An extensive description of unusual rainbows is included in Wilhelm Krebs, Atmosphärische Pracht-und Kraftentfaltung (Hamburg, 1894).

16. Benjamin Langwith, "Concerning the Appearances of Several Arches of Colours Contiguous to the Inner Edge of the Common Rainbow," Philosophical Transactions of the Royal Society of London, 1722, pp. 241-45.

17. Ibid., pp. 244-45.

18. Henry Pemberton, "Concerning the Abovementioned Appearance in the Rainbow," Philosophical Transactions, 1722, pp. 245-61.

19. Musschenbroek, Cours de physique expérimentale et mathématique

(translated by Sigaud de la Fond, 3 vols., Paris, 1769), III, 355-56.
20. Andrea Comparetti, *Observationes opticae de luce et coloribus* (Patavii, 1787), pp. 136-38.
21. Murhard, op. cit., V, 123.
22. See Gnau, op. cit.; or l'abbe Raillard, "Explication nouvelle et complète de l'arc-en-ciel," *Cosmos*, X (1856–57), 605-07; or A. Bravais, "Notice sur l'arc-en-ciel," *Annuaire Météorologique de la France pour 1849*, pp. 311-34.
23. The article on the rainbow is found on pp. 225-48 of vol. 74 of *Encyclopédie Méthodique*, published in 1793. This volume also is called vol. I of *Dictionnaire de Physique*, edited by Monge, Cassini, Bertholon, etc. There is a less extensive article on the rainbow in vol. I of the portion of the *Encyclopédie Méthodique* which is headed "Mathématiques," vol. I, Paris, 1784.
24. *Encyclopédie Méthodique*, LXXIV, 246.
25. For a recent account of his life and work see *Thomas Young, Natural Philosopher, 1773–1829*, by Alexander Wood, completed by Frank Oldham (New York, 1954).
26. *Physicomathesis* (1665), p. 190.
27. I. B. Cohen, "The First Explanation of Interference," *American Journal of Physics*, VIII (1940), 99-106.
28. *Ibid.*, p. 101.
29. *Ibid.*, p. 105.
30. See Carlos de Pedroso, *Historia de las teorias sobre la constitucion de la luz desde la antiguedad hasta la muerte de Fresnel* (Habana, 1900); or Albert Wangerin, "Anciennes théories de l'optique," *Encyclopédie des Sciences Mathématiques*, tome V, vol. IV, fasc. 1, pp. 1-104.
31. Cleveland Abbe, "Benjamin Franklin as Meteorologist," *Proceedings of the American Philosophical Society*, XLV (1906), 117-28, 183.
32. Ambrogio Fusinieri, *Ricerche meccaniche e diottriche sopra la causa della rifrazione della luce* (Venezia, 1797).
33. Thomas Young, "An Account of Some Cases of the Production of Colours, Not Hitherto Described," *Philosophical Transactions*, XCII (1802), 387-97, especially p. 387.
34. Young, "Experiments and Calculations Relative to Physical Optics," *Philosophical Transactions*, 1804, pp. 1-16, especially p. 9; or see his *Lectures on Natural Philosophy and the Mechanical Arts* (2 vols., London, 1807), I, 470, II, 643.
35. *Philosophical Transactions*, 1804, p. 11.
36. A good account of such variations is found in E. Verdet, *Leçons d'optique physique* (vols. V and VI of his *Oeuvres*, Paris, 1869–70), V, 402-23,

37. This paper was reprinted in his *Lectures on Natural Philosophy*. See II, 643.

38. Sir David Brewster, *Manuel d'optique* (translated by P. Vergnaud, 2 vols., Paris, 1833), II, 87-93.

39. Minnaert, *op. cit.*, pp. 181-82.

40. *Loc. cit.*

41. Pla, *op. cit.*, pp. 210-11.

42. Perhaps the best history of color is that by Halbertsma cited in Chapter I, note 8.

43. Quoted from Charles J. Engard, "Poetic Scientist," *Scientific Monthly*, LXVIII (1949), 305-09.

44. Wilhelm Ostwald, *Colour Science* (translated by J. Scott Taylor, 2 parts, London, 1931–33); R. C. Maclaurin, *Light* (New York, 1909).

XI

1. Henry Crew, "Thomas Young's Place in the History of the Wave Theory of Light," *Journal of the Optical Society of America*, XX (1930), 3-10.

2. Richard Potter, "Mathematical Considerations on the Problem of the Rainbow, Shewing it to Belong to Physical Optics," *Transactions of the Cambridge Philosophical Society*, VI (1838), 141-52. See also his later article, "On the Interference of Light Near a Caustic and the Phenomenon of the Rainbow," *Philosophical Magazine* (4), IX (1855), 321-26.

3. "Observations on the Absorption of Specific Rays, in Reference to the Undulatory Theory of Light," *Philosophical Magazine* (3), II (1833), 360-63.

4. *Philosophical Magazine* (3), II (1833), 276-81.

5. See reference in note 2.

6. Potter, *An Elementary Treatise on Optics* (London, 1847), p. 28.

7. They were, however, studied earlier by Euler. See *American Mathematical Monthly*, XXV (1918), 276-82.

8. Vol. VI (1838), pp. 379-403.

9. *Ibid.*, p. 391.

10. *Ibid.*, p. 392.

11. For details see Sir George Biddell Airy, *Autobiography* (Cambridge, 1896).

12. Vol. VIII, pp. 595-600.

13. G. G. Stokes, "On the Numerical Calculations of a Class of Definite Integrals and Infinite Series," in his *Mathematical and Physical*

Papers (5 vols., Cambridge, 1880–1905), II, 329-57. The paper originally appeared in the Transactions of the Cambridge Philosophical
Society, vol. IX.

14. Jacques Babinet, "Mémoires d'optique météorologique," Académie des
    Sciences, Comptes Rendus, IV (1837), 638-48.

15. Ibid., p. 648.

16. W. H. Miller's papers in Transactions of the Cambridge Philosophical
    Society, VII (1842), 277-86, and London, Edinburgh and Dublin
    Philosophical Magazine (3), XVIII (1841), 520-21.

17. See especially Poggendorff's Annalen der Physik und Chemie, LIII (1841),
    214-15, LVI (1842), 558-67.

18. See J. G. Galle, "Measurements of the Rainbow," Philosophical Maga-
    zine (3), XXVI (1845), 279-80.

19. Philosophical Magazine (3), XIII (1838), 9-12.

20. Ibid., p. 12.

21. Philosophical Magazine (4), IX (1855), 321-26.

22. See Philosophical Magazine (3), II (1833), 81-94, 161-67, 276-81. Cf. also
    pp. 191-94, 284-87.

23. Gnau, op. cit.

24. Bravais, "Sur l'arc-en-ciel blanc," Comptes Rendus, XXI (1845), 756;
    and "Notice sur l'arc-en-ciel blanc," Journal de l'École Royale Poly-
    technique, vol. XVIII, cahier 30 (1845), 97-122.

25. Bravais, op. cit.

26. James D. Forbes, A Review of the Progress of Mathematical and Physical
    Science in More Recent Times (Edinburgh, 1858), p. 101.

27. Comptes Rendus, XLIV (1857), 1142-44.

28. F. Raillard, "Nouvelle note sur l'arc-en-ciel," Comptes Rendus, LX (1865),
    1287-89.

29. J. H. Pratt, "The Supernumerary Bows in the Rainbow Arise From Inter-
    ference," Philosophical Magazine (4), V (1853), 78-86.

30. Carl Pulfrich, "Ein experimenteller Beitrag zur Theorie des Regenbogens
    und der überzählige Bogen," Annalen der Physik (3), XXXIII (1883),
    194-208.

31. Félix Billet, "Mémoire sur les dix-neuf premiers arcs-en-ciel de l'eau,"
    Annales Scientifiques de l'École Normale Supérieure, V (1868),
    67-109.

32. Billet, "Études expérimentales sur les arcs surnuméraires des onze premiers
    arcs-en-ciel de l'eau," Comptes Rendus, LVI (1863), 999-1000, LVIII
    (1864), 1064.

33. Forbes, op. cit., p. 123.

34. Pla, op. cit., p. 231.

35. *Philosophical Transactions*, 1846, pp. 1-2.
36. *Philosophical Magazine* (4), XXIII (1862), 15, 22. See also E. Taylor Jones, "The Life and Work of James Clerk Maxwell," *Proceedings of the Royal Philosophical Society*, XL (1931–32), 54f.
37. G. G. Stokes, *On Light* (London, 1887), p. 17.
38. J. W. Strutt, "On the Light from the Sky, its Polarization and Colour," *Philosophical Magazine* (4), XLI (1871), 107-20, 274-79. Cf. also pp. 447-54. A popular account of Lord Rayleigh's scientific achievements, including that on the twinkling of stars, was given by Sir Oliver Lodge in *National Review*, XXXII (1898), 89-102.
39. A good account of this phenomenon, as well as helpful explanations on the rainbow, is available in J. C. Johnson, *Physical Meteorology* (New York, 1954).
40. "Investigations in Optics," *Philosophical Magazine* (5), XII (1881), 81-101.
41. "The Incidence of Light upon a Transparent Sphere of Dimensions Comparable with the Wave Length," *Proceedings of the Royal Society of London*, series A, LXXXIV (1910), 25-46.
42. Boitel, "Sur les arcs surnuméraires qui accompagnent l'arc-en-ciel," *Comptes Rendus*, CVI (1888), 1522-24. Cf. *Philosophical Magazine*, XXVI (1888), 239.
43. Sir Joseph Larmor, *Mathematical and Physical Papers* (2 vols., Cambridge, 1929), I, 174-80.
44. Eleuthère Mascart, "Sur l'arc-en-ciel," *Comptes Rendus*, CVI (1888), 1575-77.
45. Mascart, "Sur l'arc-en-ciel blanc," *Comptes Rendus*, CXV (1892), 453-55.
46. In 3 vols., Paris, 1899–94. See especially I, 382-415.
47. J. M. Pernter, *Ein Versuch der richtigen Theorie des Regenbogens Eingang in die Mittelschulen zu verschaffen* (Wien, 1898).
48. See especially "Die Farben des Regenbogens und der weisse Regenbogen," *Sitzungsberichte der Mathematisch-Naturwissenschaftlichen Classe der Kaiserlichen Akademie der Wissenschaften in Wien*, CVI (1897), Heft 3-4, Part IIa, pp. 135-235; also "Zur Theorie des von kreisförmigen Lichtwelle erzeugte Regenbogens," and "Erklärung des fälschlich 'weisser Regenbogen' benannten Bouguer'schen Halos," in the same journal, vol. CXIV (1905), Part IIa.
49. Pernter and Exner, *Meteorologische Optik* (3 parts in 1 vol., Wien and Leipzig, 1902–10), pp. 482-558.
50. We here are following Pernter's account in the Vienna *Sitzungsberichte* for 1897, especially pp. 213f. A summary of Pernter's work is given by D. Hammer, "Airy's Theory of the Rainbow," *Journal of the*

Franklin Institute, CLVI (1903), 335-49. Cf. also Cabannes, op. cit.
51. This formula is based on one given by Minnaert, op. cit., p. 183.
52. Minnaert, op. cit., p. 184.
53. I have used the revised French edition, Traité de physique, vol. II (Paris, 1906), pp. 553ff.
54. Willy Möbius, "Die Helligkeit des klaren Himmels und die Beleuchtung durch Sonne, Himmel und Rükstrahlung," Kaiserl, Leop.-Carol. Deutschen Akademie der Naturforscher, LXXIII (1900), 1-239, XCI (1909), 81-292. See especially LXXIII, 33-94 on the rainbow.
55. Möbius, "Zur Theorie des Regenbogens an Kugeln von 1 bis 10 Lichtwellenlängen Durchmesser," Preisschriften gekrönt und herausgegeben von der Fürstlich Jablonowskischen Gesselschaft zu Leipzig, XLII (1912), 1-31.
56. James E. McDonald, "The Shape of Raindrops," Scientific American, CXC (1954), 64-68. See also Roscoe R. Braham, Jr., "How Does a Raindrop Grow?", Science, CXXIX (1959), 123-29.
57. K. Aichi and A. Tanakadate, "Theory of the Rainbow Due to a Circular Source of Light," Philosophical Magazine (6), VIII (1904), 598-610; or Journal of the College of Science, Imperial University, Tokyo, Japan, XXI (1906), article 3, pp. 1-29.
58. W. V. R. Malkus et al., "Analysis and Preliminary Design of an Optical Instrument for the Measurement of Drop Size and Free-water Content of Clouds," National Advisory Committee for Aeronautics, Technical Note No. 1622, Washington, June, 1948. See especially p. 28.
59. See Ch. E. Papanastassiou, Les théories sur la nature de la lumière de Descartes à nos jours (Paris, 1935), or Pla, op. cit., or Ronchi, op. cit.
60. Harold Jeffreys, "The Shape of the Sky," The Meteorological Magazine, LVI (1921), 173-77.
61. H. Hartridge, "Theories of Trichromatic Vision," Nature, CLIII (1944), 45-46. See also Halbertsma, op. cit.

Selected References

THE FOLLOWING BRIEF LIST INCLUDES ONLY THOSE WORKS WHICH HAVE PARTICULAR relevance or significance for the history of the theory of the rainbow. For further bibliographical sources see references in the Notes.

Of General Scope or Interest

Gehler, J. S. T.—Article on "Regenbogen," in *Physikalisches Wörterbuch* (new ed., vol. VII, Part 2, Leipzig, 1834, pp. 1318-40).
[Contains extensive historical notes, especially on the earlier periods.]
Gnau, E.—Der Regenbogen, *Humboldt*, II (1883), 226-36.
[An historical account of fundamental contributions, with special reference to the period from Fleischer to Airy.]
Halbertsma, K. T. A.—*A History of the Theory of Color* (Amsterdam, 1949).
[Perhaps the most important history of color theory, but it contains virtually nothing specifically on the rainbow.]
Just, Friedrich—*Geschichte der Theorien des Regenbogens* (Marienburg, 1863).
[Covers in seventeen pages the history from the Greeks to Airy. The account is weak on Arabic contributions.]
Kunze, Alfred—*Zur Geschichte der Theorie des Regenbogens* (Jahresbericht über das Karl-Friedrichs-Gymnasium zu Eisenach, 1870).
[Brief account with special reference to the period from the Greeks to Newton.]
Morrow, Martha Goddard—What Makes Rainbows? *Sky*, V (1941), no. 8, pp. 12-14, no. 9, pp. 9-10.
[A sketchy historical survey on an elementary level.]
Reclam, Franz—*Über den Regenbogen* (Neustettin, 1877).
[Pages 10-17 constitute a brief history from antiquity through the eighteenth century.]
Volkamer, C. T.—*Thaumantiados, sive iridis admiranda* (Noribergae, 1699).
[Almost two hundred pages on the history of the theory of the rainbow to the end of the seventeenth century. However, it is excessively discursive; and it is now, of course, quite out of date.]

355

On the Material of Chapter I

Bähr, Gerhard—El arco iris y la vía láctea en Guipúzcoa (vocables, etimologías y difusión, creencias populares), *Revista international de los Estudios Vascos*, XXII (1931), 397-414.

[An account of various names and popular beliefs associated with the rainbow.]

Hild, J. A.—Article on "Iris" in Ch. Daremberg and Edm. Saglio, *Dictionnaire des antiquités grecques et romaines* (vol. III, Part 1, Paris, 1899, pp. 573-76).

[Describes the place of Iris in classical mythology.]

Mayer, M.—Article on "Iris" in W. H. Roscher, *Ausführliches Lexikon der griechischen und römischen Mythologie* (vol. II, Part 1, Leipzig, 1890-97, columns 320-57.

[An extensive account of the role of Iris in antiquity.]

Merian, Samuel—*Die französischen Namen des Regenbogens* (Halle, 1914).

[A dissertation on some two hundred French names for the rainbow.]

Menzel, Wolfgang—*Mythologische Forschungen und Sammlungen* (Stuttgart & Tübingen, 1842).

[Pages 235-76 include a wide variety of myths clinging to the rainbow.]

Renel, Ch.—L'arc-en-ciel dans la tradition religieuse de l'antiquité, *Revue de l'histoire des religions*, XLV (1902), 58-80.

[The religious interpretation of myths and legends.]

On Chapter II

Alexander of Aphrodisias—*Alexandri Aphrodisiensis Maximi Peripatetici, in quatuor libros meteorologicorum Aristotelis commentatio lucidissimi, Alexandro Piccolomineo interprete* (Venetiis, 1540).

[One of the most important of the ancient commentaries on Aristotle's meteorology. There are copies in many American libraries.]

Aristotle—*Meteorologica* (with an English translation by H. D. P. Lee, Loeb Classical Library, Cambridge, Mass., 1952).

[A convenient English version of Aristotle's meteorology.]

Aristotle—*Meteorologica* (ed. by J. L. Ideler, 2 vols., Lipsiae, 1834-36).

[Important for scholarly apparatus and excerpts from commentaries on Aristotle's meteorology.]

Gilbert, Otto—*Die meteorologischen Theorien des griechischen Altertums* (Leipzig, 1907).

[Greek views on the rainbow are described on pp. 604-18.]

Poske, Fr.—Die Erklärung des Regenbogens bei Aristoteles, *Zeitschrift für Mathematik und Physik, Historisch-literarische Abtheilung*, XXVIII (1883), 134-38.

[Mathematical treatment of Aristotle's explanation of the rainbow.]

Sayili, Aydin M.—The Aristotelian Explanation of the Rainbow, *Isis*, XXX (1939), 65-83.

[A good general account of Aristotle's theory.]

[Seneca] Clarke, John—*Physical Science in the Time of Nero* (being a translation of the *Quaestiones naturales* of Seneca, London, 1910).
[Seneca's ideas on the rainbow are found on pp. 16-33.]

On Chapter III

[Averroës] Aristotle—*Omnia quae extant opera* (with commentary of Averroës, 11 vols. in 5, Venetiis, 1550–52).
[Commentary on the *Meteorologicorum* is in vol V. There is a copy of this work at Harvard University.]
Hellmann, Gustav, ed.—*Neudrucke von Schriften und Karten über Meteorologie und Erdmagnetismus*, vol. XV, Denkmäler mittelalterlicher Meteorologie (Berlin, 1904).
[Important for reproduction of excerpts from medieval treatises.]
Hoffmann, Immanuel—Die Anschauungen der Kirchenväter über Meteorologie, *Münchener Geographische Studien*, XXII (1907), 1-96.
[See especially pp. 80-82 for medieval views on the rainbow.]
Horten, M.—Avicennas Lehre vom Regenbogen nach seinem Werk al Schifa, *Meteorologische Zeitschrift*, XXX (1913), 533-44.
[The best account of Avicenna's views on the rainbow.]
Job of Edessa—*Encyclopaedia of philosophical and natural sciences as taught at Baghdad about A. D. 817* (Cambridge, 1935).
[For the rainbow see pp. 208-10.]
Olympiodorus—*Olympiodori Philosophi Alexandrini in Meteora Aristotelis commentarii . . . Ioanne Baptista . . . Camotio Philosopho interprete* (Venetiis, 1551).
[A copy of the rare Latin version of this important commentary is available at the University of Virginia.]
Wiedemann, Eilhard—Theorie des Regenbogens von Ibn al Haitam, *Sitzungsberichte der Physikalisch-Medizinsche Sozietät in Erlangen*, XLII (1914), 39-56.
[Alhazen's theory and commentary by Kamal al-Din.]
Würschmidt, Joseph—Die Theorie des Regenbogens und des Halo bei Ibn al Haitam und bei Dietrich von Freiberg, *Meteorologische Zeitschrift*, XXXI (1914), 484-87.
[Important for the little that is known about Alhazen's ideas on the rainbow.]

On Chapter IV

Albertus Magnus—*Opera* (21 vols., edited by Peter Jammy, Lugduni, 1651).
[For the work on the rainbow see vol. II, Book III, Tractatus IV.]
Aquinas, St. Thomas—*Aristotelis stagirite in libros meteororum, cum duplici Interpretatione antiqua et Francisci Vatabli, expositore divo Thoma Aquinate* (Venetiis, 1537).

[The explanation of the rainbow in this work attributed to St. Thomas Aquinas is on folios 42-54. There is a copy of this book at the University of Pennsylvania.]

Bacon, Roger—*The Opus majus of Roger Bacon* (translated by R. B. Burke, 2 vols., Philadelphia, 1928).
[Comments on the rainbow are scattered throughout the volumes.]

Crombie, A. C.—*Robert Grosseteste and the origins of experimental science 1100–1700* (Oxford, 1953).
[The most important of all accounts of the history of the theory of the rainbow. See especially chapters VII, VIII, IX, and X and the bibliography.]

Grosseteste, Robert—*De iride seu de iride et speculo* (edited by Ludwig Baur, in vol. IX of *Beiträge zur Geschichte der Philosophie des Mittelalters*, Münster i. W., 1912).
[This is described fully in the volume above by Crombie.]

Peckham, John—*Perspectiva communis* (Norimbergae, 1542).
[See Book III, Props. 18-21 on the rainbow. There is a copy at Columbia University.]

Sayili, Aydin M.—Al-Qarafi and His Explanation of the Rainbow, *Isis*, XXXII (1940), 16-26.

Wiedemann, Eilhard—Arabische Studien über den Regenbogen, *Archiv für die Geschichte der Naturwissenschaften und der Technik*, III (1910–12), 1-53.
[Two anonymous treatises and one by al-Qarafi.]

Witelo—*Opticae* (bound with Alhazen, *Opticae thesaurus*, Basileae, 1572).
[On the rainbow see Book X, pp. 457-74.]

On Chapter V

Crombie, A. C.—*Robert Grosseteste and the Origins of Experimental Science 1100–1700* (Oxford, 1953).
[Chapter IX, "Experimental Method and Theodoric of Freiberg's Explanation of the Rainbow," is the best account in English of the theory of Theodoric.]

Dietrich of Freiberg—See Theodoric of Freiberg.

Duns Scotus, Joannes—*Opera omnia* (new edition, 26 vols., Parisiis, 1891–95).
[See vol. IV, pp. 200-208 on the rainbow.]

Faral, E.—Jean Buridan. Notes sur les manuscrits, les éditions et le contenu de ses ouvrages, *Archives d'Histoire Doctrinale et Littéraire du Moyen Âge*, XV (1946), 1-53.
[See pp. 21-24 for Buridan's *Meteorologica*.]

Gaetan de Thiene—*Meteorologicorum . . . cum commentariis* (Venetiis, 1491).
[There is a microfilm copy at Columbia University. See Book III for the rainbow.]

Hellmann, Gustav—*Meteorologische Volksbücher* (Berlin, 1891).
[See especially for Konrad von Mengenberg.]

Themo—*In quatuor libros meteororum* (edited by George Lokert, Parisiis, 1518).
[See folios CLXXVII-CCIIII on the rainbow. There is a copy of the book at the University of Pennsylvania.]

Theodoric of Freiberg—*De iride et radialibus impressionibus* (ed. by Joseph Würschmidt in *Beiträge zur Geschichte der Philosophie des Mittelaters*, Bd. XII, Heft 5-6, Münster i. W., 1914).
[Latin text and German summary of the most important work on the rainbow in the fourteenth century.]

Thorndike, Lynn—Oresme and Fourteenth Century Commentaries on the Meteorologica, *Isis*, XLV (1954), 145-52.
[Description of a manuscript at Paris, Bibliothèque Nationale, Latin MS 15156.]

Wiedemann, Eilhard—Über das Sehen durch eine Kugel bei den Arabern, *Annalen der Physik und Chemie*, new series, XXXIX (1890), 565-76.
[Especially important for the ideas on the rainbow of Kamal al-Din.]

Wiedemann, Eilhard—Über die Brechung des Lichtes in Kugeln nach Ibn al Haitam und Kamal al Din al Farisi, *Sitzungsberichte der Physikalisch-Medizinische Sozietät in Erlangen*, XLII (1910), 15-58.
[Kamal al-Din's experiments and the theory of the rainbow.]

Wiedemann, Eilhard—Zur Optik von Kamal al Din, *Archiv für die Geschichte der Naturwissenschaften und der Technik*, III (1910-12), 161-77.
[Kamal al-Din's ideas on color and the rainbow.]

Würschmidt, J.—Über die Brennkugel, *Monatshefte für den Naturwissenschaftlichen Unterricht*, IV (1911), 98-113
[Comparison of various opinions on the burning sphere, with particular reference to Kamal al-Din.]

On Chapter VI

Ailly, Pierre d'—*Tractatus Petri de Eliaco episcopi Cameracensis: super libros metheororum: de impressionibus aeris* (Argentinae, 1504).
[Folios XVIII-XX are on the rainbow. There is a copy of this book at the University of Oklahoma.]

Cardan, Jerome—*De subtilitate* (Norimbergae, 1550).
[Pages 97-102 of Book IV are on the rainbow.]

Demerlier, Jean—*Iridis coelestis, et coronae brevis descriptio* (Parisiis, 1576).
[An unexceptional treatise of the time.]

Fleischer, Johann—*De iridibus doctrina Aristotelis et Vitellionis* (Witebergae, 1579).
[Unusually clear exposition of the ideas of Aristotle and Witelo.]

Lefevre d'Étaples, Jacques—*Totius naturalis philosophiae Aristotelis paraphrases per Iacobum Fabrum Stapulensem recognitiae iam . . . et scholijs doctissimi viri Iudoci Clichtovei illustratae* (Friburgi Brisgoviae, 1540).
[There is a copy of the first edition of 1501 at Harvard University.]

Maurolico, Francesco—*The Photismi de lumine* of Maurolycus (translated by Henry Crew, New York, 1940).

[A convenient and dependable edition of an important work in the story of the rainbow.]

Piccolomini, Alessandro—*Tractatus de iride* (in his edition of *Alexandri Aphrodisiensis . . . in quatuor libros Meteorologicorum Aristotelis commentatio lucidissima* (Venetiis, 1540).

[The *Tractatus* occupies folios 59-64. This was an important work in the sixteenth century.]

Porta, Giovanni Batista della—*De refractione optices parte libri novem* (Neapoli, 1593).

[Material on the rainbow is found on pp. 189-222. There is a copy of the book at Columbia University.]

Reisch, Gregor—*Margarita philosophica* (Friburgi, 1503).

[The material on the rainbow is found in Book IX, Chapter xxii. At the New York Public Library there are about a dozen editions of this book, including the first.]

Scaliger, Julius Caesar—*Exotericarum exercitationum lib. XV. De subtilitate ad Hieronymum Cardanum* (Francofurti, 1576).

[See Exercitatio LXXX, De iride, pp. 297-312. There is a copy of the book at the library of the New York Academy of Medicine.]

Vimercati, Francisco—*In quatuor libros Aristotelis Meteorologicorum commentarij* (Venetiis, 1565).

[An important work of the sixteenth century. There is copy at the University of Michigan.]

Vieri, Francesco de'—*Trattato nel quale si contengono i libri delle metheore* (new edition, Fiorenza, 1582).

[A completely Aristotelian commentary.]

On Chapter VII

Bacon, Francis—*Works* (ed. by James Spedding and R. L. Ellis, new edition, 14 vols., London, 1887–1902).

[See volumes IV and V for occasional references to the rainbow.]

De Dominis, Marco Antonio—*De radiis visus et lucis in vitris perspectivis et iride tractatus* (Venetiis, 1611).

[The most controverted of all works on the rainbow.]

Froidmont, Liebert—*Fromondus meteorologicorum libri sex* (Antverpiae, 1627).

[An unusually popular book, with editions also in 1639, 1656, and 1670.]

Kepler, Johann—*Opera omnia* (edited by Ch. Frisch, 8 vols., Francofurti a. M. and Erlangae, 1858–70).

[Most of the work on the rainbow is in volume II.]

Ockenden, R. E.—Marco Antonio de Dominis and his explanation of the rainbow, *Isis*, XXVI (1936), 40-49.

[An excellent evaluation of the work of De Dominis.]

Rhodius, Ambrosius—*Opticae* (Witebergae, 1611).
[The last eight pages are on the rainbow.]
Stöckl, Karl—*Bericht des Naturwissenschaftlichen Vereins zu Regensburg*, 1928–30, Heft XIX.
[This Kepler *Festschrift*, edited by Stöckl, contains nothing specifically on the rainbow, but it includes excellent papers on related optical topics.]

On Chapter VIII

Barbari, Giuseppe Antonio—*L'iride: opera fisicomatematica* (Bologna, 1678).
[One of the last books in the Aristotelian tradition. There is a copy in the Burndy Library.]
Chambre, Marin Cureau de la—*Nouvelles observations et coniectures sur l'iris* (Paris, 1650).
[An unusually diffuse and old-fashioned book of 345 pages. There is a copy at the American Philosophical Society. Harvard University has a copy of the edition of 1662, as well as of the same author's *La lumière*, also of 1662.]
Descartes, René—*Oeuvres* (ed. by Charles Adam and Paul Tannery, 12 vols. and supplement, Paris, 1897–1913).
[*Les météores* is found in volume VI.]
Duhamel, Jean Baptiste—*De meteoris et fossilibus libri duo* (Parisiis, 1660).
[This is found also in his *Operum philosophicum* (2 vols., Norimbergae, 1681). Copies of both are found at Yale University.]
Gassendi, Pierre—*Opera omnia* (6 vols., Florentiae, 1727).
[See especially vol. II, pp. 86-93.]
Gilson, E.—*Météores cartésiens et météores scholastiques, Études de Philosophie Médiévale* (Strasbourg, 1921), pp. 247-86.
[Shows the persistence of scholastic influences in Descartes.]
Kramer, P.—*Descartes und das Brechungsgesetz des Lichtes, Abhandlungen zur Geschichte der Mathematik*, IV (1882), 233-78.
[This contains nothing on the rainbow directly, but it is an excellent analysis of the related question of the discovery of the law of refraction.]
Marci, Marcus—*Thaumantias, sive liber de arcu coelesti* (Prague, 1648).
[There is a copy of this book at the University of Michigan.]

On Chapter IX

Bernoulli, Jean—*Opera omnia* (4 vols., Lausannae and Genevae, 1742).
[See IV, 197-203 for the work on the rainbow.]
Edwards, Jonathan—*Puritan Sage. Collected Writings of Jonathan Edwards* (edited by Vergilius Ferm, New York, 1953).
[On the rainbow see chapter II.]
Grimaldi, Francesco—*Physico-mathesis de lumine, coloribus, et iride* (Bononiae, 1665).

[Propositions 46-60 (pp. 420-72) are on the rainbow. There is a copy of this book at the John Crerar Library in Chicago.]

Halley, Edmund—De iride, sive de arcu coelesti, dissertatio geometrica, Royal Society of London, Philosophical Transactions, XXII (1700–01), 714-25.
[The first extensive published account of multiple rainbows.]

Hermann, Jacob—Méthode géométrique et générale de déterminer le diamètre de l'arc-en-ciel, Nouvelles de la République de Lettres, XXXII (1704), 658-71.
[Includes formulas for rainbows of higher orders.]

Mariotte, Edme—Oeuvres (new edition, La Haye, 2 vols., 1740).
[See I, 244-68 for the work on the rainbow.]

Mather, Cotton—Thoughts for the Day of Rain. In Two Essays: I. The Gospel of the Rainbow. II. The Saviour with His Rainbow (Boston, 1712).
[The pertinent material on the theory of the rainbow is included also in Mather's The Christian Philosopher (London, 1721).]

Newton, Sir Isaac—Optical Lectures (translated into English, London, 1728).
[This book, appearing in various editions, is not to be confused with the Opticks.]

Newton, Sir Isaac—Opticks (New York, 1952. Dover, reprint of the fourth edition of London, 1730, with additional material).
[This is an inexpensive edition of an important book.]

Nicolson, Marjorie Hope—Newton Demands the Muse. Newton's Opticks and the Eighteenth Century Poets (Princeton, 1946).

Rohault, Jacques—Traité de physique (2 vols. in 1, Paris, 1671).
[See Part III, pp. 288-311 on the rainbow.]

Spinoza, Benedictus—Opera (edited by J. van Vloten and J. P. N. Land, 2 vols., Hagae Comitum, 1882–83).
[See II, 507-20 for his little treatise on the rainbow.]

On Chapter X

Forbes, James D.—A Review of the Progress of Mathematical and Physical Science ... Between 1775 and 1850 (Edinburgh, 1858).
[Includes remarks and references on the rainbow.]

Kotelnikow, Simon—Phaenomenorum iridis seu arcus coelestis disquisitio, Novi Commentarii Academiae Petropolitanae, VII (1748–49, pub. 1761), 252-76.
[An extensive account of rainbows of higher orders.]

Langwith, Benjamin—Concerning the Appearances of Several Arches of Colours Contiguous to the Inner Edge of the Common Rainbow, Philosophical Transactions, 1772, 241-45.
[The first full and clear description of the supernumerary bows.]

Monge, Gaspard, et al.—Article on "Arc-en-ciel" in Encyclopédie Méthodique (vol. 74, Paris, 1793, pp. 225-48).
[Includes historical notes and accounts of multiple and supernumerary bows.]

Musschenbroek, Pieter van—*Cours de physique expérimentale et mathématique* (translated by Sigaud de la Fond, 3 vols., Paris, 1769).
[See especially III, 348 f on the rainbow.]
Noceti, Carlo—*De iride et aurora boreali carmina* (Romae, 1747).
[The most significant material on the rainbow is found in the notes by Boscovich.]
Pemberton, Henry—Concerning the Abovementioned Appearance in the Rainbow, *Philosophical Transactions*, 1722, 245-61.
[An attempted explanation of the supernumerary rainbow arcs.]
Puttkammer, J. N. van—*Dissertatio physico-mathematica de iride* (Lugduni batavorum, 1827).
[Contains historical notes and mathematical calculations on the rainbow.]
Spittler, L. T. von, and C. M. T. Breunlin—*Dissertatio physica de iride* (Tubingae, 1772).
[A clarification of the explanation of the rainbow given in Segner's *Physica*.]
Young, Thomas—*A Course of Lectures on Natural Philosophy and the Mechanical Arts* (2 vols., London, 1807).
[See I, 470 f and II, 643 f for work on the rainbow.]
Young, Thomas—Experiments and Calculations Relative to Physical Optics, *Philosophical Transactions*, 1804, 1-16.
[The important Bakerian Lecture of 1803.]

On Chapter XI

Aderholt, A. E.—*Die Theorie des Regenbogens in fasslicher Darstellung* (Jena, 1858).
[Especially good for diagrams of the caustics associated with the rainbow.]
Aichi, K., and A. Tanakadate—Theory of the Rainbow Due to a Circular Source of Light, *Philosophical Magazine* (6), VIII (1904), 598-610.
[This material, amplified, is found also in *Journal of the College of Science, Imperial University, Tokyo*, vol. XXI (1906), article 3.]
Airy, Sir George Biddell—On the Intensity of Light in the Neighbourhood of a Caustic, *Transactions of the Cambridge Philosophical Society*, VI (1838), 379-403, VIII (1849), 595-600.
[Airy's classic paper on the rainbow.]
Babinet, Jacques—Mémoires d'optique météorologique, *Comptes Rendus*, IV (1837), 638-48.
[Describes his method for observing supernumerary arcs experimentally.]
Billet, Félix—Études expérimentales sur les arcs supernuméraires des onze premiers arcs-en-ciel de l'eau, *Comptes Rendus*, LVI (1863), 999-1000, LVIII (1864), 1046.
[Brief reports on his extension of the work of Babinet and Miller.]
Billet, Félix—Mémoire sur les dix-neuf premiers arcs-en-ciel de l'eau, *Annales Scientifiques de l'École Normale Supérieure*, V (1868), 67-109.
[The most extensive work on rainbows of high order.]

364

Selected References

Bravais, A.—Notice sur l'arc-en-ciel, *Annuaire Météorologique de la France pour 1849*, pp. 311-34.
[On supernumerary bows and the theory of Airy.]

Bravais, A.—Notice sur l'arc-en-ciel blanc, *Journal de l'École Royale Polytechnique*, Tome XVIII, Cahier 30 (1845), 97-122.
[Defends his thesis of vesicular drops. In Cahier 31 of this *Journal* Bravais published one of the most extensive of all accounts of "Les halos et les phénomènes optiques qui les accompagnent."]

Cabannes, J.—L'explication scientifique de l'arc-en-ciel, *La Science Moderne*, VIII (1931), 217-26.
[Contains historical remarks on the modern period.]

Chwolson, O. D.—*Traité de physique* (translated by E. Davaux, revised edition, vol. II, Paris, 1906).
[Contains a good account of Pernter's work on the relation between the supernumerary arcs and the size of the drops.]

Hammer, D.—Airy's Theory of the Rainbow, *Journal of the Franklin Institute*, CLVI (1903), 345-49.
[On Airy's theory and its experimental verification.]

Humphreys, W. J.—*Physics of the Air* (second edition, New York, 1929).
[Contains a good account of the modern theory (pp. 458-82).]

Johnson, John C.—*Physical Meteorology* (New York, 1954).
[Includes accurate values for the first three rainbows (pp. 175-85).]

Krebs, Wilhelm—*Atmosphärische Pracht- und Kraftentfaltung* (Hamburg, 1894).
[Contains an elaborate classification of unusual rainbows.]

Larmor, Sir Joseph—*Mathematical and Physical Papers* (2 vols., Cambridge, 1929).
[Includes a formula for the radii of supernumerary bows. This is reprinted from *Proceedings of the Cambridge Philosophical Society*, VI (1888), 280-86.]

Malkus, W. V. R., et al.—Analysis and Preliminary Design of an Optical Instrument for the Measurement of Drop Size and Free-water Content of Clouds, *National Advisory Committee for Aeronautics, Technical Note no. 1622* (Washington, June, 1948).
[Describes the use of rainbows in determining the constitution of clouds. Includes a new calculation of the rainbow.]

Mascart, Éleuthère—*Traité d'optique* (3 vols., Paris, 1889-94).
[This contains an unusually extensive account (I, 382-415) of the rainbow. See also his notes in *Comptes Rendus*, CVI (1888), 1575-77, and CXV (1892), 453-55.]

Miller, W. H.—On Spurious Rainbows, *Transactions of the Cambridge Philosophical Society*, VII (1842), 277-86.
[Important for the experimental confirmation of Airy's theory.]

Minnaert, M.—*The Nature of Light and Colour in the Open Air* (translated by H. M. Kremer-Priest and revised by K. E. Brian Jay, London, 1940).
[A charming book which includes much interesting information on the

rainbow (but no history). An inexpensive reprint (New York: Dover, 1954) is available.]

Möbius, Willy—Zur Theorie des Regenbogens an Kugeln von 1 bis 10 Lichtwellen-längen Durchmesser, *Preisschriften gekrönt und herausgegeben von der Fürst-lich Jablonowskischen Gesellschaft zu Leipzig*, XLII (1912), 1-31.
[A highly technical contribution filling the gap between the calculations of Rayleigh and those of Mascart.]

Möbius, Willy—*Zur Theorie des Regenbogens und ihrer experimentellen Prüfung* (Leipzig, 1907).
[A review, highly specialized, of the theory of the rainbow since Airy. A doctoral dissertation.]

Pernter, J. M.—Die Farben des Regenbogens und der weisse Regenbogen, *Sitz-ungsberichte der Mathematisch- Naturwissenschaftlichen Classe der Kaiser-lichen Akademie der Wissenschaften*, CVI (1897), Heft 3-4, Part II.a, 135-235.
[An important study of the relations between drop size and the colors of the supernumerary rainbow arcs.]

Pernter, J. M., and F. M. Exner—*Meteorologische Optik* (4 parts in 1, Wien and Leipzig, 1902–10).
[Contains an unusually thorough treatment of the rainbow (pp. 482-558).]

Pernter, J. M.—*Zur Theorie des von einer kreisförmigen Lichtquelle erzeugten Regenbogens* (Wien, 1905).
[A recalculation of intensities of illumination, making allowance for the breadth of the sun. Reprinted from *Sitzungsberichte der Kaiserlichen Akademie der Wissenschaften in Wien, Math.- Naturw. Klasse*, CXIV, Abt. II a, June, 1905.]

Pernter, J. M.—*Ein Versuch der richtigen Theorie des Regenbogens Eingang in die Mittelschulen zu verschaffen* (Wien, 1898).
[A good account of the advantages of the theory of Airy over that of Des-cartes. Reprinted from *Kaiser-Jubiläums-Hefte der Zeitschrift für die Oester-reichischen Gymnasien*, 1898.]

Potter, Richard—Mathematical Considerations on the Problem of the Rainbow, Shewing It to Belong to Physical Optics, *Transactions of the Cambridge Phil-osophical Society*, VI (1838), 141-52.
[Important for the calculation of caustics and wave-fronts associated with the rainbow.]

Potter, Richard—On the Radii and Distance of the Primary and Secondary Rain-bows, as Found by Observation, and on a Comparison of Their Values with Those Given by Theory, *Philosophical Magazine* (3), XIII (1838), 9-12.
[Expresses scepticism of the correctness of the wave theory of light and of Airy's theory of the rainbow.]

Potter, Richard—On the Interference of Light Near a Caustic, *Philosophical Maga-zine* (4), IX (1855), 321-26.
[Questions the correctness of Airy's theory of the rainbow.]

Pulfrich, Carl—Ein experimenteller Beitrag zur Theorie des Regenbogens und der überzählige Bogen, Annalen der Physik (3), XXXIII (1883), 194-208.
[Experimental confirmation of Airy's theory.]

Raillard, F.—Explication nouvelle et complète de l'arc-en-ciel, Comptes Rendus, XLIV (1857), 1142-44.
[A plea for the theory of Airy rather than that of Descartes. See also Cosmos, X (1856-57), 605-07.]

Raillard, F.—Nouvelle note sur l'arc-en-ciel, Comptes Rendus, LX (1865), 1287-89.
[A renewal of the plea for the theory of Airy.]

Rayleigh, Lord—The Incidence of Light Upon a Transparent Sphere, Royal Society of London, Proceedings, Series A, LXXXIV (1910), 25-46.
[A highly technical account of the intensity of light on the basis of the electromagnetic theory of light.]

Stevens, W. Le Conte—Theory of the Rainbow, United States Weather Bureau, Monthly Weather Review and Annual Summary, XXXIV (1906), 170-73.
[A good comparison of the theories of Descartes and Airy.]

Stokes, G. G.—On the Numerical Calculation of a Class of Definite Integrals and Infinite Series, Transactions of the Cambridge Philosophical Society, IX, part I.
[A more expeditious device for calculating the values of Airy's rainbow integral. This is reprinted in Stokes' Mathematical and Physical Papers (5 vols., Cambridge, 1880–1905), II, 329-57.]

Tyndall, John—Six Lectures on Light (New York, 1895).
[A charming account of some of the important developments in optics.]

Verdet, É.—Leçons d'optique physique (vols. 5 and 6 of his Oeuvres, Paris, 1869–70).
[Contains a good account (V, 402-23) and bibliography of the theory of the rainbow in the nineteenth century through the work of Airy.]

Index

Abbe, Cleveland, 350
Acta Eruditorum, 248
Adam de Bokefeld, 87
Aderholt, A. E., 363
Aeneid, 58
Aetius, 56
Agrippina, 58
Aichi, Keiichi, 317f, 354
Ailly, Pierre d', see Pierre d'Ailly
Airy, George Biddle, 301-310; 294, 313-318, 321, 351, 353, 355, 363ff; Airy's rainbow integral, 302
Akenside, Mark, 268
Albert of Saxony, 131-132; 138, 337f
Albertus Magnus, 94-99; 36, 88, 101ff, 109, 111f, 133, 140ff, 147f, 150, 152, 154, 169f, 173, 189, 198, 357; De Meteoris, 94-97, 130; Opera, 357
Al-Biruni, 189
Albumasar, 169
Alchemy, 28
Alexander of Aphrodisias, 62-65; 67, 69, 71, 76, 86, 119, 122, 148f, 154ff, 167, 170, 232, 328f, 339, 356, 360; Alexander's band, 64, 119, 122, 215
Alexander the Great, 56, 74, 202, 276
Alexandrian Age, 56-65
Alfred of Sareshel, 87f, 331
Alhazen, 80-83; 69, 85, 87, 90, 102f, 105, 107, 111f, 125ff, 133, 138, 156f, 178, 186f, 192, 204f, 330, 336, 358f; Treasury of Optics, 80, 82, 90, 103, 111f, 125ff, 157, 166, 192, 330, 333, 358; Alhazen's problem, 80
Al-Qarafi, 125ff, 358
Al-Shirazi, see Qutb al-Din al-Shirazi
Ambrose, St., 66
Ammonius, 154
Anaxagoras, 35ff, 42, 80, 259, 320
Anaximander, 34, 38
Anaximenes, 34, 47, 229, 327

Antoniadi, E.-M., 327f
Aphrodisian paradox, 64f, 119; see also Alexander of Aphrodisias
Apollonius, 45f, 208, 275f
Aquinas, St. Thomas, 98-99; 148, 263, 333, 357f
Arafat, W., 335
Arago, François, 305
Archimedes, 55, 61, 82, 143
Aristarchus, 57, 197
Aristophanes, 23, 37, 48, 61, 323
Aristotle, 36-53 and passim; Meteorologica, 38ff, 325ff, 328f, 338f, 356f, and passim; Posterior Analytics, 91f; Problemata, 40, 326; theory of color, 47-49, 53, 55, 57f, 328; theory of vision, 50
Artemidorus, see Parianus Artemidorus
Athens, Chapter II, 66, 75
Australia, 26f
Autrecourt, Nicolaus of, see Nicolaus of Autrecourt
Averroes, 83-84; 87, 91, 108, 131, 163, 331f, 357
Avicenna, 76-80; 82f, 85, 87f, 91f, 94, 99, 111, 126ff, 173, 330, 357
Aymeric de Plaisance, 110

Babinet, Jacques, 304f, 352, 363
Babylon, see Mesopotamia and Islamic tradition
Bachtold-Stäubli, Hanns, 324
Bacon, Francis, 195-196; 113, 204, 234, 342, 360; Advancement of Learning, 195; Historia Ventorum, 195; Novum Organum, 204; Sylva Sylvarum, 204; Works, 360; Baconian method, 208
Bacon, Roger, 99-102; 87, 97, 106, 109, 113, 115, 120f, 139, 196, 201, 204, 273, 331, 358; Opus Majus, 89, 99f, 333, 358
Bähr, Gerhard, 324, 356

367